1 MONTH OF
FREE
READING

at
www.ForgottenBooks.com

By purchasing this book you are eligible for one month membership to ForgottenBooks.com, giving you unlimited access to our entire collection of over 1,000,000 titles via our web site and mobile apps.

To claim your free month visit:
www.forgottenbooks.com/free55775

ISBN 978-1-5282-6378-8
PIBN 10055775

A

TREATISE ON CHEMISTR

BY

SIR H. E. ROSCOE F.R.S. AND C. SCHORLEMMER F.R.S

VOLUME III

THE CHEMISTRY OF THE HYDROCARBONS AND THEIR DERIVATIVES

OR

ORGANIC CHEMISTRY

PART III

NEW AND THOROUGHLY REVISED EDITION

" *Chymia, -alias Alchemia et Spagirica, est ars corpora vel mixta, vel composita vel -aggregata etiam in principia sua resolvendi, aut ex principiis in talia combinandi.*"—STAHL, 1723

London

MACMILLAN AND CO.

AND NEW YORK

1891

The Right of Translation and Reproduction is Reserved

RICHARD CLAY AND SONS, LIMITED,
LONDON AND BUNGAY.

First Edition 1886.   Reprinted 1889.
New Edition 1891.

PREFACE TO FIRST EDITION

This part of the work commences the consideration of the complicated but most important series of bodies known as the Aromatic Compounds.

It contains, after an introduction, a description of the mode of formation, and of the properties, of the Aromatic Hydrocarbons and their derivatives, together with an historical discussion of their isomeric modifications. The constitution of Benzene and the characteristic reactions of its di-substitution products are then fully explained. Next follows a review of all the important Benzene Derivatives, so that the part now published forms a complete chapter of the ever-increasing volume of Aromatic Chemistry.

H. E. R.
C. S.

January, 1886.

PREFACE TO SECOND EDITION

In consequence of the rapid progress of the Chemistry of the Aromatic Compounds during the last five years, the authors have found it necessary to prepare an entirely new edition of this part, including all the more important work published up to the date of going to press. In addition to the general revision of the whole subject-matter, they would draw special attention to the renewed discussion of the constitution of Benzene, and to the researches of Nietzki and his co-workers on the higher substitution-products of Benzene, which have explained the constitution of the remarkable substances derived from the explosive compound of potassium and carbonic oxide. The authors desire particularly to express their obligation to their friend Dr. H. G. Colman, by whom the new edition has been prepared.

H. E. R.

C. S.

January, 1891.

CONTRACTIONS EMPLOYED IN THIS PART

adj. = adjacent.
as. = asymmetric.
m. = meta.
n. = normal.
o. = ortho.
p. = para.
s. = symmetric.

Am. Chem. Journ. = American Chemical Journal.
Annalen = Liebig's Annalen der Chemie.
Ann. Chim. Phys. = Annales de Chimie et Physique.
Ber. = Berichte der Deutschen Chemischen Gesellschaft.
Bull. Soc. Chim. = Bulletin de la Société Chimique de Paris.
Chem. Centr. = Chemisches Centralblatt.
Chem. Zeit. = Chemiker Zeitung.
Compt. Rend. = Comptes Rendus de l'Académie des Sciences.
Gazzetta = Gazzetta Chimica Italiana.
Journ. Chem. Soc. = Journal of the Chemical Society.
J. Pr. Chem. = Journal für Praktische Chemie.
Journ. Soc. Chem. Ind. = Journal of the Society of Chemical Industry.
Monatsh. = Monatshefte für Chemie.
Phil. Mag. = Philosophical Magazine.
Pogg. Ann. = Annalen der Physik und Chemie (Poggendorf).
Proc. Roy. Soc. = Proceedings of the Royal Society.
Rec. Trav. Chim. = Recueil des Travaux Chimiques des Pays-Bas.
Wied. Ann. = Annalen der Physik (Wiedemann).
Zeit. analyt. Chem. = Zeitschrift für analytische Chemie.
Zeit. phys. Chem. = Zeitschrift für physikalische Chemie.
Zeit. angew. Chem. = Zeitschrift für angewandte Chemie.
Zeit. physiol. Chem. = Zeitschrift für physiologische Chemie.

CONTENTS

ORGANIC CHEMISTRY

ORGANIC CHEMISTRY

OR THE CHEMISTRY OF THE HYDROCARBONS AND THEIR
DERIVATIVES

PART III

AROMATIC COMPOUNDS

926 THE term " Aromatic compounds " was formerly employed
to denote a small number of naturally occurring substances which
were distinguished by an aromatic taste and smell. It was at
that time customary to classify chemical compounds according
to their physical properties, and it is indeed still usual in some
of our manuals to have chapters on " colouring matters," "bitter
principles," &c., in which those compounds are described whose
constitution is not yet known, and which cannot therefore be classified in any other manner. To these compounds, termed by
Gerhardt " *corps à série*," the greater portion of the older works
on organic chemistry was devoted, but their number is decreasing
from day to day, and with the advance of science will finally
disappear.

The original group of aromatic compounds has long ago
undergone this change, closer investigation having shown, that
in addition to their aromatic smell and taste, they were also
closely related to one another in their chemical properties. The
first step in this direction was made by Wöhler and Liebig in
1832, in their classical " Research on the Radical of Benzoic
Acid." [1] In their introduction they remark, " We may con-
gratulate ourselves if we succeed in finding in this dark realm
of organic nature, one single ray of light, which may serve to

[1] *Annalen*, **3**, 249.

lead us to a starting point, from which we may possibly succeed in discovering the true way for the exploration of this wide field, though we know well its illimitable extent. Nor indeed can we here reasonably expect at first to arrive at any very profound or wide generalisations, owing to the absence of previous investigations, as well as the difficulty of obtaining the necessary materials. Under these circumstances the experiments which are described in the sequel must be regarded, as far as their extent and connection with other branches of investigation are concerned, only as the opening out of a wide and fruitful field for future research."

This wide field embraces the whole of the aromatic compounds, and it has proved to be even more fruitful than Liebig and Wöhler could have dreamed. Their first observation was that oil of bitter almonds is distinguished from similar compounds by its power, first noticed by Stange, of absorbing oxygen, with formation of benzoic acid—a compound known so long ago as the seventeenth century, and prepared from the aromatic gum benzoïn. The two chemists then showed that both these compounds contain the same radical, benzoyl, C_7H_5O, and that the oil of bitter almonds must be regarded as benzoyl hydride, C_7H_6O which on oxidation passes into benzoic acid, $C_7H_6O_2$. By replacing an atom of hydrogen in benzoyl hydride by the elements of the chlorine group, by cyanogen, or by sulphur, other compounds of this same radical are obtained.

Mitscherlich prepared the hydrocarbon, C_6H_6, by distilling benzoic acid with slaked lime, it being formed by the separation of carbon dioxide from benzoic acid, and to this body he gave the name of benzin, which was altered by Liebig to benzol. English chemists now give to this hydrocarbon the name of benzene. Mitscherlich also found that this hydrocarbon is converted by the action of nitric acid into nitrobenzene, $C_6H_5NO_2$. Zinin next showed that the oxygen in this body can be replaced by hydrogen, and that a basic oil is thus obtained which was soon proved to be identical with aniline, a compound previously obtained by Fritsche by distilling indigo with caustic potash. The Russian chemist had also observed that indigo when treated with caustic potash yields an acid termed anthranilic acid, $C_7H_7NO_2$, and that this is decomposed on distillation into carbon dioxide and aniline.

Gerhardt afterwards noticed that when nitrous acid acts upon anthranilic acid, this latter is converted into salicylic acid,

$C_7H_6O_3$, a body first obtained from the ethereal oil of *Spiræa Ulmaria*, together with salicyl hydride, $C_7H_6O_2$. These two compounds stand in the same relation to one another as benzoyl hydride does to benzoic acid; just as the latter is converted into benzene by the separation of carbon dioxide, so from salicylic acid, carbolic acid, C_6H_6O, is obtained. This latter substance is also found in coal-tar, together with benzene and toluene, C_7H_8, a body whose name is derived from the fact that it was first obtained by distilling the aromatic tolu-resin. This compound closely resembles benzene; thus, by the action of nitric acid it is converted into nitrotoluene, and from this the base toluidine can readily be prepared. These investigations proved that an intimate relation exists between the two series of compounds, containing respectively six and seven atoms of carbon :—

Benzene.	Nitrobenzene.	Aniline.	Carbolic Acid.
C_6H_6.	$C_6H_5NO_2$.	C_6H_7N.	C_6H_6O.
Toluene.	Nitrotoluene.	Toluidine.	—
C_7H_8.	$C_7H_7NO_2$.	C_7H_9N.	
Benzoyl. Hydride.			Salicyl Hydride.
C_7H_6O.			$C_7H_6O_2$.
Benzoic Acid.			Salicylic Acid.
$C_7H_6O_2$.			$C_7H_6O_3$.

In course of time the existence of many similar groups was discovered, and to all these the general title of "aromatic compounds" was given, although among them many are found which either possess no smell, or whose odour is anything but aromatic. The designation of "aromatic" is applied in the same sense as that of "fatty bodies," in which we now class almost all the organic substances not belonging to the aromatic series. (R. Meyer.)

It was at first difficult to find a sharp distinction between aromatic compounds and other groups, as many bodies were known which, according to their products of decomposition as well as their mode of combination, may equally well be placed either in the one or in the other division. Thus it came about that many compounds which clearly belong to the aromatic group were not classed under this head. On the other hand, however, the fact was not then recognised that bodies properly classed as aromatic could yield compounds belonging either to the group of saturated or unsaturated fatty bodies. Nor was it known that the reverse action could take place (Kekulé.)

927 *Kekulé's Theory of the Constitution of Aromatic Bodies.*— No expression of opinion is to be found respecting the constitution of the aromatic compounds until Kekulé proposed a theory which shed an unexpected light on these hitherto neglected bodies.[1] The effect of the general recognition of the truth of Kekulé's theory has been, that this special field of organic chemistry has since been, and still is, most industriously cultivated, perhaps to the undeserved neglect of other branches.

In justification, however, it must be remembered that the study of the aromatic compounds has an especial attraction, inasmuch as not only all the natural, but also all the artificial colouring matters, as well as the most valuable medicines and most potent poisons, belong to this class of bodies.

Kekulé's views are best given in his own words : " If we wish to form an idea of the constitution of the aromatic compounds we are bound to explain the following facts :—(1st) Even the simplest of aromatic compounds are relatively richer in carbon than the compounds belonging to the fatty group; (2nd) Amongst the aromatic compounds as well as amongst the fatty bodies a large number of homologous substances exist; (3rd) The simplest of aromatic compounds contain at least six atoms of carbon; (4th) The derivatives of these substances all exhibit certain family relationships to one another and to the compounds from which they are obtained, and therefore also belong to the group of aromatic compounds. In the more vigorous reactions, a part of the carbon is, it is true, frequently eliminated, but the chief product contains at least six atoms of carbon, the decomposition ceasing with the formation of these products, unless the action is so vigorous that the organic group is completely broken up.

" These facts justify the supposition that one and the same group of atoms occurs in all the aromatic compounds, or that they all contain a common nucleus of six atoms of carbon. Within this ring the carbon atoms are more intimately united, or lie closer to one another, and this is the cause of the relatively large quantity of carbon contained in all aromatic compounds.

" Other atoms of carbon may attach themselves to this nucleus in the same manner, and according to the same laws, as is the

[1] *Bull. Soc. Chim.* 1865, **1**, 98.

case in the group of fatty bodies, and in this way the existence of homologous compounds is readily explained." [1]

According to Kekulé, the six atoms of carbon in the aromatic nucleus are united together into a closed chain or ring. Of the remaining twelve valencies of the carbon atoms, six are saturated

by combination with hydrogen or other simple or compound radicals, whilst the six remaining valencies are supposed by Kekulé to saturate one another in the manner shown in the following graphic formula, the six carbon atoms being alternately connected by single and double linkage.

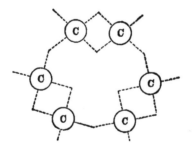

The last-named hypothesis of Kekulé's has been warmly disputed during the past twenty-five years, and the question cannot be even now regarded as settled. It will be more fully discussed later on. (See p. 58 *et seq.*) The other hypotheses have, however, found almost universal acceptance, and for the discussion of the questions on isomerism, &c., the formula

[1] Kekulé, *Lehrbuch*, **2**, 195.

which embodies the first two hypotheses and leaves the question of the disposition of these six valencies undetermined, is quite sufficient.

928 *Isomerism in the Aromatic Group.*—The simplest aromatic compound is that in which the aromatic nucleus is combined with six atoms of hydrogen, the compound formed being benzene, or

$$\begin{array}{c}
\text{CH} \\
\text{HC} \diagup \quad \diagdown \text{CH} \\
\text{HC} \diagdown \quad \diagup \text{CH} \\
\text{CH}
\end{array}$$

From this all other aromatic compounds may be derived in the same way as the fatty bodies are derived from methane. Hence, benzene may be regarded as the marsh gas of the aromatic series, and this series is frequently defined as the group of the benzene derivatives. Kekulé's formula not only explains the above-named facts, but also the following, which will be mentioned in detail in the sequel :

(1) Among the mono-substitution products of benzene no case of isomerism occurs, as all the six atoms of hydrogen of this hydrocarbon are of equal value, and, therefore, indistinguishable one from the other.

(2) If two atoms of hydrogen be replaced by elements or radicals, termed "side chains," three isomeric compounds may be formed whether the entering elements or side chains be identical or different, and this conclusion is borne out by experiment. This isomerism is due to the fact that the bodies replacing hydrogen take up different positions in the benzene ring. If we represent benzene by a hexagon, at each of whose angles a carbon atom is placed, and number these 1 to 6, it is clear that three different dibromobenzenes, $C_6H_4Br_2$, or nitrobromobenzenes, $C_6H_4(NO_2)Br$, can exist in which bromine or nitroxyl takes the following positions :—

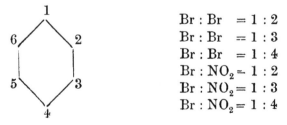

$$\begin{array}{ll}
\text{Br : Br} & = 1 : 2 \\
\text{Br : Br} & = 1 : 3 \\
\text{Br : Br} & = 1 : 4 \\
\text{Br : NO}_2 = 1 : 2 \\
\text{Br : NO}_2 = 1 : 3 \\
\text{Br : NO}_2 = 1 : 4
\end{array}$$

Every other position is identical with one of the above; hence, these may be distinguished as adjacent, alternate, and opposite. Tri-substitution products of benzene containing identical elements or side chains can also occur, but only in three modifications :—

$$1 : 2 : 3$$
$$1 : 2 : 4$$
$$1 : 3 : 5$$

These are distinguished as adjacent, asymmetric, and symmetric. It is also clear that if four atoms of hydrogen are replaced by some element or radical three isomeric compounds can occur.

Supposing, however, that the entering elements or radicals are not identical, it is plain that the number of isomeric bodies will be increased; thus, for example, we are acquainted with six nitrodibromobenzenes, $C_6H_3(NO_2)Br_2$, in which, if the nitroxyl occupies position 1, the bromine atoms may occupy the following positions :—

$$2 : 3$$
$$2 : 4$$
$$2 : 5$$
$$2 : 6$$
$$3 : 4$$
$$3 : 5$$

If all three side chains or even a greater number are different, the number of possible isomerides will of course be correspondingly larger.

929 *Characteristic Properties of Aromatic Bodies.*—By the action of chlorine and bromine on benzene, substitution products are formed, but under certain conditions additive products may also be produced, inasmuch as benzene combines with two, four, or six atoms of the halogen, but not with more. In support of his theory Kekulé adds : " Some few chemists incline to the view that benzene and hydrocarbons homologous with it are derived from hydrocarbons belonging to the class of fatty bodies by simple loss of hydrogen, and the consequent more intimate union of the carbon atoms. I do not share this view, but rather believe that the hydrocarbons of the formula C_6H_6 prepared from C_6H_{14} by subtraction of hydrogen, will turn out to be only isomeric, and not identical with benzene." Such a hydrocarbon is in fact now known, and is termed dipropinyl, $CH \equiv C.CH_2.CH_2.C \equiv CH$. (See **Vol**. III. Part II. p. 550.) This compound combines energetically with bromine to form a tetrabromide, $C_6H_6Br_4$, and on warming with bromine it yields an octobromide, $C_6H_6Br_8$. On the other hand, its isomeride benzene exhibits totally different reactions, inasmuch as it combines with bromine and chlorine only slowly, and moreover is incapable of taking up more than six atoms.

The substitution products of benzene react quite differently from those of the paraffins. By the action of chlorine, monochlorobenzene, C_6H_5Cl, is formed, and this was formerly supposed to be the chloride of a monad alcohol radical and termed *phenyl chloride*. It does not, however, exhibit any of the properties of an alcoholic chloride, inasmuch as it holds its chlorine much more firmly, and is not attacked when heated with caustic potash solution, silver salts, ammonia, &c.

The action of concentrated nitric acid serves as another characteristic test for the aromatic compounds, as these readily form nitro-compounds, bodies which in the fatty series are only obtained by indirect processes; thus benzene yields nitrobenzene, $C_6H_5NO_2$, and this passes easily by reduction into aniline, $C_6H_5NH_2$, a body which was formerly called *phenylamine*, but is now termed *amidobenzene;* for, although in some respects it possesses analogies with the compound ammonias, it differs widely from them in other respects, especially in not possessing any ammoniacal smell and having a perfectly neutral reaction.

Another characteristic property of the aromatic compounds is the formation of sulphonic acids when the hydrocarbons are treated with strong sulphuric acid; thus, for example, benzene yields benzenesulphonic acid, $C_6H_5SO_3H$.

If an atom of hydrogen in benzene be replaced by hydroxyl, a substance is obtained which is in many respects analogous to the alcohols of the fatty series, and was therefore formerly termed *phenyl alcohol*. It is however now termed phenol, as it has a number of properties which are not possessed by the fatty alcohols. For example, this substance and its homologues are extremely stable compounds which can only be oxidized with difficulty in acid solution. Alcohol is converted by concentrated nitric acid into a nitrate, and by strong sulphuric acid into a sulphate; phenol, on the other hand, treated in like manner, yields nitrophenol, $C_6H_4(NO_2)OH$, and phenolsulphonic acid, $C_6H_4OH.SO_3H$. These examples suffice to show that well-defined differences are observable between the members of the fatty and those of the aromatic group of substances.

AROMATIC HYDROCARBONS.

930 The homologues of benzene are formed by the replacement of hydrogen by an alcoholic radical, such as methyl, ethyl, propyl, &c.; the compounds thus obtained act on the one hand as aromatic, and on the other, as fatty bodies. When the hydrogen in the nucleus of methylbenzene or toluene, $C_6H_5.CH_3$, is replaced by chlorine, hydroxyl or nitroxyl, bodies are formed which exhibit the closest similarity to the corresponding benzene derivatives. If, however, the replacement takes place in the methyl group, compounds of the alcohol radical, phenylmethyl or benzyl are produced, such as benzyl chloride, $C_6H_5.CH_2Cl$, a body which is converted by the action of ammonia into strongly alkaline benzylamine; similarly, the same body can be converted by oxidation into benzaldehyde and benzoic acid. Thus the two following series are obtained :—

Toluene, Methylbenzene or Phenylmethane.
$$C_6H_5.CH_3.$$

Monochlorotoluene.	Benzyl Chloride.
$C_6H_4Cl.CH_3.$	$C_6H_5.CH_2Cl.$
Toluidine or Amidotoluene.	Benzylamine.
$C_6H_4(NH_2)CH_3.$	$C_6H_5.CH_2.NH_2.$
Cresol or Hydroxytoluene.	Benzyl Alcohol.
$C_6H_4(OH)CH_3.$	$C_6H_5.CH_2.OH.$
Nitrocresol.	Benzyl Nitrate.
$C_6H_3(NO_2)(OH)CH_3.$	$C_6H_5.CH_2O.NO_2.$
—	Benzaldehyde.
	$C_6H_5.COH.$
	Benzoic Acid.
	$C_6H_5.CO.OH.$

Here we find several different isomeric compounds, but their number is really still greater, for each mono-substitution product of toluene exists in three isomeric forms.

Moreover, substitution can take place at the same time in the aromatic nucleus and in the alcohol radical, whereby bodies are obtained which act both as aromatic substitution-products and as compounds of alcohol radicals. Such a body is chlorobenzyl chloride, $C_6H_4Cl.CH_2Cl$, which can be easily converted into chlorobenzylamine, chlorobenzyl alcohol, &c.

Again, not only can one atom of hydrogen in benzene be replaced by an alcohol radical, but all six can be consecutively substituted. If this take place with methyl, a series of homologous bodies is obtained whose members are isomeric with monosubstituted benzenes which contain higher alcohol radicals, thus :

$$\text{Benzene} \quad . \quad . \quad . \quad C_6H_6$$
$$\text{Methylbenzene} \quad . \quad C_6H_5.CH_3$$

Ethylbenzene .	$C_6H_5.C_2H_5$	Dimethylbenzene . .	$C_6H_4(CH_3)_2.$
Propylbenzene.	$C_6H_5.C_3H_7$	Trimethylbenzene .	$C_6H_3(CH_3)_3.$
Butylbenzene .	$C_6H_5.C_4H_9$	Tetramethylbenzene .	$C_6H_2(CH_3)_4.$
Pentylbenzene.	$C_6H_5.C_5H_{11}$	Pentamethylbenzene .	$C_6H(CH_3)_5.$
Hexylbenzene.	$C_6H_5.C_6H_{13}$	Hexmethylbenzene .	$C_6(CH_3)_6.$

Here we find two kinds of homologous bodies, the isomerism of the first series depending on the lengthening of the side chain, and the number of its members is only limited by the number of alcohol radicals which is known, whilst the homology of the second series depends upon an increase in the number of side chains, so that here hexmethylbenzene is the last member of the series.

With the monosubstituted benzenes we find additional isomerides as soon as propylbenzene is reached, the number being dependent on the number of isomerides of the alcohol radical. Thus we are acquainted with a normal propylbenzene, $C_6H_5.CH_2.CH_2.CH_3$, and isopropylbenzene, $C_6H_5.CH.(CH_3)_2$. With the second series the isomerism can, as already explained, occur only in the case of the di- tri- and tetra-substituted compounds, provided the radical is in all cases the same. If, however, the hydrogen atoms in benzene be replaced by different alkyl groups, the number of possible isomerides is greatly increased.

In all these hydrocarbons the hydrogen atoms in the aromatic nucleus, as well as those in the side chain, can again be replaced, and from this it will be seen that the number of benzene derivatives may be extremely large, and the possibility of a further increase is evident when we remember that the hydrogen of the benzene is not merely capable of replacement by a monad alcohol radical, C_nH_{2n+1}, but also by groups containing less hydrogen. Thus for example we are acquainted with the following hydrocarbons :

Styrolene or phenylethylene . . . $C_6H_5.CH{=}CH_2$.
Allylbenzene or phenylpropylene . . $C_6H_5.CH{=}CH.CH_3$.
Crotonylbenzene or phenylbutylene . $C_6H_5.CH_2.CH_2.CH{=}CH_2$.

And derivatives of these, such as :

Cinnamic alcohol or phenylallyl alcohol, $C_6H_5.CH{=}CH.CH_2.OH$.
Cinnamic aldehyde or phenylacrylaldehyde,$C_6H_5.CH{=}CH.COH$.
Cinnamic acid or phenylacrylic acid, $C_6H_5.CH{=}CH.COOH$
Hydroxyphenylpropylene, $C_6H_4(OH)CH{=}CH.CH_3$.

These compounds exhibit the general properties of the olefines or of the compounds of the allyl series ; they unite with hydrogen, the elements of the chlorine group and their hydracids, &c., forming saturated compounds.

The number of side chains which is contained in an aromatic hydrocarbon, or in its derivatives obtained by substitution in the side chains, may easily be ascertained by oxidation with chromic acid solution. If only one side chain be present as in toluene, ethylbenzene, propylbenzene, amylbenzene, phenylethylene, or cinnamic alcohol, &c., the monobasic benzoic acid is produced. The constitution of the side chain in such hydrocarbons may also be ascertained in the same way; thus amylbenzene, $C_6H_5.CH_2.CH_2.CH(CH_3)_2$, in addition to benzoic acid, yields isobutyric acid.[1]

If two side chains be present, one of the three phthalic acids or benzenedicarboxylic acids, $C_6H_4(CO_2H)_2$, is formed, and if three such side chains be present a tribasic benzenetricarboxylic acid, $C_6H_3(CO_2H)_3$, is formed. All the carboxylic acids of benzene yield, when distilled with slaked lime, carbon dioxide and benzene.

When dilute nitric acid or potassium permanganate is employed as the oxidizing agent, only one side chain is at first oxidized to carboxyl, and, if these side chains are different, the longest of them is attacked first. The dimethylbenzenes, $C_6H_4(CH_3)_2$, ethylmethylbenzenes, $C_6H_4\left\{\begin{smallmatrix}CH_3\\C_2H_5\end{smallmatrix}\right.$, and propylmethylbenzenes, $C_6H_4\left\{\begin{smallmatrix}CH_3\\C_3H_7\end{smallmatrix}\right.$, yield monobasic toluic acids, $C_6H_4\left\{\begin{smallmatrix}CH_3\\CO_2H\end{smallmatrix}\right.$, and, of

[1] Popow and Zincke, *Ber.* **5**, 384.

the last class, that which contains normal propyl yields in addition acetic acid.[1] One trimethylbenzene, $C_6H_3(CH_3)_3$, treated in a similar way, first yields a monobasic acid, $C_6H_3 \left\{ \begin{array}{l} (CH_3)_2, \\ CO_2H \end{array} \right.$ and then a dibasic acid, $C_6H_3 \left\{ \begin{array}{l} CH_3 \\ (CO_2H)_2 \end{array} \right.$; the first of these yields dimethylbenzene when distilled with lime, and the second, like the toluic acids, yields toluene. In certain cases it has been observed that when the oxidation is carried on very slowly, a long side chain is not at once oxidized to carboxyl, but that intermediate products are formed; thus, for example, by oxidizing ethylbenzene, $C_6H_5CH_2CH_3$, some methylphenylketone, $C_6H_5.CO.CH_3$, is obtained,[2] whilst from triethylbenzene, $C_6H_5(C_2H_5)_3$, together with trimesitic acid, $C_6H_3(CO_2H)_3$, the tribasic isophthalacetic acid, $C_6H_3 \left\{ \begin{array}{l} (CO_2H)_2 \\ CH_2CO_2H \end{array} \right.$, is obtained.[3]

931 *Formation of Aromatic Hydrocarbons.*—In addition to the above-mentioned mode of formation from the acids, a large number of the hydrocarbons can be obtained by simple synthesis, as is the case with paraffins, and by this means their constitution can be ascertained.

1st. A mixture of a brominated hydrocarbon and the iodide or bromide of an alcohol radical is treated with sodium thus:[4]—

$$C_6H_5Br + CH_3I + 2Na = C_6H_5.CH_3 + NaBr + NaI.$$
$$C_6H_4Br.CH_3 + C_2H_5Br + 2Na = C_6H_4 \left\{ \begin{array}{l} CH_3 \\ C_2H_5 \end{array} \right. + 2NaBr.$$
$$C_6H_4Br_2 + 2CH_3I + 4Na = C_6H_4(CH_3)_2 + 2NaBr + 2NaI.$$

2nd. The zinc compound of the alcohol radical is allowed to act on benzyl chloride or benzylene chloride thus:

$$2C_6H_5.CH_2Cl + Zn(C_2H_5)_2 = C_6H_5.C_3H_7 + ZnCl_2.$$
$$C_6H_5.CHCl_2 + Zn(CH_3)_2 = C_6H_5.CH(CH_3)_2 + ZnCl_2.$$

3rd. Friedel and Craffts reaction. This method depends upon the fact that when an aromatic hydrocarbon is mixed with the haloid salt of an alcohol radical and aluminium chloride, hydrochloric acid is evolved and a higher homologue of the

[1] Dittmar and Kekulé, *Annalen,* **162,** 337.
[2] Bahlson, *Bull. Soc. Chim.* **32,** 615.
[3] *Ibid.* **34,** 625.
[4] Fittig and Tollens, *Annalen,* **131,** 303.

aromatic hydrocarbon formed. Thus when methyl chloride is passed into a mixture of benzene and aluminium chloride the chief product is durene or tetramethylbenzene,[1] hydrochloric acid being also evolved.

$$C_6H_6 + 4CH_3Cl = C_6H_2(CH_3)_4 + 4HCl.$$

The part played by the aluminium chloride in the reaction is not thoroughly understood, but it is certain that it is not merely a so-called " contact action."

This reaction is also capable of a very much wider application than that above mentioned, for not only the alkyl chlorides, but also acid chlorides, and many other halogen compounds act on the aromatic hydrocarbons in the same manner in the presence of aluminium chloride, and further, the hydrocarbons may be replaced by phenol ethers such as anisoïl. These various reactions will be more fully discussed under the compounds themselves.

Under certain circumstances the reverse reaction may take place, and indeed generally takes place concurrently with the chief reaction. Thus if hexmethylbenzene is heated with aluminium chloride methyl chloride is evolved, and if a current of hydrogen chloride be passed through at the same time the methyl groups may be completely eliminated.[2]

4th. Aromatic hydrocarbons are also formed by the condensation of hydrocarbons of the acetylene series. Berthelot found that if acetylene be heated to a temperature at which glass softens, a considerable quantity of benzene is produced.

At the same time styrolene, C_8H_8, naphthalene, $C_{10}H_8$, and other hydrocarbons are formed.

If methylacetylene be dissolved in sulphuric acid and the

[1] Friedel and Craffts, _Ann. Chim. Phys._ VI. **1**, 449 ; **14**, 443.
[2] Friedel and Craffts, _Brit. Ass. Reports_, **1883**, 468 ; Jacobsen, _Ber._ **18**, 338 ; Anschütz and Immendorf, _ibid._ **18**, 657.

mixture distilled, mesitylene or symmetric trimethylbenzene
is formed :

$$3CH\equiv C.CH_3 = C_6H_3(CH_3)_3.$$

Dimethylacetylene, $CH_3C\equiv C.CH_3$, when shaken with sulphuric
acid, is converted into hexmethylbenzene, $C_6(CH_3)_6$. (See **Vol.**
III. Part II. pp. 524-529.)

5th. When a ketone containing the group $CH_3.CO$ is distilled
with sulphuric acid it is converted into an aromatic hydrocarbon,
three molecules of the ketone condensing with elimination of
three molecules of water. The three alcohol radicals occupy
the symmetric position.

$$3CH_3.CO.CH_3 = C_6H_3(CH_3)_3 + 3H_2O.$$
$$3CH_3.CO.C_2H_5 = C_6H_3(C_2H_5)_3 + 3H_2O.$$
$$3CH_3.CO.C_3H_7 = C_6H_3(C_3H_7)_3 + 3H_2O.$$

6th. Aromatic hydrocarbons occur in many balsams and ethereal
oils, as well as in certain petroleums. They are formed by the
dry distillation of organic substances, and hence they are found
in wood-tar and in larger quantities in coal-tar, in which we find
benzene, toluene, dimethylbenzenes, trimethylbenzenes, and a
large number of other aromatic compounds.

In a memoir on the products obtained in the preparation of
illuminating gas from resin, Pelletier and Walter so early as
the year 1838 remark, that chemical operations conducted on
the large scale offer opportunities for observing phenomena, in-
vestigating laws, and preparing new products which do not present
themselves in laboratory experiments. Chemical industry, largely
indebted to theory for its progress, repays the chemist whose
assistance she needs, by presenting him with new compounds for
investigation, and thus science is widened and industry developed.
In this way the manufacture of coal-gas has enriched organic
chemistry with many new compounds, whose investigation has
proved of the greatest interest. But for coal-gas Faraday's
liquid hydrocarbons, butylene and benzene, Kidd's naphthalene
and Dumas and Laurent's paranaphthalene (anthracene) would
still have to be discovered.[1]

Pelletier and Walter discovered toluene in the oil obtained
from distillation of resin. This body, together with its homologues, benzene, naphthalene, anthracene, is now obtained from

[1] *Ann. Chim. Phys.* **67,** 269.

coal-tar on the large scale. This thick black, and unpleasant smelling substance, an essential product of the coal-gas manufacture, and formerly a noxious and useless article, has now become not only of the greatest importance to the colour manufacturer but most valuable to the scientific chemist, as a means of carrying on the investigation of interesting and important bodies to an extent which would have been impossible but for the introduction of the gas manufacture. For this coal-tar has proved a source, as yet inexhaustible, of new aromatic compounds, from which valuable materials, such as the colouring matters of madder and indigo, bodies previously only known as natural products, can now be artificially prepared.

Additive Products of the Aromatic Compounds.—When benzene and its homologues are heated to 280° with fuming hydriodic acid they combine with six atoms of hydrogen and give rise to a homologous series whose members are isomeric with the olefines.[1]

$$\text{Hexhydrobenzene} \quad . \quad . \quad . \quad C_6H_{12}.$$
$$\text{Hexhydromethylbenzene} \quad . \quad C_7H_{14}.$$
$$\text{Hexhydrodimethylbenzene} \quad C_8H_{16}., \&c.$$

These compounds resemble the benzene derivatives inasmuch as they contain a closed chain of six carbon atoms, but all the valencies are saturated with hydrogen. Hexhydrobenzene is therefore identical with hexamethylene.

In their properties, these compounds resemble the saturated fatty hydrocarbons much more closely than the aromatic substances from which they are derived. A number of derivatives of hexamethylene, which have been synthetically obtained from fatty compounds have already been described (vol. iii. Part II. p. 730), together with the analogous tri- tetra- and. pentamethylene derivatives. These hydrocarbons are found in Baku petroleum,[2] and also probably, together with paraffins and with hydrocarbons of the benzene series, in that of Galicia.[3]

[1] Wreden, *Annalen*, **187**, 168.
[2] Beilstein and Kurbatow, *Ber.* **13**, 1818 ; **14**, 1620.
[3] Lachowicz, *Annalen*, **220**, 188.

According to Markownikow and Oglobin, the hydrocarbons which occur in Caucasian petroleum are not hydrogen addition products of benzene and its homologues, but consist of a peculiar group of bodies, to which they gave the name of naphthenes.[1]

Ethereal oils contain hydrocarbons having the common formulæ $C_{10}H_{16}$ and termed terpenes; these stand in close connection with cymene or propylmethylbenzene, $C_{10}H_{14}$, and behave as hydrogen additive-products of this body.

DERIVATIVES OF AROMATIC HYDROCARBONS.

932 *Halogen Substitution Products.*—Derivatives may be obtained from aromatic hydrocarbons by replacing hydrogen either in the side chain or in the aromatic nucleus. We have already seen that in the first case chlorine is very intimately combined, inasmuch as it cannot be removed by alkalis, silver salts, sodium sulphite, ammonia, etc.; on the other hand, the compounds obtained by substitution in the side chain react like the chlorides of radicals of the fatty group.

When benzene is treated by chlorine alone no simple reaction takes place, but if a small quantity of iodine be added the substitution takes place quietly and step by step, the iodine acting as a carrier of the chlorine in consequence of the alternate formation and decomposition of iodine chloride.[2] In this reaction some iodobenzene is always formed. In order to replace the last atoms of hydrogen, antimony chloride is advantageously employed as a carrier of chlorine (H. Müller). The reaction takes place even more easily and regularly if one per cent. of molybdenum chloride be employed instead of iodine.[3]

Chlorine reacts upon the homologues of benzene in a peculiar way; if this gas be passed through the liquid in the cold substitution only takes place in the nucleus, but in presence of iodine and antimony chloride this same substitution occurs whether the saturation takes place in the cold or at the boiling point; if, however, free chlorine be allowed to act by itself upon the boiling hydrocarbon, a replacement of hydrogen in the side chain is alone effected. If the mixture be allowed to cool,

[1] *Ber.* **16**, 1873.
[2] Hugo Müller, *Journ. Chem. Soc.* **15**, 41.
[3] Aronheim, *Ber.* **8**, 1400.

substitution again occurs in the nucleus; on heating a second time the chlorine reverts to the side chain. Thus we obtain as final products from toluene, tetrachlorobenzenyl trichloride, $C_6HCl_4.CCl_3$, and pentachlorobenzidene dichloride, $C_6Cl_5.CHCl_2$. If an attempt be made further to chlorinate these bodies in presence of iodine, a decomposition into hexchlorobenzene, C_6Cl_6, and tetrachloromethane, CCl_4, takes place.[1] The homologues of toluene exhibit quite a similar behaviour, and when thoroughly chlorinated yield the two chlorides of carbon as final products.[2]

Bromine acts in the same manner as chlorine, but less quickly. In the cold, or in presence of iodine, the bromine takes its place in the nucleus; at the boiling point it appears in the side chain. As commercial bromine frequently contains iodine, it must be purified by distillation with potassium bromide when a substitution in the side chain has to be effected.[3] If bromine containing iodine be heated with benzene homologues, the formation of hexbromobenzene and tetrabromomethane is observed.

Iodine cannot effect a direct substitution, but if benzene be heated with iodine and iodic acid to a temperature of 200—240° the following reaction occurs:

$$5C_6H_6 + HIO_3 + 2I_2 = 5C_6H_5I + 3H_2O.$$

In this case part of the benzene is oxidized by the iodic acid to carbon dioxide and water.[4] The iodine substitution products of the aromatic hydrocarbons are far more easily obtained by the action of hydriodic acid on the diazo-compounds (p. 26). They are as stable as the chlorine and bromine derivatives, which may also be obtained by means of the diazo-reaction. Halogen substitution products of the aromatic hydrocarbons are also formed by action of phosphorus compounds of the halogens on phenol, or the aromatic alcohols; they may likewise be obtained from haloid substitution products of the acids, by heating them with lime or baryta.

Nitro-Substitution Products are formed by the action of concentrated nitric acid on hydrocarbons, when nitroxyl, NO_2, replaces the hydrogen in the aromatic nucleus but not in the

[1] Beilstein and Geitner, *Annalen*, **139**, 331.
[2] Krafft and Merz, *Ber.* **8**, 1296.
[3] Thorpe, *Proc. Roy. Soc.* **18**, 123.
[4] Kekulé, *Annalen*, **137**, 161.

side chain.[1] When the action takes place in the cold, the mono-nitro compound is easily formed; when heated, or when a mixture of nitric and sulphuric acids is employed, higher substitution products are obtained. The larger the number of side chains which a compound contains, the more readily does the nitration take place, and the action may be moderated by previously dissolving the substance in glacial acetic acid.

The halogen substitution products are in like manner converted into nitro-compounds by the action of nitric acid; but if, on the other hand, the nitrated hydrocarbons be treated with chlorine or bromine, the reactions take place only on warming, and then the halogen takes the place of the nitroxyl group. The substitution is aided by the presence of iodine, and, as in the case of chlorine, the action takes place more readily in presence of antimony chloride.

933 *Sulphonic acids.*—By acting with concentrated or fuming sulphuric acid on the hydrocarbons, or on their halogen substitution products, &c., monosulphonic or disulphonic acids are formed, the substitution taking place in the nucleus:

$$C_6H_6 + SO_2(OH)_2 = C_6H_5SO_2.OH + H_2O.$$
$$C_6H_5.CH_3 + 2SO_2(OH)_2 = C_6H_3(SO_2.OH)_2CH_3 + 2H_2O.$$

Trisulphonic acids are formed by heating with sulphuric acid and phosphorus pentoxide. Chlorosulphonic acid, $SO_2Cl(OH)$, acts in a similar way to sulphuric acid;[2] it may also be employed for the preparation of trisulphonic acids.[3]

The sulphonic acids of the aromatic group closely resemble those of the members of the fatty group and, like these, are easily soluble in water. On heating, they split up into sulphur trioxide and the original hydrocarbon, and this reaction may be employed for separating aromatic hydrocarbons from those containing more hydrogen, in cases, such as that of coal-tar, when the two series occur together.[4] This reaction depends on the fact that hydrocarbons rich in hydrogen are either, like the paraffins, unattacked by sulphuric acid or, like the olefines, converted into the acid sulphates of the alcohol radicals or polymerised. When the sulphonic acids undergo dry distillation

---

[1] Some compounds containing two carbon atoms in double linkage in the side chain form an exception, such as styrolene or phenylethylene, $C_6H_5.CH{=}CH_2$, which is converted by nitric acid into phenylnitroethylene, $C_6H_5.CH{=}CHNO_2$.

[2] Beckurts and Otto, *Ber.* **11**, 2061.   [3] Claesson, *ibid.* **14**, 307.

[4] Beilstein, *Annalen*, **133**, 34.

they always yield by-products, which diminish the yield of the hydrocarbon; better results are obtained when the sulphonic acid is dissolved in concentrated sulphuric acid and the mixture then treated with superheated steam.[1]

According to Friedel and Craffts a better yield of the hydrocarbon is obtained by adding to the sodium or potassium salt an excess of a phosphoric acid solution of 60° B, and then passing steam.[2] When acted upon by phosphorus pentachloride the sulphonic acids yield sulpho-chlorides and these are converted by zinc into sulphinic acids such as $C_6H_5SO_2H$; these are also readily formed by leading sulphur dioxide into a mixture of a hydrocarbon and aluminium chloride.[3]

$$C_6H_6 + SO_2 = C_6H_5SO_2H.$$

They readily take up oxygen again and are converted into sulphonic acids.

934 *Phenols.*—This name is applied to compounds containing one or more hydroxyls in the nucleus. They are obtained by fusing a sulphonic acid with caustic potash thus:

$$C_6H_5.SO_3K + HOK = C_6H_5OH + SO_3K_2,$$

They are also formed from amido-compounds by the action of nitrous acid:

$$C_6H_5.NH_2 + HO.NO = C_6H_5.OH + H_2O + N_2.$$

In this case, however, the diazo-compound is formed as an intermediate product. They may be further obtained by acting on the hydrocarbons with oxygen in presence of aluminium chloride.[4]

Aromatic hydroxyacids which contain a hydroxyl group in the nucleus decompose when heated alone or with baryta into carbon dioxide and a phenol; this same decomposition takes place on heating with hydrobromic or hydriodic acids; thus salicylic acid readily yields common phenol or monohydroxybenzene:

$$C_6H_4 \begin{cases} OH \\ CO_2H \end{cases} = C_6H_5.OH + CO_2.$$

This substance is found, together with other phenols and their methyl ethers, in wood-tar and coal-tar. Thymol, $C_6H_3(OH)$

[1] Armstrong and Miller, *Journ. Chem. Soc.* 1884, **2**, 148.
[2] *Compt. Rend.* **109**, 95.
[3] Friedel and Craffts, *Ann. Chim. Phys.* VI. **14**, 442.
[4] Friedel and Craffts, *ibid.* VI. **14**, 436.

$CH_3(C_3H_7)$, is a phenol which occurs as a constituent of a large number of ethereal oils. The phenols exhibit in certain relations strong analogies with the alcohols, and indeed were formerly classed amongst them.

In other respects, however, they differ from them very distinctly—*e.g.* the hydrogen of the hydroxyl is very easily replaced by metals. In order to prepare sodium ethylate, C_2H_5ONa, alcohol must be treated with sodium, whereas sodium phenate is formed at once by action of caustic soda upon phenol. A very characteristic reaction of the phenols is their behaviour with nitric and sulphuric acids. They do not form any acid ethers (or ethereal salts) with these acids, but yield substitution products, and these are more readily formed than is the case with the hydrocarbons; thus we obtain nitrophenol, $C_6H_4(NO_2)OH$, phenolsulphonic acid, $C_6H_4(OH)SO_3H$, &c. Whilst the nitro-hydrocarbons are only attacked by chlorine and bromine with difficulty, the nitrophenols readily yield substitution products. The phenols are moreover distinguished from the alcohols by the fact that they withstand the action of acid oxidizing agents, though in alkaline solutions they absorb oxygen, and those which contain many hydroxyls readily reduce the salts of the noble metals. These and similar reactions are of a complicated character, and differ greatly from the simple oxidation of the alcohols.

Phenol Ethers are obtained heating the phenols with caustic potash and the alcoholic iodides:

$$C_6H_5OK + CH_3I = C_6H_5OCH_3 + KI.$$

By heating with the hydracids these bodies are converted into phenol and a haloid ether. They are acted upon by the elements of the chlorine group, nitric acid, sulphuric acid, and chromic acid, exactly as the hydrocarbons are; thus cresolmethyl ether, $C_6H_4\left\{\begin{array}{l} OCH_3 \\ CH_3, \end{array}\right.$ yields on oxidation methoxybenzoic acid, $C_6H_4\left\{\begin{array}{l} OCH_3 \\ CO_2H. \end{array}\right.$

The acid phenol ethers are obtained by the action of chlorides or oxides of the acid radicals on the phenols :[1]

$$C_6H_5.OH + (C_2H_3O)_2O = C_6H_5.O.C_2H_3O + (C_2H_3O)OH.$$

They are very readily saponified by alkalis.

---

[1] Baeyer, *Annalen*, **140**, 295.

The phenols when heated with zinc-dust are reduced to hydrocarbons:

$$C_6H_5OH + Zn = C_6H_6 + ZnO.$$

Phenol ethers, on the other hand, may be distilled over zinc-dust without undergoing change. Neutral ferric chloride colours aqueous solutions of the phenols violet, blue, red, or green, but not those of their ethers. Many hydroxyacids which are at the time phenols likewise exhibit this reaction.[1]

If a phenol be shaken with a solution of potassium nitrite in concentrated sulphuric acid (Liebermann's reagent) the dark coloured solution becomes cherry red on addition of water, and this changes to a light blue when an excess of caustic soda is added to the dilute solution.[2] Potassium nitrite may be replaced by any nitroso-compound, and the reaction is frequently used conversely for the detection of the latter class of compounds.

When the hydrogen in the nucleus of a phenol is substituted by a halogen or by nitroxyl, or other electro-negative radicals, its acid character becomes more marked; thus trinitrophenol or picric acid, $C_6H_2(NO_2)_3OH$, possesses all the properties of a powerful monobasic acid.

In addition to phenols containing one hydroxyl, others are known containing two and three hydroxyls; thus we are acquainted with three dihydroxybenzenes, $C_6H_4(OH)_2$, viz. catechol, resorcinol, and quinol.

Thiophenols are obtained by the action of phosphorus penta-sulphide on phenols, by heating sulphur with a mixture of an aromatic hydrocarbon and aluminium chloride, and still more readily by treating the sulphonic chlorides with zinc and dilute sulphuric acid. Like the fatty mercaptans these are un-pleasantly-smelling bodies, readily forming metallic salts or mercaptides. On oxidation they are easily converted into sulphides and disulphides; thus ordinary thiophenol, C_6H_5SH, yields $(C_6H_5)_2S$ and $(C_6H_5)_2S_2$.

935 *Sulphones.* The monosulphides are readily oxidized by chromic acid to sulphones; thus phenyl sulphide yields diphenyl sulphone $(C_6H_5)_2SO_2$. They are also formed by the action of chlorosulphonic acid or sulphur trioxide on the hydrocarbons, and by acting on a sulphonic chloride with a hydrocarbon in presence of aluminium chloride.

---

[1] H. Schiff, *Annalen*, **169**, 164.
[2] Liebermann, *Ber.* **7**, 248, 1098.

We are acquainted with a number of sulphones containing both fatty and aromatic residues, such as phenyl-sulphone-acetic acid, $C_6H_5SO_2.CH_2.COOH$. All these compounds have a great resemblance in their properties to the corresponding ketones, and it would therefore appear that the carbonyl and sulphuryl groups exert an analogous influence on the compounds in which they occur.

936 *Amido-Compounds* are formed by the action of nascent hydrogen on nitro-compounds; thus from nitrobenzene we obtain aniline or amidobenzene :

$$C_6H_5N{\Large\langle}{\overset{O}{\underset{O}{|}}} + 6H = C_6H_5.N{\Large\langle}{\overset{H}{\underset{H}{}}} + 2H_2O.$$

Aniline and its homologues, such as toluidine or amidotoluene, $C_6H_4(CH_3)NH_2$, were formerly classed as amines. They are, however, as different from these as phenols are from alcohols, and hence Griess proposed to call them amido-compounds.[1]

They are only weak bases, although they combine readily with acids to form salts which easily crystallize but exhibit an acid reaction. On the other hand benzylamine, $C_6H_5.CH_2NH_2$, an isomeride of toluidine containing the amido-group in the side chain, is a strongly caustic liquid, miscible with water, possessing a powerful alkaline reaction, and rapidly absorbing carbon dioxide from the air. Aniline and its homologues, on the contrary, are only slightly soluble in water, possess an aromatic smell, exhibit no alkaline reaction, and do not form carbonates.

These bodies resemble the amines, however, inasmuch as the hydrogen of the amido-group can be replaced by alcohol radicals by the action of the haloid salts of the latter. The action of nitrous acid on the amido-bases is characteristic. The primary amido-compounds such as aniline are thus converted into phenols, as the primary amines are into alcohols. The secondary bases when treated with nitrous acid also behave in a similar manner to the secondary fatty amines, forming nitroso-compounds; thus methylaniline $C_6H_5.NH.CH_3$ yields methylphenylnitrosamine $C_6H_5.N(NO).CH_3$. There is, however, a great difference between the action of nitrous acid on the tertiary amido-derivatives of the fatty and aromatic series, for whilst the former remain unaltered, the latter are at once attacked, the nitroso-group replacing the hydrogen in the nucleus standing in the para-

[1] *Annalen*, **121**, 258.

position to the amido-group; thus from dimethylaniline we obtain *p-nitrosodimethylaniline,* $C_6H_4(NO).N(CH_3)_2$. The tertiary bases combine readily with the iodides of the alcohol radicals to form ammonium iodides and these are decomposed by moist silver oxide, with formation of strongly alkaline and caustic hydroxides.

Secondary and tertiary bases are also formed when aniline hydrochloride is heated with methyl alcohol or other alcohols to from 280—300°. In this case the chloride of the alcohol radical is, doubtless, first produced. When wood-spirit is employed, other bases in which the hydrogen in the nucleus is also replaced by methyl are produced in addition to methylaniline and dimethylaniline. Thus Hofmann and Martius found in the material which is prepared on the large scale for the manufacture of colours, dimethyltoluidine, $C_6H_4(CH_3)N(CH_3)_2$, dimethylxylidine, $C_6H_3(CH_3)_2N(CH_3)_2$ and dimethylcumidine, $C_6H_2(CH_3)_3N(CH_3)_2$.[1] In addition to these a non-volatile diacid base, $C_{19}H_{26}N_2$, occurs together with hexmethylbenzene.[2]

When trimethylphenyl-ammonium iodide, $C_6H_5N(CH_3)_3I$, is heated to from 220—230°, two dimethyltoluidines are formed, together with a methylxylidine and dimethylxylidine; whilst at higher temperatures such as 335°, or the melting-point of lead, cumidine, $C_6H_2(CH_3)_3NH_2$,[3] is chiefly formed. At this temperature methylaniline hydrochloride passes into toluidine,[4] whilst ethylaniline, $C_6H_5NH(C_2H_5)$, is converted into amido-ethylbenzene, $C_6H_4(C_2H_5)NH_2$, and amylaniline, $C_6H_5NH(C_5H_{11})$, into amido-amylbenzene, $C_6H_4(C_5H_{11})NH_2$.[5] Xylidine hydrochloride when heated with methyl alcohol to from 250—300° is converted into cumidine.[6]

The hydrogen of the amido-group in the primary and secondary bases can be replaced by treatment with the chlorides or nitrates of the alcohol radicals. In the case of the tertiary compounds, no action of this kind can of course occur, and hence this reaction is employed for the separation of secondary and tertiary compounds which are often formed together.

The hydrogen in the nucleus of the amido-compounds can also be replaced by the halogen and nitroxyl groups, when the basic character of the original compound is weakened. This also occurs when the hydrogen of the amido-group in aniline is replaced by phenyl; thus diphenylamine, $(C_6H_5)_2NH$, yields salts

[1] *Ber.* **4**, 742. [2] *Ibid.* **6**, 345. [3] Hofmann, *ibid.* **5**, 704.
[4] *Ibid.* **5**, 720. [5] *Ibid.* **7**, 526. [6] *Ibid.* **13**, 1730

which are decomposed by water, whilst triphenylamine, $(C_6H_5)_3N$, does not combine with acids.

The dinitro-substitution products of the aromatic hydro-carbons are converted by nascent hydrogen into diacid bases, such as diamidobenzene or phenylenediamine, $C_6H_4(NH_2)_2$, which exists in three modifications. The nitrophenols are thus converted into amidophenols, which do not combine with bases to form salts, but unite with acids.

937 *Azo-Compounds* are formed under certain conditions by the reduction of nitro-substitution products; thus azobenzene, $C_{12}H_{10}N_2$, is obtained from nitrobenzene by acting upon it with caustic soda and zinc-dust, or by treating it in alcoholic solution with sodium amalgam:

$$\begin{matrix} C_6H_5.NO_2 \\ C_6H_5.NO_2 \end{matrix} + 8H = \begin{matrix} C_6H_5.N. \\ \| \\ C_6H_5.N \end{matrix} + 4H_2O.$$

The same body is also produced when aniline undergoes moderate oxidation:

$$\begin{matrix} C_6H_5.NH_2 \\ C_6H_5.NH_2 \end{matrix} + 2O = \begin{matrix} C_6H_5.N \\ \| \\ C_6H_5.N \end{matrix} + 2H_2O.$$

This compound unites with nascent hydrogen to form hydrazo-benzene, $(C_6H_5NH)_2$, which by further reduction yields aniline.

The hydrazo-compounds are colourless, whilst the azo-compounds have a yellow or red colour; but they do not possess the properties of colouring matters if they only contain hydrogen in the nucleus. If, however, this is replaced by hydroxyl, by an amido-group, &c., the *azo-colours* are formed, many of which are now made on the large scale, and concerning which more will be said in the sequel.

938 *Diazo-Compounds.*—These bodies form one of the most characteristic and important series of the aromatic compounds, and possess a great theoretical as well as practical interest. They were discovered by Griess, who obtained them by the action of nitrous acid on the amido-compounds. The action of this acid on ammonia is as follows:

$$NH_3 + NO.OH = N_2 + 2H_2O.$$

Now Piria found in 1849 that by the action of nitrous acid, asparagine, or amidosuccinamic acid, is converted into malic acid :

$$C_2H_3(NH_2)\left\{\begin{array}{l}CO.NH_2\\CO.OH\end{array}\right. + 2NO.OH = 2N_2 + C_2H_3(OH)\left\{\begin{array}{l}CO.OH\\CO.OH\end{array}\right.$$
$$+ 2H_2O.$$

Soon after this, Strecker in like manner obtained glycollic acid from amidoacetic acid, and we now know that all amido-compounds and amines are acted upon in a similar manner.

Analogous reactions had also been observed in the aromatic group. Thus Hunt in 1849 found that nitrous acid converted aniline into phenol.[1] This latter substance is, however, more easily formed if aniline hydrochloride be treated with silver nitrite :[2]

$$C_6H_5.NH_2 + NO.OH = N_2 + C_6H_5OH + H_2O.$$

In a similar way Gerland in 1853 prepared hydroxybenzoic acid, $C_6H_4(OH)CO_2H$, from amidobenzoic acid, $C_6H_4(NH_2)CO_2H$.[3]

Griess then showed in a series of classical researches, that when nitrous acid acts upon the aromatic amido-compounds no nitrogen is evolved, but that the resulting compounds contain the nitrogen both of the nitrous acid and of the amido-group, and that they very readily part with this nitrogen, frequently with explosive violence, though when treated with certain reagents they are converted into compounds, the formation of which had been previously observed.[4]

By acting with nitrous acid on an aqueous solution of aniline nitrate, Griess obtained diazobenzene nitrate, its formation being represented as follows :

$$C_6H_7N.HNO_3 + HNO_2 = C_6H_4N_2.HNO_3 + 2H_2O.$$

According to Kekulé's view, which has been generally accepted by chemists, the group N_2 has the constitution $- N = N -$. The formation of diazobenzene nitrate is thus represented in the following manner :

$$C_6H_5.NH_2.HNO_3 + HNO_2 = C_6H_5.N{=}N.NO_3 + 2H_2O.$$

Many chemists, however, upheld the view that diazo-compounds contain pentad nitrogen,[5] according to which diazobenzene nitrate

[1] *Jahresber.* **1849**, 391 ; *Annalen*, **76**, 285.　　[2] Hofmann, *ibid.* **75**, 359.

[3] *Ibid.* **86**, 143.　　[4] Kekulé, *Lehrb.* **2**, 703.

[5] Strecker, *Ber.* **4**, 786 ; Erlenmeyer, *ibid.* **7**, 1110 ; Blomstrand, *ibid.* **8**, 51.

would have the formula, $C_6H_5N\begin{smallmatrix} \diagup N \\ \diagdown NO_3 \end{smallmatrix}$. The formation of hydrazine derivatives by the reduction of the diazo-compounds, renders this view highly improbable, and is in full agreement with Kekulé's view.

By acting with sulphuric acid on the nitrate, acid diazobenzene sulphate, $C_6H_5.N_2.SO_4H$, is obtained, and this is better suited for most reactions, inasmuch as nitric acid is apt to yield nitro-products. In many cases, however, it is not necessary to use the pure salt, but the amido-compound may be dissolved in the requisite quantity of dilute sulphuric acid and the theoretical amount of sodium nitrite gradually added in aqueous solution.[1] The product is then boiled with water, when a phenol is formed:

$$C_6H_5.N_2.SO_4H + H_2O = C_6H_5OH + N_2 + SO_4H_2.$$

Hydriodic acid decomposes the salts of diazobenzene easily with formation of iodine substitution products, which are best obtained in this way:

$$C_6H_5.N_2.SO_4H + HI = C_6H_5I + N_2 + SO_4H_2.$$

Hydrobromic acid does not act so readily. If the diazo-salt be treated at the same time with hydrobromic acid and bromine, a diazobenzene perbromide is formed, which is easily decomposed on heating with alcohol:

$$C_6H_5.NBr - NBr_2 + C_2H_6O = C_6H_5Br + 2HBr + N_2 + C_2H_4O.$$

Hydrochloric acid acts with still greater difficulty on the diazo-salt, but if hydrochloric acid, platinic chloride and alcohol be added to its solution, a platinochloride is precipitated, which on heating with caustic soda decomposes as follows:

$$(C_6H_5N_2)_2PtCl_6 = 2C_6H_5Cl + N_2 + Pt + 2Cl_2$$

If the diazo-salt be boiled with absolute alcohol the corresponding hydrocarbon is produced:

$$C_6H_5.N_2.SO_4H + C_2H_6O = C_6H_6 + N_2 + SO_4H_2 + C_2H_4O.$$

[1] Meyer and Ambühl, *Ber.* **8**, 1074.

Very frequently, however, the reaction proceeds in quite a different manner, phenyl ethyl ether or phenetoïl being formed.[1]

$$C_6H_5.N_2.SO_4H + C_2H_5OH = C_2H_5.O.C_6H_5 + N_2 + H_2SO_4.$$

A better method of replacing the diazo-group by hydrogen is to pour a solution of stannous chloride in caustic soda into an ice-cold solution of a diazo-salt, to which caustic soda has likewise been added.[2]

The diazo-group can also be readily replaced by the halogens or by cyanogen by Sandmeyer's reaction, which consists in treating the diazo-compound with the corresponding cuprous salt. Thus cuprous chloride converts diazobenzene chloride into chlorobenzene.[3] By treating the diazo-salt with moist cuprous oxide and sodium nitrite, the diazo-group is replaced by NO_2.[4]

Instead of a cuprous salt, finely-divided copper and an alkali-salt containing the residue it is desired to introduce, may be employed. Thus, when a solution of potassium bromide is mixed with a solution of diazobenzene sulphate, and finely-divided copper added, bromobenzene is formed. This reaction takes place in the cold.[5] If potassium cyanate be substituted for the bromide, the cyanates of the aromatic radicals are obtained.

Instead of converting the amido-compounds into diazo-salts, it is often possible to replace the amido-group directly by hydrogen. This is accomplished by heating the base with alcohol saturated with nitrogen trioxide and therefore containing ethyl nitrite, when the diazo-compound is first formed, but afterwards decomposed by the alcohol.

939 *Diazoamido-Compounds* are formed when nitrous acid acts upon an excess of a base, such as aniline, especially in alcoholic solution, and also when a base is brought in contact with a diazo-salt in aqueous or alcoholic solution:

$$C_6H_5N\!=\!NCl + NH_2.C_6H_5 = C_6H_5N\!=\!N.NH(C_6H_5) + HCl.$$

A large number of diazoamido-compounds can be prepared from other bases and diazo-salts. These are not colourless like the diazo-salts but possess a yellow tint; they do not combine with acids but form platinochlorides. They are converted into diazo-compounds by the further action of nitrous acid; when

[1] Haller, *Ber.* **17**, 1887 ; Hofmann, *ibid.* **17**, 1917 ; Remsen and Orndorff, *Am. Chem. Journ.* **9**, 317 ; **10**, 368.   [2] Friedländer, *Ber.* **22**, 587.
[3] *Ber.* **17**, 2652.   [4] *Ibid.* **20**, 1494.   [5] Gattermann, *Ber.* **23**, 1222.

heated with water or acids they undergo a similar decompo-
sitiou to the diazo-salts, amido-bases being formed at the
same time :

$$C_6H_5N_2.NH.C_6H_5 + H_2O = C_6H_5.OH + NH_2.C_6H_5 + N_2.$$

Some of these bodies readily pass into the isomeric amidoazo-
compounds. In the case of diazoamidobenzene this change
occurs when its alcoholic solution is allowed to stand for a few
days; the reaction takes place more readily if some aniline
hydrochloride be added to the solution.[1] In this reaction the
aniline residue $C_6H_4.NH_2$ takes the place of $NH.C_6H_5$; but as
aniline hydrochloride is formed in the reaction, a small quantity
of this salt is sufficient to convert a large quantity of diazo-
amidobenzene into amidoazobenzene :

$$C_6H_5.N{=}N.NH.C_6H_5 + (C_6H_5)NH_2.ClH =$$
$$C_6H_5.N{=}N.C_6H_4.NH_2 + NH_2(C_6H_5).ClH.$$

This molecular change is only possible when the para-position
in the original substance is unoccupied. Thus the diazo-amido-
compound obtained from p-toluidine does not undergo this
change, whereas it readily occurs with the compounds obtained
from o- and m- toluidine.

By the action of diazobenzene chloride on p-toluidine and
of p-diazotoluene on aniline, one would expect the formation
of two different diazo-amido-compounds, having the formulæ,
(a) $C_6H_5.N = N.NH.C_7H_7$ and (b) $C_6H_5.NH.N = N.C_7H_7$ respectively. Experiment has, however, shown that in this and
all analogous cases, one and the same compound is obtained.
This fact may be readily explained if we assume that the
diazo-salts, at any rate in solution, have a constitution corresponding to the formula, $C_6H_5 . N . N . Cl$; the reactions in the
above cases, would be then represented as follows :

(1) $C_6H_5.N . N . Cl + H_2N.C_7H_7 = C_6H_5.N . N . N . C_7H_7 + HCl.$

(2) $C_6H_5.NH_2 + Cl.N . N . C_7H_7 = C_6H_5.N . N . N . C_7H_7 + HCl.$

[1] *Zeitschr. Chem.* **1866,** 689.

The compound thus obtained then loses water and passes into the diazo-amido compound.[1]

In order to ascertain which of the above formulæ the compound really has, Goldschmidt and Holm recommend that it should be treated with phenyl cyanate, which unites with it, forming an additive compound. This must have the formula,

$$C_6H_5NH.CO.N{\Large<}{\small\begin{matrix}N_2.C_6H_5 \\ C_7H_7\end{matrix}} \quad \text{or} \quad C_6H_5.NH.CO.N{\Large<}{\small\begin{matrix}N_2.C_7H_7 \\ C_6H_5\end{matrix}}$$

according as the formula (a) or (b) is correct. If the first formula be correct the addition compound must on treatment with hydrochloric acid yield p-tolylphenyl carbamide, whilst according to (b) it would yield diphenylcarbamide. The experiment in this instance has decided in favour of the first formula.

When phenol acts upon diazobenzene sulphate, diphenyl oxide is produced,

$$C_6H_5.N_2.SO_4H + HO.C_6H_5 = N_2 + SO_4H_2 + C_6H_5.O.C_6H_5.$$

If, however, potassium phenate be used, hydroxyazobenzene is obtained :

$$C_6H_5.N_2.NO_3 + KO.C_6H_5 = KNO_3 + C_6H_5.N_2.C_6H_4OH.$$

A series of hydroxyazo-compounds, also used as colours, has been obtained from diazo-salts and alkaline solutions of phenols or their sulphonic acids.

Azylines are compounds formed by the action of nitric oxide on tertiary amido-compounds, which are then partly oxidized :

$$2C_6H_5.(CH_3)_2N + 2NO = (CH_3)_2NC_6H_4.N{=}N.C_6H_4.N(CH_3)_2. + H_2O + O.$$

They are strong bases and form salts which crystallize well.

Diazo-amines are formed when a diazo-salt acts on primary and secondary amines.[2]

Diazobenzene-ethylamine.
$$C_6H_5.N_2.NO_3 + H_2N.C_2H_5 = C_6H_5.N_2.NH.C_2H_5. + NO_3H.$$

Diazobenzene-dimethylamine.
$$C_6H_5.N_2.NO_3 + HN(CH_3)_2 = C_6H_5.N_2.N(CH_3)_2 + NO_3H.$$

These cannot be converted into the isomeric amidoazo-compounds, but corresponding mixed nitroazo-compounds are known

[1] Goldschmidt and Holm, and Molinari, **21**, 1016 ; **21**, 2557.
[2] Baeyer and Jaeger, *Ber.* **8**, 148.

which are formed when the sodium compound of a nitro-paraffin is brought in contact with a diazo-salt.[1]

Azophenylnitroethyl.

$$C_6H_5.N_2.NO_3 + NaC_2H_4.NO_2 = C_6H_5.N_2C_2H_4.NO_2 + NaNO_3.$$

The reactions of these bodies resemble those of the other azo-compounds; they are yellow colouring matters.

940 *Hydrazines* are formed by the action of an excess of acid potassium sulphite on a diazo-salt:[2]

$$C_6H_5.N{=}N.NO_3 + 3KHSO_3 =$$
$$C_6H_5.NH{-}NHSO_3K + KHSO_4 + KNO_3 + SO_2.$$

In this reaction the potassium salt of phenylhydrazinesulphonic acid is first formed, and this converted by boiling with hydrochloric acid into phenylhydrazine hydrochloride:

$$C_6H_5.N_2H_2.SO_3K + HCl + H_2O = C_6H_5.N_2H_3.ClH + KHSO_4.$$

In place of acid potassium sulphite, hydrochloric acid and stannous chloride may frequently be employed with advantage.[3] Hydrazines are also formed when a diazoamido-compound is treated with zinc-dust and acetic acid:

$$C_6H_5.N{=}N.NH.C_6H_5 + 4H = C_6H_5NH - NH_2 + H_2N.C_6H_5.$$

Secondary hydrazines are formed in a similar way from nitroso-bases; thus nitrosodiphenylamine yields diphenylhydrazine:

$$(C_6H_5)_2N - NO + 4H = (C_6H_5)_2N - NH_2 + H_2O.$$

Asymmetric ethylphenylhydrazine, $C_6H_5N(C_2H_5)-NH_2$, is formed from nitrosoethylaniline. The same compound is also formed together with symmetric ethylphenylhydrazine, $(C_6H_5)NH-NH(C_2H_5)$, by the action of ethylbromide on phenyl-hydrazine. This latter compound is oxidized by mercuric oxide to azophenylethyl, $C_6H_5N = NC_2H_5$, and this compound unites with hydrogen to form the original compound, which may there-fore be called hydrazophenylethyl.[4] The asymmetric compound

[1] Meyer and Ambühl, *Ber.* **8**, 751, 1073.
[2] E. Fischer, *Annalen*, **190**, 67.
[3] V. Meyer and Lecco, *Ber.* **16**, 2976.
[4] Fischer and Ehrhard, *Annalen*, **199**, 325.

yields on oxidation with mercuric oxide, diethyldiphenyltetrazone, a body having the following composition:

$$\begin{array}{c} \text{N}-\text{N}\!\!<\!\!\begin{array}{l} \text{C}_6\text{H}_5 \\ \text{C}_2\text{H}_5 \\ \text{C}_6\text{H}_5 \\ \text{C}_2\text{H}_5. \end{array} \\ \text{N}-\text{N}\!\!< \end{array}$$

The aromatic hydrazines resemble in their general properties the corresponding fatty derivatives, being stable towards reducing agents, but readily undergoing oxidation. They are, however, only *monacid* bases.

The mono-substituted, and asymmetric di-substituted hydrazines unite with aldehydes and ketones ·with elimination of water, forming characteristic, mostly crystalline compounds, termed *hydrazones*. Thus acetone reacts with phenylhydrazine in the following manner:

$$(CH_3)_2CO + H_2N.NHC_6H_5 = CH_3.C\!=\!\!N.NH.C_6H_5 + H_2O.$$

Diacetyl unites with two molecules of phenylhydrazine, forming a compound of the constitution

$$\begin{array}{c} CH_3\,C = N.NHC_6H_5 \\ | \\ CH_3.C = N.NHC_6H_5. \end{array}$$

To all such compounds containing two hydrazone groups combined with adjacent carbon atoms, the name *osazones* has been given. Phenylhydrazine also combines with substances containing the group — CH.OH — CO — forming osazones, and these are therefore yielded by the sugars (cf. Vol. III., Part II., p. 584).

The hydrazones and osazones have proved of the greatest possible value in separating and identifying ketones and aldehydes, and have made possible the separation of the sugars recently·obtained by synthesis (Part II., p. 595).

941 *Aromatic Alcohols and Acids.*—Aromatic alcohols are formed by replacement of hydrogen in the side chain of the hydrocarbon by hydroxyl; their mode of preparation corresponds exactly to that employed in the preparation of the alcohols of the fatty group, and, as in this latter case, primary, secondary and tertiary aromatic alcohols are known as well as those containing divalent and polyvalent radicals.

Phenylethyl alcohol.　　　　　Phenyldimethyl carbinol.
$C_6H_5.CH_2.CH_2.OH.$　　　　$C_6H_5.C(OH)(CH_3)_2.$

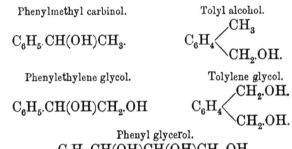

Phenylmethyl carbinol.

Tolyl alcohol.

$$C_6H_5.CH(OH)CH_3.$$

$$C_6H_4\Big\langle{}^{CH_3}_{CH_2.OH.}$$

Phenylethylene glycol.

Tolylene glycol.

$$C_6H_5.CH(OH)CH_2.OH$$

$$C_6H_4\Big\langle{}^{CH_2.OH.}_{CH_2.OH.}$$

Phenyl glycerol.

$$C_6H_5.CH(OH)CH(OH)CH_2.OH.$$

In these alcohols the hydrogen of the nucleus can be replaced by other elements or radicals, and a series of compounds thus obtained, which, on the one hand, act as fatty bodies, and on the other, like aromatic substances. Thus, some are known which are at once phenols and alcohols; to this class belongs o-hydroxy-benzyl alcohol, $C_6H_4\Big\langle{}^{OH}_{CH_2OH.}$

Towards oxidizing agents the aromatic alcohols act in the same manner as the fatty compounds to which they correspond.

942 *Aromatic Acids.*—These primary alcohols yield aldehydes and then acids, and these latter can be obtained according to other reactions characteristic of the bodies of the aromatic-group, viz. :—

1st. They may be obtained from the hydrocarbons not only by oxidation of the side chain, but also synthetically by combination with carbon dioxide in presence of aluminium chloride.[1] Thus from benzene we obtain benzoic acid :

$$C_6H_6 + CO_2 = C_6H_5.CO_2H.$$

and if carbonyl chloride be employed in place of carbon dioxide benzoyl chloride is formed :

$$C_6H_6 + COCl_2 = C_6H_5.COCl + HCl.$$

This cannot however be obtained in large quantity, as it acts further on the benzene with formation of diphenylketone,[2] $CO(C_6H_5)_2$.

2nd. Acids are obtained from monosubstitution products of the hydrocarbons by the simultaneous action of sodium and carbon dioxide :[3]

$$C_6H_5Br + CO_2 + 2Na = C_6H_5.CO_2Na + NaBr.$$

[1] Friedel and Crafts, *Ann. Chim. Phys.* VI. **14**, 441.
[2] Friedel, Crafts, and Ador, *Ber.* **10**, 1854.   [3] Kekulé, *Annalen*, **137**, 178.

3rd. In place of carbon dioxide and sodium, ethyl chloroformate and sodium amalgam may be used, when the ethyl salt is formed :[1]

$$C_6H_5Br + ClCO_2C_2H_5 + 2Na = C_6H_5.CO_2C_2H_5 + NaBr + NaCl.$$

4th. Monobasic acids are obtained from the sulphonic acids of hydrocarbons by heating them with sodium formate :

$$C_6H_5SO_3Na + HCO_2Na = C_6H_5CO_2Na + HNaSO_3.$$

If sulphobenzoic acid be employed, dibasic isophthalic acid is formed :[2]

$$C_6H_4\begin{matrix} CO_2Na \\ \\ SO_3Na \end{matrix} + HCO_2Na = C_6H_4\begin{matrix} CO_2Na \\ \\ CO_2Na \end{matrix} + HNaSO_3.$$

Isophthalic acid is also prepared by heating sodium bromobenzoate with sodium formate :[3]

$$C_6H_4Br.CO_2Na + 2CHO_2Na = C_6H_4(CO_2Na)_2 + NaBr + CH_2O_2.$$

5th. When an amido-compound is heated with oxalic acid a formamide is obtained; thus aniline yields phenylformamide or formanilide, which is converted on heating with hydrochloric acid into benzonitrile,[4] the carbamine which is at first formed undergoing an intramolecular change :

$$C_6H_5N(COH)H = C_6H_5.CN + H_2O.$$

6th. Amido-compounds can also be converted by Weith's method into acids, by preparing the thiocarbimide or mustard oil, and heating this with finely-divided copper :[5]

$$C_6H_5NCS + 2Cu = C_6H_5.CN + Cu_2S.$$

Aromatic nitriles are also formed when a sulphonic acid salt is heated with potassium cyanide,[6] as well as by the reactions employed for the preparation of the nitriles of the fatty series. Like this latter class of bodies, they are converted into the corresponding acids by heating with alkalis or strong acids.

[1] Wurtz, *Annalen*, Suppl. **7**, 125.   [2] V. Meyer, *Ber.* **3**, 112.
[3] Ador and Meyer, *ibid.* **4**, 259.   [4] Hofmann, *Annalen*, **142**, 121.
[5] *Ber.* **6**, 210.   [6] Merz, *Zeitschr. Chem.* **1868**, 33.

7th. Aromatic nitriles are much more readily obtained from the amido-compounds by Sandmeyer's method (p. 29). Thus aniline, after diazotizing, yields large quantities of benzonitril on treatment with a mixture of potassium cyanide and copper sulphate solutions, which has been warmed till the precipitate first formed has been redissolved.[1] A solution of potassium cyanide and finely-divided copper acts in a similar manner.[2]

8th. The amides of the aromatic acids are readily obtained by acting on aromatic hydrocarbons with chloroformamide, $NH_2.CO.Cl$, in presence of aluminium chloride.[3] The latter may be replaced by a mixture of the two substances into which it dissociates on heating, viz., hydrochloric and cyanic acids.[4]

Aromatic acids can also be obtained by means of the aceto-acetic and malonic syntheses.

Aromatic acids containing unsaturated side-chains, and resembling, therefore, in constitution the members of the acrylic series, are also known. They are obtained by heating an aromatic aldehyde with the sodium salt of a fatty acid and a dehydrating agent such as acetic anhydride:

Cinnamic Acid.

$$C_6H_5.CHO + CH_3.COOH = C_6H_5.CH{=}CH.COOH + H_2O.$$

This reaction is generally known, from the name of its discoverer, as Perkin's reaction.

It takes place in reality in two stages, in the first of which the benzaldehyde and sodium acetate combine to form *sodium phenyl β-lactate*, $C_6H_5.CH(OH).CH_2.CO_2Na$, a condensation similar to that of acetaldehyde to aldol. (Vol. III., Part II., p. 188.) On further heating this loses water forming sodium cinnamate.

In place of the salts of the fatty acids, those of malonic acid or of its homologues of analogous constitution may be employed, as these decompose on heating into monocarboxylic acids and carbon dioxide.

943 *Aromatic hydroxyacids* which are at the same time phenols, can be prepared from the latter in various ways.

1st. Kolbe and Lautemann showed that when carbon dioxide and sodium act simultaneously upon phenol, salicylic acid or orthohydroxybenzoic acid is formed.[5] Kolbe then proved that

[1] Sandmeyer, *Ber.* **17**, 2653.   [2] Gattermann, *ibid.* **23**, 1218.
[3] Gattermann, *Annalen*, **244**, 29.   [4] Gattermann, *Ber.* **23**, 1190.
[5] *Annalen*, **115**, 201.

this acid may be easily prepared by heating sodium phenate in a current of carbon dioxide to 180°:[1]

$$2C_6H_5.ONa + CO_2 + C_6H_4{<}^{ONa}_{CO_2.Na} + C_6H_5.OH.$$

2nd. Its *ethyl* salt is formed by the action of sodium on a mixture of phenol and ethyl chloroformate:[2]

$$C_6H_5ONa + ClCO_2.C_2H_5 = C_6H_4{<}^{OH}_{CO_2.C_2H_5} + NaCl.$$

3rd. The aldehydes of these hydroxyacids are formed when a phenol is heated with caustic soda and chloroform:[3]

$$C_6H_5.ONa + 3NaOH + CHCl_3 = C_6H_4{<}^{ONa}_{COH} + 3NaCl + 2H_2O.$$

4th. If tetrachloromethane be employed instead of chloroform the acid itself is obtained:[4]

$$C_6H_5.ONa + 5NaOH + CCl_4 = C_6H_4{<}^{ONa}_{CO_2Na} + 4NaCl + 3H_2O.$$

5th. When a hydroxyacid is heated with caustic soda and chloroform an aldehydo-acid is produced, and this is converted on oxidation into a dibasic hydroxyacid.[5]

$$C_6H_4{<}^{ONa}_{CO_2Na} + 3NaOH + CHCl_3 = C_6H_3{<}^{COH}_{ONa}{-}_{CO_2Na} + 3NaCl + 2H_2O.$$

6th. Hydroxyacids are also formed when a sulphonic acid, such as sulphobenzoic acid, is fused with caustic potash.

7th. Certain dihydroxybenzenes are converted into dihydroxy-acids when heated with ammonium carbonate and water to 110°. Thus resorcinol gives rise to a mixture of isomeric dihydroxybenzoic acids:[6]

$$C_6H_4{<}^{OH}_{OH} + CO_2 = C_6H_3{<}^{OH}_{OH}{-}_{CO_2H}.$$

[1] *J. Pr. Chem.* II. **10**, 93.   [2] Wilm and Wischin, *Zeitschr. Chem.* **1868**, 6.
[3] Reimer and Tiemann, *Ber.* **9**, 824.
[4] *Ibid.* 1285.   [5] *Ibid.* 1268.
[6] Senhofer and Brunner, *ibid.* **13**, 930.

In addition to these, hydroxyacids are known containing the hydroxyl group in the side-chain, such as phenylglycollic acid, $C_6H_5 - CH.OH - COOH$. These resemble the fatty hydroxyacids in their properties.

ISOMERISM IN THE AROMATIC GROUP.

944 It has been already stated that experiment has shown that no isomeric mono-substitution products of benzene exist. It was indeed at one time thought that such bodies had been discovered, but careful investigation proved that in these cases the observed differences were due to impurities.

Di-substitution products, on the other hand, can theoretically occur in three isomeric forms, and this conclusion has been completely verified by experiment. In this case also, some chemists assumed the existence of more than three modifications, but here, as with the monosubstitution products, it was found that the supposed isomerides were either impure or mixtures of the isomeric compounds, or that the observed differences were due to physical isomerism, such as is frequently observed in both the fatty and inorganic compounds, and known as dimorphism and trimorphism.[1]

When Kekulé proposed his theory, many substitution products of benzene were known, and amongst them several di-substitution products, such as certain derivatives of benzoic acid, which could be arranged in three classes according to their genetic relations. Kekulé gives the following arrangement, observing that in the condition of our knowledge at that time it was scarcely possible to determine with any degree of certainty the positions which the elements or radicals replacing the hydrogen in benzene occupy.[2]

	(1)	(2)	(3)
$C_6H_4\begin{cases}Cl\\CO_2H\end{cases}$ —	Chlorodracylic acid.	Chlorosalylic acid.	Chlorobenzoic-acid.
$C_6H_4\begin{cases}OH\\CO_2H\end{cases}$ —	Para-oxybenzoic acid.	Salicylic acid.	Oxybenzoic acid.
$C_6H_4\begin{cases}NH_2\\CO_2H\end{cases}$ —	Amidodracylic acid.	Anthranilic acid.	Amidobenzoic acid.

The members of the last series are direct derivatives of benzoic acid; by the action of nitric acid, nitrobenzoic acid,

[1] The history of the di-substitution products is fully given in Richard Meyer's work entitled *Einleitung in das Studium der aromatischen Substanzen.*

[2] *Lehrbuch,* **2,** 517 (1866).

$C_6H_4(NO_2)CO_2H$, is obtained, and this on reduction is converted into amidobenzoic acid: this again is converted into oxybenzoic acid by the action of nitrous acid, while chlorobenzoic acid is obtained from it by the diazo-reaction. It has already been stated that anthranilic acid is obtained by heating indigo with caustic potash; by the action of nitrous acid this is converted into the earlier known salicylic acid, and the latter substance when acted upon with phosphorus pentachloride and water yields chlorosalylic acid. The members of the first series are obtained from toluene, $C_6H_5CH_3$, by conversion first by nitric acid into nitrotoluene, $C_6H_4(NO_2)CH_3$, and then by oxidation into nitrodracylic acid, from which chlorodracylic acid is obtained in the same way as chlorobenzoic acid from nitrobenzoic acid. Para-oxybenzoic acid was previously known, and thus termed because it was found to be isomeric with oxybenzoic acid.

Bromotoluene can be obtained directly from toluene like nitrotoluene, and hence Kekulé assumed that both bodies have an analogous constitution. Bromotoluene when acted upon by sodium and methyl iodide, yields a dimethylbenzene (paraxylene), and when treated with sodium and carbon dioxide forms toluic acid, $C_6H_4(CH_3)CO_2H$, both derivatives being converted by oxidation into terephthalic acid, $C_6H_4(CO_2H)_2$. For this reason Kekulé supposed that both these bodies belong to the first series. Later investigation has, however, shown that an analogous mode of formation does not always give rise to similarly constituted compounds, but in the above cases Kekulé's view was found to be correct.

Cases of isomerism, similar to that of the substituted benzoic acids, were then observed in other compounds: thus, for example, two series of substituted anilines were known. To those which were obtained directly from aniline no special designation was given, whilst the others were termed para-compounds, although there was no intention of connecting this series in any way with the series of para-oxybenzoic acid. Various other compounds were obtained from these substituted anilines, such as halogen substitution products of benzene, substituted phenols, &c.

When Körner discovered a third iodophenol, he termed it meta-iodophenol and suggested for the first series the name of ortho-compounds. The same method of designation was then applied to the derivatives of benzoic acid obtained by direct substitution, and, as a para-series already existed, bodies connected with salicylic acid were termed meta-compounds. This led to much confusion, as it was not unreasonably supposed that a similar mode of designation indicated an analogous constitution in the compounds of both groups.

The following table, taken with some abbreviation from Kekulé's *Lehrbuch*,[1] shows the state of knowledge at the time that work was published (1866).

	Ortho-series.	Para-series.	Meta-series.
$C_6H_4Cl_2$	Dichlorobenzene.	—	—
$C_6H_4Br_2$	Dibromobenzene.		—
$C_6H_4I_2$	Di-iodobenzene.	-	—
$C_6H_4(NO_2)_2$	—	Dinitrobenzene.	—
$C_6H_4(NO_2)Cl$	Nitrochlorobenzene	Paranitrochlorobenzene.	—
$C_6H_4(NO_2)Br$	Nitrobromobenzene	Paranitrobromobenzene.	—
$C_6H_4(NO_2)I$	Nitro-iodobenzene.	Paranitro-iodobenzene.	—
$C_6H_4(NH_2)Cl$	Chloraniline.	Parachloraniline.	—
$C_6H_4(NH_2)Br$	Bromaniline.	Parabromaniline.	—
$C_6H_4(NH_2)I$	Iodaniline.	Paraiodaniline.	—
$C_6H_4(NH_2)NO_2$	Nitraniline.	Paranitraniline.	—
$C_6H_4(NH_2)_2$	Phenylenediamine.	Paraphenylenediamine.	—
$C_6H_4(OH)I$	Iodophenol. ·	Para-iodophenol.	Meta-iodophenol
$C_6H_4(OH)_2$	Hydroquinone.	Resorcin.	Pyrocatechin.

It has already been stated that the above ortho-compounds are those which are obtained from aniline by direct substitution, or by means of the diazo-reaction.

The starting-point for the para-compounds was dinitrobenzene, which is easily converted by partial reduction into paranitraniline. This yields the nitro-halogen compounds by the diazo-reaction, and from these again the substituted anilines and phenols can be obtained. The phenylenediamines or diamidobenzenes are obtained by further reduction of the nitranilines. The ortho-compound of the iodophenols, in addition to its production from iodaniline, was prepared by Körner by the action of iodine and iodic acid on phenol, a small quantity of meta-iodophenol being produced at the same time. The earlier known dihydroxy-benzenes were obtained by Körner by fusing the iodophenols with caustic potash.

Hydroquinone on oxidation forms quinone, $C_6H_4O_2$, and this may also be obtained by the oxidation of phenylenediamine, a further proof that hydroquinone belongs to the first series.

945 As soon as the fact was recognised that the di-substitution products exist in three isomeric forms, and that they may be arranged in series whose members can be converted by simple reactions into the others, it became necessary to investigate the orientation or the relative position of the substituted radicals or elements. This indeed Kekulé attempted to effect, and he was of opinion that when two similar elements or radicals replace the hydrogen in benzene, they take up positions in the molecule removed as far as possible from one another, inasmuch as all the atoms lying within the sphere of action of the first bromine atom would have their affinity for bromine weakened.

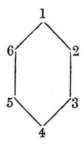

Let us suppose that in monobromobenzene, C_6H_5Br, the bromine takes up position 1, then in dibromobenzene the second atom of bromine will occupy position 4.[1]

Baeyer opposed this view, pointing out that when ethyl chloride is chlorinated, ethidine dichloride, $CH_3.CHCl_2$, and not ethylene dichloride, $CH_2Cl.CH_2Cl$, is formed, so that the second chlorine atom instead of being repelled by the first, is rather attracted to it.[2] In the above case, however, both views are correct; for when bromobenzene is brominated, 1, 4, dibromobenzene is the chief product, but a certain quantity of 1, 2, dibromobenzene is also formed. This was, however, not known at the time, and these considerations did not exert any influence on the problem. On the other hand, the views held by Baeyer respecting the constitution of mesitylene, C_9H_{12}, exerted considerable influence.[3]

[1] *Lehrbuch*, **2**, 553.   [2] *Annalen* Suppl. **5**, 84.
[3] *Ibid.* **140**, 306.

This hydrocarbon is formed when acetone is heated with sulphuric acid :

$$3C_3H_6O = C_9H_{12} + 3H_2O.$$

Fittig had previously found that mesitylene yields on oxidation trimesitic acid, $C_6H_3(CO_2H)_3$, and that this when heated with lime yields benzene and carbon dioxide. Hence it is proved that mesitylene is trimethylbenzene, $C_6H_3(CH_3)_3$. The formation of this substance was explained by Baeyer by supposing that each molecule of acetone loses the elements of water, forming the residue—$CH\!=\!\!CH—CH_3$, three of which then combine forming symmetric trimethylbenzene, the methyl groups occupying the positions 1, 3, 5.

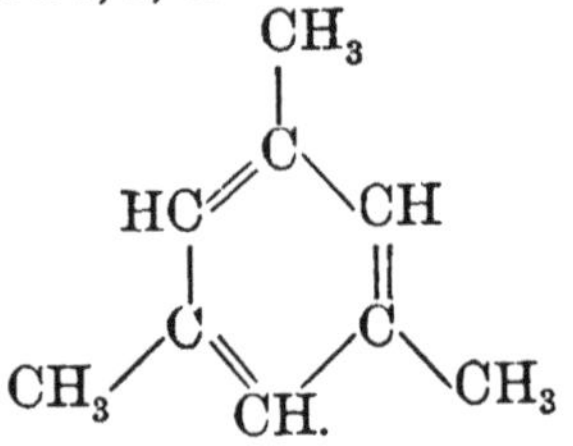

This view, though probable, was not proved to be correct, any more than another idea which was at one time entertained, namely, that trichlorobenzene derived from benzene hexchloride $C_6H_6Cl_6$, by removal of three molecules of hydrogen chloride, contains the chlorine atoms symmetrically arranged. Ladenburg, however, afterwards proved the truth of Baeyer's hypothesis, and it now stands as one of the most important methods of effecting orientation (see p. 51).

Mesitylene yields on moderate oxidation a monobasic and a dibasic acid.

Mesitylenic Acid.

$$C_6H_3 \begin{cases} (CH_3)_2 \\ CO_2H. \end{cases}$$

Uvitic Acid.

$$C_6H_3 \begin{cases} CH_3 \\ (CO_2H)_2. \end{cases}$$

The former of these when distilled with lime yields a dimethylbenzene which is identical with the metaxylene contained in coal-tar oils, and yields on oxidation the dibasic isophthalic acid $C_6H_4(CO_2H)_2$, isomeric with phthalic and terephthalic acids.

Assuming the above constitution of mesitylene, it follows that the side chains of metaxylene and of isophthalic acid occupy alternate positions :

$$1 : 3 = 1 : 5 = 3 : 5.$$

946 Phthalic acid differs from terephthalic acid, and, as was afterwards found, also from isophthalic acid, inasmuch as when heated, it is readily resolved into water and an anhydride :

$$C_6H_4\begin{matrix}CO.OH\\CO.OH\end{matrix} = C_6H_4\begin{matrix}CO\\CO\end{matrix}O + H_2O.$$

From this reaction Gräbe concluded that the carboxyls are adjacent or occupy the positions 1, 2,[1] and this conclusion he corroborated in his valuable research on naphthalene, $C_{10}H_8$.

Erlenmeyer had previously expressed the opinion that this latter hydrocarbon contains two benzene rings, having two carbon atoms in common. These two rings we may designate as a and b :

$$\begin{matrix} & CH & & CH & \\ HC & & C & & CH \\ & a & & b & \\ HC & & C & & CH \\ & CH & & CH & \end{matrix}$$

Now from Gräbe's investigations it appears very probable that this formula does represent the constitution of naphthalene ; and as this hydrocarbon yields phthalic acid on oxidation, the two carboxyls must occupy the positions 1 and 2.[3] Gräbe's proof will be given hereafter under naphthalene. Nölting and Reverdin[4] have, however, since given a much simpler proof, as follows: nitronaphthalene, $C_{10}H_7NO_2$, yields on oxidation nitrophthalic acid, $C_6H_3(NO_2)(CO_2H)_2$. If we now assume that the nitroxyl is contained in the nucleus a, it follows that the nucleus b has been destroyed. By reduction nitronaphthalene is converted into amidonaphthalene, $C_{10}H_7NH_2$, and this on oxidation does not yield amidophthalic acid, or an oxidation product of it, but phthalic acid. Hence it follows that in this latter case, the nucleus a has been oxidized to carboxyl. Now as we are acquainted with the constitution of phthalic and isophthalic acids, it is clear that that of terephthalic acid is also known, as well as that of the bromotoluene which can be converted into

[1] Gräbe and Born, *Annalen*, **142**, 330.
[2] *Ibid.* **137**, 346.   [3] *Ibid.* **149**, 1.
[4] *Constitution of Naphthalene and its Derivatives*, 1880.

toluic acid, and of the dimethylbenzene which on oxidation yields terephthalic acid.

947 Hydroquinone, $C_6H_4(OH)_2$, is distinguished from its isomerides inasmuch as it readily passes by oxidation into quinone, $C_6H_4O_2$, and this can again be easily reduced to hydroquinone. From these facts Gräbe concluded that in the formation of quinone the two oxygen atoms combined together:

$$C_6H_4 \Big\langle {}^{OH}_{OH} + O = C_6H_4 \Big\langle {}^{O}_{O} \Big\rangle + H_2O.$$

This peculiar reaction of hydroquinone was explained by Gräbe, by the supposition that in this body the hydroxyls occupy the adjacent positions of 1 and 2. But as hydroquinone is formed by fusing ortho-iodophenol with potash, it was concluded that the substitution in ortho-compounds takes place in the adjacent position.

At the end of his memoir on naphthalene Gräbe gives the following table. The expressions, ortho, meta, para, are here employed in the sense we now use them.

	Ortho-series. 1 : 2.	Meta-series. 1 : 3.	Para-series. 1 : 4.
$C_6H_4 \{ {}^{OH}_{OH}$	Hydroquinone.	Pyrocatechin.	Resorcin.
$C_6H_4 \{ {}^{OH}_{CO_2H}$	Oxybenzoic acid.	Salicylic acid.	Para-oxybenzoic acid.
$C_6H_4 \{ {}^{CO_2H}_{CO_2H}$	Phthalic acid.	Isophthalic acid.	Terephthalic acid.

It has already been stated that Kekulé assumed that para-oxybenzoic and terephthalic acids belong to the same series; the former of these was obtained from nitrotoluene, which on reduction yields a toluidine, from which Hofmann obtained the same toluic acid which is formed by Kekulé's reaction from bromotoluene, and which on oxidation yields terephthalic acid, thus corroborating Kekulé's supposition. The connection between bromotoluene and nitrotoluene is further shown, inasmuch as the former is converted by oxidation into a bromobenzoic acid, which is identical with that obtained by the diazo-reaction from amidodracylic acid; hence all these compounds belong to the para-series.

A connection between resorcin and terephthalic acid was not at that time proved, but Irelan [1] and Garrick [2] soon after showed that bromobenzenesulphonic acid and benzenedisulphonic acid yield resorcin when fused with caustic potash, whilst when they are heated with potassium cyanide and the products decomposed by caustic potash, terephthalic acid is obtained. But if hydroquinone is an ortho-compound and resorcin a para-compound, then pyrocatechin must belong to the meta-series.

The insertion of oxybenzoic and salicylic acids was, on the other hand, purely arbitrary.

948 The relations existing between the di-substitution products of benzene and the mono-substitution products of benzoic acid were at that time only determined in so far as that both were placed in the para-series. The assumption that the same relative position of the substituents held good for the ortho-and meta-compounds of the two groups was in no way proved, as was pointed out by Victor Meyer, who then drew attention to the fact that in the meta-derivatives of benzoic acid the adjacent position of the side chains must be assumed with the same degree of probability as in the case of phthalic acid, inasmuch as, of the three isomeric oxybenzoic acids, only salicylic acid yields an anhydride on heating.

$$C_6H_4\Big\langle{}^{OH}_{CO.OH} = C_6H_4\Big\langle{}^{O}_{CO}\Big\rangle + H_2O.$$

It therefore appeared to Meyer more probable that oxybenzoic acid belongs to the series 1, 3, and of this he discovered the proof; for Barth had shown that sulphobenzoic acid when fused with caustic potash only yields oxybenzoic acid,[3] whilst on heating it with sodium formate, Meyer obtained isophthalic acid,[4] which is also formed when ordinary bromobenzoic acid is employed in place of the sulphonic acid.[5]

He also proved that the crystallized dibromobenzene belongs to the para-series, inasmuch as when acted upon by sodium and methyl iodide it is converted into paradimethylbenzene, which yields terephthalic acid when oxidized.

Of the three known nitrotoluenes the solid one yields on oxidation paranitrobenzoic acid (nitrodracylic acid). The second, liquid one, yields common or metanitrobenzoic acid, and the

[1] Zeitschr. Chem. 1869, 164.   [2] Ibid. 549.
[3] Annalen, 148, 30.   [4] Ibid. 156, 268.
[5] Ader and Meyer, ibid. 159, 1.

third, also a liquid body, is decomposed completely, a property characteristic of the ortho-compounds, which are either scarcely attacked by oxidizing agents, or else completely burned. Hence there can no longer be any doubt respecting the classification of the nitrobenzenes and of the toluidines which are formed from them.

The three dimethylbenzenes or xylenes were also known at the time referred to. It has been already stated that one yields on oxidation terephthalic acid, whilst the second is converted into isophthalic acid, the third, or orthoxylene, being entirely decomposed. Finally, by the dry distillation of two diamido-benzoic acids, Griess obtained the third diamidobenzene, which up to that time had not been prepared, and this was naturally considered to be a meta-compound. The following table exhibits the views respecting the constitution of the di-substitution products which were held about the year 1870; only some of the compounds contained in the table on p. 35 being mentioned :—

	Ortho-series. 1 : 2	Meta-series. 1 : 3	Para-series. 1 : 4
$C_6H_4Br_2$	—	—	Dibromobenzene, M.P. 89°.
$C_6H_4(NO_2)_2$	—	—	Dinitrobenzene.
$C_6H_4(NO_2)Br$	Bromonitrobenzene M.P. 126°.	Bromonitrobenzene, M.P. 41°·5.	Bromonitrobenzene, M.P. 56°.
$C_6H_4(NH_2)Br$	Bromaniline, from aniline.	Metabromaniline, from bromonitro-benzene.	Parabromaniline, from dinitroben-zene.
$C_6H_4(NH_2)_2$	Phenylenediamine, M.P. 147°.	Phenylenediamine, M.P. 99°.	Paraphenylene-diamine, M.P. 63°.
$C_6H_4I(OH)$	Ortho-iodophenol, from phenol.	Meta-iodophenol, from phenol.	Para-iodophenol, from dinitrobenzene.
$C_6H_4(OH)_2$	Hydroquinone.	Pyrocatechin.	Resorcin.
$C_6H_4Br(CH_3)$	—	Metabromotoluene.	Parabromotoluene.
$C_6H_4(NO_2)CH_3$	Nitrotoluene, from toluene (liquid).	Metanitrotoluene.	Paranitrotoluene, from toluene (solid).
$C_6H_4(NH_2)CH_3$	Orthotoluidine (liquid).	Metatoluidine (liquid).	Paratoluidine (solid).
$C_6H_4Cl.CO_2H$	Chlorosalicylic acid.	Chlorobenzoic acid.	Chlorodracylic acid.
$C_6H_4BrCO_2H$	—	Bromobenzoic acid.	Bromodracylic acid.
$C_6H_4(NH_2)CO_2H$	Anthranilic acid.	Amidobenzoic acid.	Amidodracylic acid.
$C_6H_4(OH)CO_2H$	Salicylic acid.	Oxybenzoic acid.	Para-oxybenzoic acid.
$C_6H_4(CH_3)_2$	Orthoxylene.	Metaxylene.	Paraxylene.
$C_6H_4(CO_2H)_2$	Phthalic acid.	Isophthalic acid.	Terephthalic acid.

949 The arrangement of the compounds mentioned in this table was chiefly founded on experimental evidence, and those compounds for which this was not the case were classified from analogy. But, notwithstanding this, many emendations were soon found necessary. Thus the para-derivatives of benzene, with the exception of paradibromobenzene and resorcin, had not been brought into close connection with the para-derivatives of toluene or of benzoic acid. Petersen then showed that the two nitrotoluenes, of which the one yields on oxidation orthonitrobenzoic acid and the other paranitrobenzoic acid, yield on further nitration the same dinitrotoluene, and for this reason the nitroxyls must occupy the meta-position:

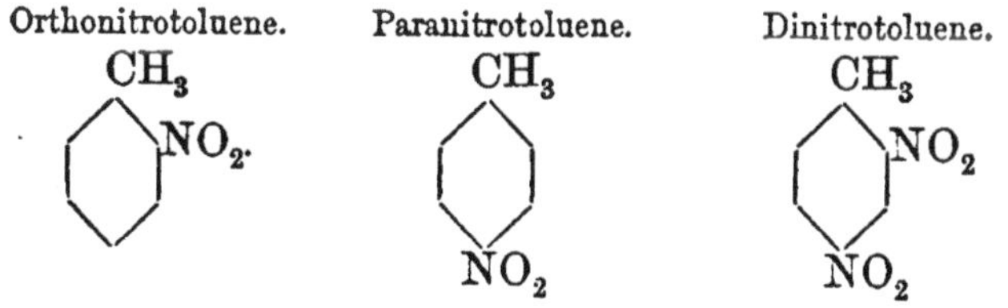

Orthonitrotoluene.   Paranitrotoluene.   Dinitrotoluene.

Now, dinitrobenzene so closely resembles dinitrotoluene that an analagous position of the nitroxyls may well be assumed in both compounds. If this be correct, dinitrobenzene and the compounds connected with it must also belong to the meta-series. The same chemist then proved that the phenylene-diamine melting at 99°, is not a meta-compound but belongs to the ortho-series.

Anisic acid, or paramethoxybenzoic acid, $C_6H_4(OCH_3)CO_2H$, yields nitranisic acid, in which the nitroxyl can only occupy the position 2 or 3 = 6 or 5.

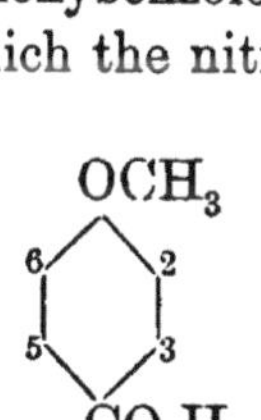

By heating nitranisic acid with ammonia Salkowski obtained a nitro-amidobenzoic acid:

$$C_6H_3(NO_2)(OCH_3)CO_2H + NH_3 =$$
$$C_6H_3(NO_2)(NH_2)CO_2H + CH_3OH.$$

By means of the diazo-reaction the amido-group can be replaced by hydrogen; in this way metanitrobenzoic acid is obtained;

and from this it follows that the above-mentioned nitro-amido-benzoic acid has the following constitution:

$$
\begin{array}{c}
NH_2 \\
\bigcirc NO_2 \\
CO_2H.
\end{array}
$$

By reducing this, diamidobenzoic acid, $C_6H_3(NH_2)_2CO_2H$, is formed, and if carbon dioxide be then eliminated, orthodiamido-benzene or phenylenediamine is obtained, a body melting at 99°. Mcyer and Wurster, as well as Salkowski, had shortly before come to the conclusion that this belongs to the series 1 : 2. The same substance is also obtained from the bromonitrobenzene, $C_6H_4(NO_2)Br$, melting at 41·5°, by heating it with ammonia, when it is converted into nitraniline, and reducing this. The same nitrobromobenzene when heated with caustic potash yields a nitrophenol melting at 45°; this may be converted into amido-phenol and afterwards into chlorophenol, which latter on fusion with caustic potash yields pyrocatechin, $C_6H_4(OH)_2$, which therefore is *ortho-* and not *meta-*dioxybenzene.

950 The nitroxyl in the bromonitrobenzene melting at 126°, can be replaced by bromine, paradibromobenzene being formed; hence bromonitrobenzene is a para-compound. When heated with caustic potash it gives a nitrophenol which melts at 115°, yielding on reduction an amidophenol, which, like phenylene-diamine, yields quinone on oxidation. This same paranitro-benzene can also be obtained from the nitraniline, which melts at 146°, and which, according to Hofmann, also yields a large quantity of quinone. Hence, then, Petersen concluded that all these compounds, and therefore hydroquinone, belong to the series 1 : 4.

From this and former statements it follows. that resorcin ought to be classed in the 1 : 3 series, for it can be obtained from dinitrobenzene, which, according to Petersen, is a meta-compound. Now, resorcin is also formed by fusing benzene-disulphonic acid with caustic potash; but this acid yields on heating with potassium cyanide the nitrile of terephthalic acid, which latter is undoubtedly a para-compound. Petersen suggested that, on fusing the sulphonic acid with caustic potash intra-molecular changes may take place; and this idea is borne out in many cases by experiment, especially where the reaction requires a high temperature.

Wurster next prepared the diamidobenzene melting at 63° by oxidizing the above-named dinitrotoluene to form a dinitrobenzoic acid, which on reduction and elimination of carbondioxide yielded this diamidobenzene, so that there can be no doubt that it is a meta-compound.

Moreover, Wurster and Ambühl showed that this body, like orthodiamidobenzene, can be obtained from two different diamidobenzoic acids, whilst the para-compound

$$NH_2$$

$$NH_2$$

corresponds to only one such acid.

Salkowski also concluded that the common phenylenediamine is metadiamidobenzene, and he proved that hydroquinone belongs to the same series as para-oxybenzoic acid. He reduced the methyl ether of the nitrophenol melting at 115°, or nitro-anisol, $C_6H_4(NO_2)OCH_3$, to amido-anisol, $C_6H_4(NH_2)OCH_3$, substituted in this, the amido-group by bromine, and then, by means of sodium and methyl iodide, replaced this bromine by methyl, thus obtaining a cresol methyl ether, $C_6H_4(CH_3)OCH_3$, which on oxidation yielded anisic acid. He then converted the amido-anisol into a diazo-compound, and decomposed this by water, when methyl was eliminated, and hydroquinone formed. It follows from this that the hydroxyls of hydroquinone and therefore the oxygen atoms of quinone occupy the positions 1 : 4. The constitution of this latter body is, therefore, expressed by one of the following formulæ:

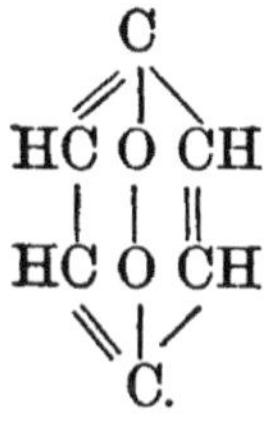

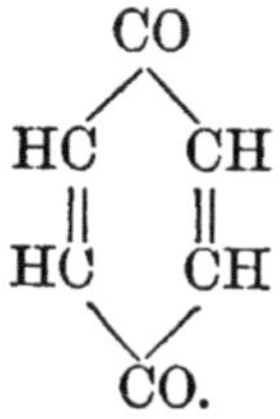

As will be shown later on it is probable that quinone can exist in both these forms, which must be regarded as tautomeric.

From the results of these and similar investigations it follows that the former table must be revised as below, only a portion being here given and the lacunæ being filled up:—

$C_6H_4Br_2$	Orthodibromobenzene, B.P. 224°.	Metadibromobenzene, B.P. 219°.	Paradibromobenzene, M.P. 89°.
$C_6H_4(NO_2)_2$	Orthodinitrobenzene, M.P. 118°.	Metadinitrobenzene, M.P. 89°·8.	Paradinitrobenzene, M.P. 172°.
$C_6H_4(NO_2)Br$	Orthobromonitrobenzene, M.P. 41°·5.	Metabromonitrobenzene, M.P. 56°·4.	Parabromonitrobenzene, M.P. 126°.
$C_5H_4(NH_2)_2$	Orthodiamidobenzene, M.P. 99°.	Metadiamidobenzene, M.P. 63°.	Paradiamidobenzene, M.P. 147°.
$C_6H_4(OH)_2$	Pyrocatechin, M.P. 104°.	Resorcin, M.P. 110°.	Hydroquinone, M.P. 169°.
$C_6H_4Br(CH_3)$	Orthobromotoluene, B.P. 182°.	Metabromotoluene, B.P. 184°.	Parabromotoluene, M.P. 28°·3.
$C_6H_4(NO_2)CH_3$	Orthonitrotoluene, B.P. 223°.	Metanitrotoluene, M.P. 16°.	Paranitrotoluene, M.P. 54°.
$C_6H_4(NH_2)CH_3$	Orthotoluidine, B.P. 199°·5.	Metatoluidine, B.P. 197°.	Paratoluidine, M.P. 45°.
$C_6H_4(OH)CO_2H$	Salicylic acid, M.P. 156°.	Meta-oxybenzoic acid, M.P. 200°.	Para-oxybenzoic acid, M.P. 210°.
$C_6H_4(CH_3)_2$	Orthoxylene, B.P. 140°.	Metaxylene, B.P. 137°.	Paraxylene, B.P 137°.
$C_6H_4(CO_2H)_2$	Phthalic acid.	Isophthalic acid.	Terephthalic acid.

951 So far all attempts at orientation had been directed towards placing the compounds in genetic connection with one of the phthalic acids. We have already seen that the constitution of terephthalic acid and of the para-series connected with it had been placed on a sure foundation.

On the other hand, the views concerning the constitution of phthalic and isophthalic acids rested on hypotheses which had not been verified. As common phthalic acid is the only one which yields an anhydride, it was assumed that in this the carboxyls occupy the adjacent position. But a similar argument was employed to show that the hydroxyls in hydroquinone also occupy the same position, and this conclusion was proved to be erroneous; hence the argument lost weight in the case of phthalic acid. The conclusions based upon the constitution of naphthalene stood on a firmer basis, but it still was a question whether they were final.

The case of isophthalic acid is a similar one : its constitution depends upon that of mesitylene, in which we know that the three methyls are contained in symmetrical position, or, what is the same thing, that the three hydrogen atoms in direct combination with the nucleus are of equal value. Ladenburg proved this as follows :—" Let us indicate the three hydrogen atoms by the letters a, b and c, and assume that in dinitromesitylene, $C_6(CH_3)_3\overset{a}{N}O_2(\overset{b}{N}O_2)H$, a and b are replaced by nitroxyl, and that by partial reduction b is converted into an amido-group; we then obtain the formula $\overset{a}{N}O_2\overset{b}{N}H_2\overset{c}{H}$ for amidonitromesitylene. Now, the hydrogen atom marked c in this may be replaced by nitroxyl, and the amido-group by hydrogen, a dinitromesitylene, having the formula $\overset{a}{N}O_2\overset{b}{H}\overset{c}{N}O_2$, being thus obtained, and as this is identical with the former one, it is clear that the hydrogen atoms b and c are of equal value.

Supposing now that the amido-group in amidonitromesitylene be replaced by hydrogen, we obtain nitromesitylene, which on reduction is converted into amidomesitylene, $\overset{a}{N}H_2\overset{b}{H}\overset{c}{H}$. If this body be nitrated only one amidonitromesitylene, $\overset{a}{N}H_2\overset{b}{N}O_2\overset{c}{H}=\overset{a}{N}H_2\overset{b}{H}\overset{c}{N}O_2$, can be formed, as both b and c are of equal value. The compound thus obtained proves, however, to be identical with the amidonitromesitylene, having the following formula : $\overset{a}{N}O_2\overset{b}{N}H_2\overset{c}{H}$. Hence the hydrogen atoms a and b are also of equal value, or the three hydrogen atoms belonging to the aromatic nucleus of mesitylene are of equal value."

952 Before Ladenburg's memoir appeared, Körner published a valuable research on the isomerism of the aromatic compounds containing six atoms of carbon.[1] In this he criticised the methods of orientation then in vogue, founded on the constitution of the three phthalic acids, of mesitylene and of naphthalene, &c. He expressed doubt whether the constitution of these compounds was ascertained with sufficient certainty to serve as a foundation for orientation and he, therefore, doubted the conclusions and experiments by which the acids containing eight atoms of carbon were brought into connection with the disubstitution products of benzene obtained by direct methods. Körner prepared 126 new compounds, and in addition revised

[1] *Gazetta* **1874**, 305 ; *Journ. Chem. Soc.* 1876, **1**, 204.

all the researches which had hitherto been published, especially those concerning the simplest benzene derivatives, in the preparation of which it frequently happened that the formation of two or three isomerides had been overlooked. Assuming the equal value of the hydrogen atoms of benzene, he employed a method of orientation depending on the following principles:

An ortho-compound (1 : 2) with similar substituents can yield two tri-derivatives, when a third hydrogen atom is replaced. Under the same conditions a meta-compound (1 : 3) yields three tri-derivatives; whereas a para-compound can only yield one. In this case, where the substituted element or radical is identical with the two already present, only three substitution products can be formed; but when a different element or radical is introduced, six may be obtained, as is seen from the following table:

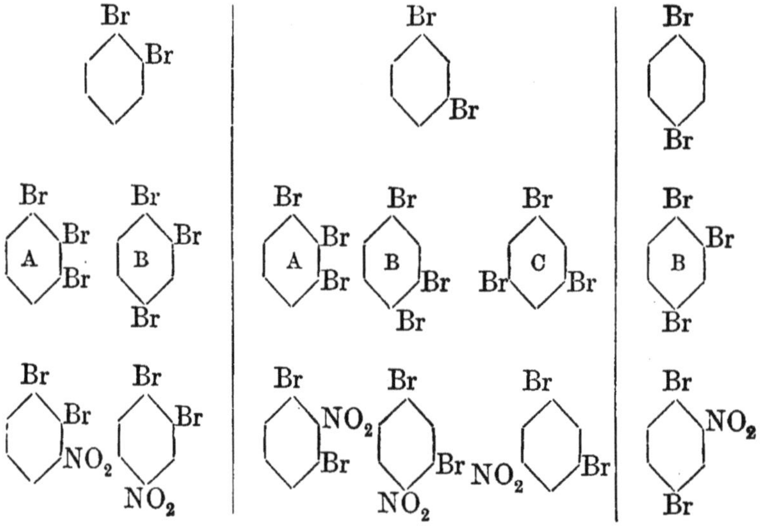

From this it follows that the dibromobenzene which yields three tribromobenzenes or three nitrodibromobenzenes, and which on the other hand, is formed from three dibromanilines, by elimination of the amido-group, contains the bromine atoms in the position 1 : 3.

Orthodibromobenzene having the position 1 : 2 can only yield two tribromobenzenes or two nitrodibromobenzenes, and can be obtained only from two dibromanilines. Lastly paradibromo-benzene (1 : 4) can only be obtained from one dibromaniline and yields only one tribromobenzene, and only one nitrodibromo-benzene.

Now, the solid dibromobenzene, melting at 89°, yields only one nitro-derivative;[1] the liquid compound obtained together with the former by the action of bromine on benzene[2] yields, on the other hand, two; and the third, obtained by Meyer and Stüber from dibromaniline,[3] forms three nitro-dibromobenzenes.

Körner showed, moreover, that when the dibromobenzenes are further brominated, the above conditions are fulfilled, so that the solid corresponds to one; the liquid, formed at the same time, to two; and the third to three dibromanilines. Solid dibromo-benzene is, therefore, (1 : 4); that obtained by Riese, (1 : 2); and that prepared by Meyer and Stüber, (1 : 3). No. 1 corresponds to hydroquinone, No. 2 to pyrocatechin, and No. 3 to resorcin.

953 In a similar way Körner has determined the constitution of a large number of compounds, and this without reference to the views respecting the positions of the carboxyls in the three phthalic acids. Almost at the same time Griess employed the same principle to ascertain the constitution of the three diamido-benzenes, by preparing them from the six diamidobenzoic acids by elimination of carbon dioxide :

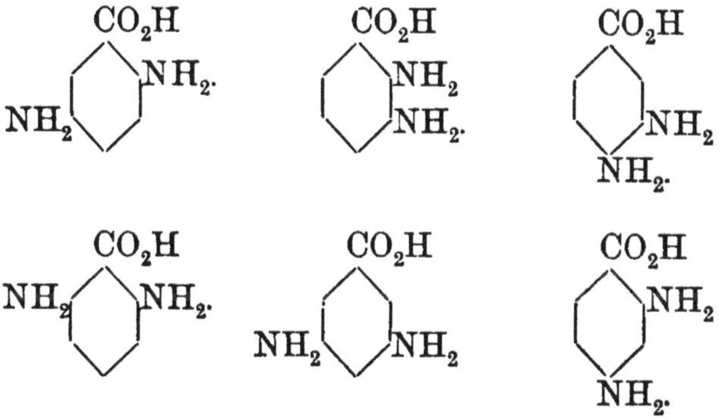

$$CO_2H,\ NH_2,\ NH_2 \qquad CO_2H,\ NH_2,\ NH_2 \qquad CO_2H,\ NH_2,\ NH_2$$

$$CO_2H,\ NH_2,\ NH_2 \qquad CO_2H,\ NH_2,\ NH_2 \qquad CO_2H,\ NH_2,\ NH_2$$

From the first of these he obtained paradiamidobenzene melting at 147°, the next two yielded orthodiamidobenzene melting at 99°, and the last three gave rise to common metadiamidobenzene melting at 63°.[4]

In close connection with these researches Nölting has given

[1] Zincke and Sintenis, *Ber.* **6**, 123.
[2] Riese, *ibid.* **2**, 61.
[3] Meyer and Stüber, *ibid.* **5**, 52.
[4] *Ibid.* **7**, 1226.

a direct proof of the constitution of phthalic acid.[1] We have already seen that dinitrotoluene has the following constitution:

$$CH_3$$

This is deduced from the facts that it has been obtained by the further nitration of the two mononitrotoluenes obtained directly from toluene, of which the solid one yields paranitrobenzoic acid on oxidation, and that it was converted by Wurster, by reduction and elimination of methyl, into ordinary metadiamidobenzene. From this it follows that the liquid nitrotoluene, and therefore the orthotoluidine obtained from it, contain the side chains in the position 1 : 2. This latter substance, however, Weith converted by means of the mustard-oil reaction into orthotoluic acid, $C_6H_4(CH_3)CO_2H$, which on oxidation with potassium permanganate yields phthalic acid; this substance must therefore also contain the carboxyls in the position 1 : 2.

Kekulé had previously converted orthotoluidine into ortho-iodotoluene and obtained orthotoluic acid from it by Wurtz's reaction, while by oxidizing this iodotoluene he obtained iodobenzoic acid, $C_6H_4I(CO_2H)$, which when fused with caustic potash yielded salicylic acid.[2]

Another very simple proof of the constitution of the phthalic acids has been given by Nölting,[3] using the method of Körner and Griess (see page 52). These acids are readily obtained by oxidizing the corresponding xylenes with potassium permanganate. Now isoxylene yields three mononitroxylenes, orthoxylene two, and paraxylene only one. It follows from this that the side chains in isoxylene, and, therefore, in isophthalic acid, have the position 1 : 3, in orthoxylene and orthophthalic acid 1 : 2, and in paraxylene and terephthalic acid, 1 : 4.

954 Adopting the principles of orientation above explained many chemists have extended their researches in this direction, so that the constitution of all the more important di-substitution products of benzene is now known with a great degree of probability. As, however, higher substitution products can be obtained from

[1] R. Meyer, *Einl. Arom. Verb.* 73.
[2] *Ber.* **7,** 1006.   [3] *Ibid.* **18,** 2687.

.these by simple reactions, and as these again can easily be reconverted into di-substitution products, in the manner shown above, it is clear that the constitution of a large number of aromatic compounds containing three or more side chains is also known.

For the further consideration of this point we must refer to the special description of the compounds.

In the foregoing historical account the old names of the substances have been employed. In many cases these have now been altered in order to systematize the nomenclature. Thus the term "hydroxy" is now employed instead of "oxy" when a hydrogen atom has been replaced by hydroxyl; the compounds pyrocatechin, resorcin, and hydroquinone, are now, in accordance with the usual terminal nomenclature of phenols, known as catechol, resorcinol, and quinol respectively. The prefixes *ortho*, *meta*, and *para*, are usually represented by their initial letter, *e.g.* orthodiamidobenzene becomes *o*-diamidobenzene.

955 In the case of penta-substitution products which contain identical elements or radicals, as in that of mono-substitution derivatives, no isomerides occur. From this we may conclude that the six hydrogen atoms of benzene are of equal value. This conclusion is, however, founded on negative evidence only, and it is of importance that it should be proved directly. For this purpose Ladenburg adopted the same principle as that which he employed for proving the equality of the three hydrogen atoms of mesitylene (p. 51).[1]

In order to prove this point, Ladenburg first makes the two following hypotheses :—(1) In benzene there are two pairs of hydrogen atoms symmetrically situated with regard to a fifth. (2) Benzene contains at least four hydrogen atoms equal in value. The proof of these two hypotheses taken together would conclusively show that *all* six atoms of hydrogen were equal in value.

As regards the first proposition, Hübner and Petermann found that when ordinary *m*-bromobenzoic acid is nitrated, two isomeric bromonitrobenzoic acids are formed, in which, therefore, the nitroxyl groups must occupy different positions, which may be denoted by *a* and *b*.

$$C_6H_3.Br.NO_2.CO_2H. \qquad C_6H_3.Br.NO_2.CO_2H.$$
$$a \qquad\qquad\qquad b$$

[1] *Ber.* **2,** 140.

Both bodies are converted by the same substance, viz. hydrogen, into anthranilic acid, bromine being replaced by hydrogen, and the nitroxyl reduced to the amido-group. Hence it follows that both a and b occupy a symmetrical position with regard to the carboxyl group.

There is, however, a second pair of hydrogen atoms symmetrically situated with regard to the carbon atom to which the carboxyl group is attached, as has been shown in the following manner.

Wroblewsky[1] found that p-toluidine is acted on by bromine with formation of bromo-p-toluidine, the amido-group of which may be replaced by hydrogen, by means of the diazo-reaction. The monobromotoluene thus obtained is converted on oxidation into m-bromobenzoic acid.

This same bromotoluidine also yields a nitrobromotoluidine, which by the diazo-reaction is converted into a nitrobromotoluene, and this on reduction passes into the corresponding amidobromotoluene. If the bromine in the latter be replaced by hydrogen, a toluidine is obtained which on replacing the amido-group by bromine in the usual manner, yields the same bromotoluene which was obtained from p-toluidine direct, and which of course on oxidation yields metabromobenzoic acid.

The following formulæ, in which the different groups are marked a, b, c, d, e, f, will render the meaning of the above series of changes more conspicuous.

Bromoparatoluidine.

$$C_6.H.H.H.Br.CH_3.NH_2.$$
$$\;\;\;\; a \;\;\; b \;\;\; c \;\;\; d \;\;\; e \;\;\;\;\;\; f$$

Metabromotoluene.

$$C_6.H.H.H.Br.CH_3.H.$$
$$\;\;\;\; a \;\; b \;\; c \;\; d \;\;\; e \;\;\;\; f$$

Nitrobromotoluidine.

$$C_6.H.H.NO_2.Br.CH_3.NH_2.$$
$$\;\;\;\; a \;\; b \;\;\; c \;\;\;\; d \;\;\; e \;\;\;\; f$$

Metabromobenzoic Acid.

$$C_6.H.H.H.Br.COOH.H.$$
$$\;\;\;\; a \;\; b \;\; c \;\;\; d \;\;\;\; e \;\;\;\; f$$

Nitrobromotoluene.

$$C_6.H.H.NO_2.Br.CH_3.H.$$
$$\;\;\;\; a \;\; b \;\;\; c \;\;\;\; d \;\;\; e \;\;\; f$$

Amidobromotoluene.

$$C_6.H.H.NH_2.Br.CH_3.H.$$
$$\;\;\;\; a \;\; b \;\;\; c \;\;\;\; d \;\;\; e \;\;\; f$$

Amidotoluene.

$$C_6.H.H.NH_2.H.CH_3.H.$$
$$\;\;\;\; a \;\; b \;\;\; c \;\;\;\; d \;\;\; e \;\;\; f$$

Metabromotoluene.

$$C_6.H.H.Br.H.CH_3.H.$$
$$\;\;\;\; a \;\; b \;\;\; c \;\; d \;\;\; e \;\;\;\; f$$

Metabromobenzoic Acid.

$$C_6.H.H.Br.H.COOH.H.$$
$$\;\;\;\; a \;\; b \;\;\; c \;\; d \;\;\; e \;\;\;\;\; f$$

[1] *Annalen*, **168**, 153; **192**, 196.

The identity of the bromobenzoic acids obtained in these two different ways, proves conclusively that the positions c and d are symmetrical with regard to e.

Hübner and Petermann in their experiments above described started with the same bromobenzoic acid that Wroblewsky obtained by the oxidation of his monobromotoluenes. As, however, we have in the two nitrobenzoic acids two nitroxyl groups symmetrically situated with regard to the carboxyl, whilst in Wroblewsky's compounds we have two bromine atoms similarly situated, it follows that *in benzene two pairs of hydrogen atoms exist which have a symmetrical position with regard to a fifth, and that therefore the sixth occupies an isolated position with regard to the same atom.*

As Hübner and Petermann obtained *ortho*-amidobenzoic acids, and Wroblewsky finally *meta*-bromobenzoic acid, it further follows that two of the two equivalent positions represent the ortho, and the remaining two the meta-compounds, whereas the para-position can only occur once.

Ladenburg has given another proof of the above statement, for which reference must be made to the original paper.[1]

956 The second of the propositions given above has been proved in the following manner. Phenol, when acted on by phosphorus pentabromide, yields bromobenzene, which by the action of sodium and carbon dioxide is converted into benzoic acid. The hydroxyl has therefore the same position in phenol as the carboxyl in benzoic acid. Let this be denoted by a. Now three different atoms of hydrogen in benzoic acid may be replaced by hydroxyl with formation of the three isomeric hydroxybenzoic acids, the formulæ of which may be written in the following manner:

$$C_6.\underset{a}{COOH}.\underset{b}{OH}.\underset{c}{H}.\underset{d}{H}.\underset{e}{H}.\underset{f}{H}.$$

$$C_6.\underset{a}{COOH}.\underset{b}{H}.\underset{c}{OH}.\underset{d}{H}.\underset{e}{H}.\underset{f}{H}.$$

$$C_6.\underset{a}{COOH}.\underset{b}{H}.\underset{c}{H}.\underset{d}{OH}.\underset{e}{H}.\underset{f}{H}.$$

All three of these compounds on heating with lime, lose carbon dioxide, and are converted into one and the same compound, viz. phenol, hence the positions b, c, and d are of equal value. But it has already been shown that the same

[1] *Ber.* **5**, 322; **8**, 1666; *Theorie. Arom. Verb.* (Vieweg and Sohn), 1876.

compound is formed when the hydroxyl group occupies the position *a*. Hence the four positions *a*, *b*, *c*, *d*, are equal in value.

The truth of these two propositions having been established, it follows that all six hydrogen atoms are of equal value, and further that one hydrogen atom corresponds to two symmetrical pairs; all of which is indicated in the hexagonal formula of benzene:

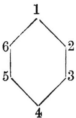

The symmetrical pairs to (1) are 2 : 6, and 3 : 5 ; to (2) 1 : 3, and 4 : 6, &c. With reference to (1) the positions which occur twice are :—

$$1 : 2 = 1 : 6$$
$$1 : 3 = 1 : 5$$

The position 1 : 4 only occurs once.

THE CONSTITUTION OF BENZENE.

957 A constitutional formula for benzene must satisfy the following conditions:

(1) The six atoms of hydrogen contained in it are of equal value.

(2) There exist three and only three isomeric di-substitution products of benzene.

(3) In the additive compounds of benzene and its derivatives substances can only be obtained to which 2, 4, or 6 monad radicals have been added.

As already mentioned (p. 7) Kekulé expressed the constitution of benzene by the following formula:

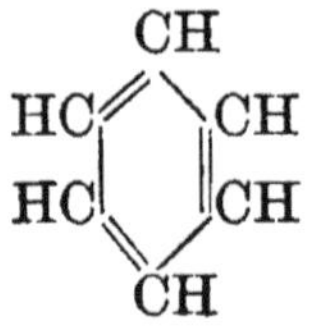

$$
\begin{array}{c}
\text{CH} \\
\text{HC} \diagup \quad \diagdown \text{CH} \\
\text{HC} \diagdown \quad \diagup \text{CH} \\
\text{CH}
\end{array}
$$

Very soon after the publication of Kekulé's paper various chemists gave their views as to the constitution of this compound, and expressed these by means of graphic formulæ, of which the following may be mentioned.

Claus[1] proposed the formulæ illustrated in Figs. A and B, giving preference to the former of these. Körner proposed a

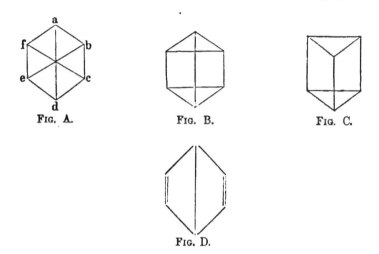

Fig. A.   Fig. B.   Fig. C.

Fig. D.

similar formula, which cannot however be represented on a plane surface, but may be readily portrayed by Kekulé's glyptic models. According to Körner's view, and in accordance with Claus's first formula, each carbon atom is combined directly with three others, but instead of the carbon atoms being situated in the same plane, the atoms *a c e*, and *b d f* lie in two different parallel planes. Claus's second formula is identical with that proposed by Ladenburg (Fig. C), except that the latter assumes a special configuration of the six carbon atoms, regarding them as situated at the six corners of a three-sided prism.[2] The formula given in Fig. D was proposed by Dewar, but as, according to it, two isomeric ortho- and para-derivatives should exist, it has not found many adherents.

According to Claus's first formula, it appears that only two isomeric di-substitution products can exist, and on this account this formula did not receive much consideration. If, however, certain other assumptions be made, which will be discussed later, this difficulty may be avoided.

[1] *Theoret. Betracht. System. Org. Chem.* Freiburg, 1867.
[2] *Ber.* **2**, 141, 272.

Ladenburg raised an objection to Kekulé's formula inasmuch as it requires, not three, but four isomeric di-substitution products and two mono-substitution products, the positions 1 : 2 and 1 : 6 not being perfectly identical; for it is clear that two substituents occupying the position 1 : 2 are combined

with two carbon atoms which are connected together by single linkage, whereas between the positions 1 : 6 the carbon atoms are doubly linked. The question whether 1 : 3 is equal to 1 : 5 is a simpler one and may be answered in the affirmative.

Victor Meyer opposed Ladenburg's view, insisting that the difference between the positions 1 : 2 and 1 : 6 was not brought about by a variation in the position of the atoms, but only by a varying arrangement of the combining units of the carbon, and that it was of so very subtle a character that it was doubtful whether such minute differences could exert a perceptible influence on the properties of the compound.[1]

Ladenburg further added that if the first two positions are identical, the two crotonic acids, which were then supposed to have the formula

$$CH_3-CH=CH.CO_2H,$$
$$CH_2=CH-CH_2.CO_2H,$$

must also be identical; in reply however it was pointed out that in the latter case the differences observed are not due merely to the manner in which the carbon atoms are linked together, but also to the distribution of the hydrogen among the carbon atoms.

958 Kekulé[2] further stated that the apparent difference between the positions 1 : 2 and 1 : 6, originates rather from the form in which these ideas are represented, than from the idea itself, which is only imperfectly represented by this form. In order to show that no such difference exists, he brought forward the following hypothesis.[3]

"The atoms in the systems which we call molecules must be

[1] *Annalen,* **156,** 265 ; **159,** 24.
[2] *Ibid,* **162,** 77 ; *Journ. Chem. Soc.* **1872,** 612.
[3] *Annalen,* **162,** 86.

assumed to be continually in motion, but hitherto no explanation as to the nature of this intramolecular motion has been given, which of course must be in accordance with the law of the linking of atoms. A planetary motion seems, therefore, inadmissible; the movement must be of such a kind, that all the atoms forming the system, retain the same relative arrangement; in other words, that they return to a mean position of equilibrium. The most probable assumption, and one which is in accordance with the view held by physicists, is that the motion of the atoms takes place in straight lines and that on striking each other, they recoil like elastic bodies. What we call valency would then be nothing but the number of contacts experienced by one atom with other atoms in the unit of time. In the same time that the monad atoms of a diatomic molecule like H_2 strike each other once, the dyad atoms of a diatomic molecule come in contact with each other twice, the temperature in both cases being the same. In a molecule consisting of one dyad and two monads, as H_2O, the number of contacts in the unit of time is 2 for the former and 1 for each of the latter.

" If two atoms of carbon are linked together by one combining unit of each, they strike against each other once in the unit of time, or in the same time in which monad hydrogen makes a complete vibration; during the same time they encounter three other atoms. Carbon atoms linked together by two bonds of each come in contact twice in the unit of time, and encounter during the same time two other atoms.

" Applying these views to benzene, the formula proposed by me then becomes the expression of the following ideas: Each carbon atom strikes against two others in the unit of time, once against the one and twice against the other. In the same unit of time it comes once in contact with the hydrogen atom, which during the same period makes a complete vibration. Accordingly in the graphic formula

$$HC=CH$$

the contacts of the atom 1 in the unit of time may be thus represented, if h stands for hydrogen:

$$1 : 2, 6, h, 2.$$

In the second unit of time they are:

$$2 : 6, 2, h, 6,$$

which would be represented by the following formula:

$$HC-CH$$

The same carbon atom is therefore, during the first unit of time linked to one of the adjoining ones by one, and during the second unit of time by two of its combining units; and *vice versâ* with regard to the other adjoining carbon atom.

" The variation of these contacts undergone by a carbon atom exhausts itself in these two units of time; its sum, viz:

$$2, 6, h, 2, 6, 2, h, 6$$

represents the whole series of possible contacts under these circumstances, and therefore repeats itself continually. From this it is evident that the carbon atom strikes against the two others, with which it is directly combined, an equal number of times, *i.e.* that it bears the same relation to both. The ordinary graphic formula only represents the contacts made during the first unit of time, and thus the view has sprung up that in the di-substitution products, the positions 1, 2, and 1, 6, must produce different compounds. If the above, or some similar conception be correct, it follows that no real difference exists."

959 Michaelis opposed Kekulé's views, which, according to him possess, from physical considerations, little probability, especially from the point of view of the kinetic theory of gases, as according to this, the square of the velocity of the atoms at equal temperatures is, in a perfect gas, inversely proportional to the atomic weights. Hence it follows that hydrogen atoms have the greatest velocity, but according to Kekulé they impinge less often than the oxygen atoms for example, and hence they must describe a much larger orbit than these; and further, their molecules must be larger than those of the oxygen. But, according to the kinetic theory of gases, the hydrogen molecules are only half as large as those of oxygen and are probably, the smallest of all the gaseous molecules. Kekulé's ideas on atomic motion cannot,

according to Michaelis, be regarded as representing a stable condition of dynamical equilibrium, and are moreover improbable because then benzene should readily decompose into three molecules of acetylene.[1]

In comparing the formulæ of Kekulé and Ladenburg, we find that both explain equally well the formation of benzene from acetylene, and of mesitylene from acetone, as well as the formation of addition-products; in both cases a closed ring of six carbon atoms remains in which each atom is connected with two others by a single bond.

Ladenburg's formula also explains the equal value of the six hydrogen atoms, besides showing as clearly as Kekulé's the existence of the three isomeric di-substitution products. If we number the six angles of the prism or the corresponding plane formula:

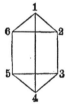

we have:

$$1 : 2 = 1 : 6 = \text{Meta.}$$
$$1 : 3 = 1 : 5 = \text{Ortho.}$$
$$1 : 4 = \text{Para.}$$

This is also easily shown by employing the proof given by Körner.

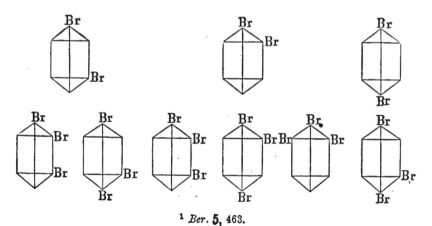

[1] *Ber.* **5**, 463.

In the case of the triderivatives we obtain the following results, similar to those previously obtained :

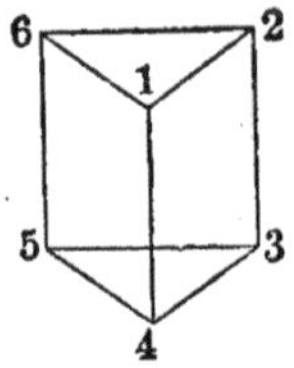

1 : 2 : 3 adjacent,
1 : 3 : 4 asymmetrical,
1 : 2 : 6 symmetrical.

If we assume that one of the triangular faces of the prism be revolved in its plane through 180°, whilst the other remains in its original position, and at the same time the figure thus obtained be projected on the surface whose position has been unchanged, the following graphic formula is obtained :

This form possesses the advantage that the numbers have the same meaning as in Kekulé's formula, and otherwise expresses all that the prism formula can.

960 *Physical Methods.*—Attempts have also been made to determine the constitution of benzene by physical methods. Brühl has investigated the refractive power of compounds containing carbon in single, double, or treble linkage, and has arrived at the conclusion that benzene contains three doubly-linked carbon atoms; hence he considers the accuracy of Kekulé's views to be proved.[1]

Lossen and Zander arrived at the same conclusion from investigations on the specific volume of certain hydrocarbons,[2] whilst Julius Thomsen, in determining the heat of combustion of similar bodies, arrived at the opposite conclusion; for according to him benzene contains nine single linkages, and, therefore, Ladenburg's formula is correct.[3]   The results drawn from physical methods are therefore diametrically opposed to one another.[4]

---

[1] *Annalen,* **200**, 93 ; *Ber.* **20**, 2288.   [2] *Annalen,* **225**, 119.
[3] *Ber.* **13**, 1808.   [4] See also Horstmann, *ibid.* **21**, 2211.

Another interesting observation which seemed likely to afford satisfactory proof has also failed. An acid termed carboxytartronic acid, $C_4H_4O_7$, was obtained by the action of nitrous acid on protocatechuic acid, $C_6H_3(OH)_2CO_2H$,[1] and on catechol, $C_6H_4(OH)_2$.[2] This acid is very unstable in the free state, quickly decomposing into carbon dioxide and tartronic acid, and for this reason Gruber and Barth proposed for it the following constitution:

$$\begin{array}{c} CO_2H \\ | \\ HO-C-CO_2H \\ | \\ CO_2H. \end{array}$$

According to these chemists it contains one carbon atom in direct linkage with three others, and as it is formed by a simple reaction from a benzene derivative it appeared very probable that this form of linkage also occurred in the aromatic nucleus, as expressed by the prism formula.

Kekulé, however, showed that the correct formula of this acid is $C_4H_6O_8$, and that it is not a tribasic but a dibasic acid, which on reduction yields racemic acid together with inactive tartaric acid. Kekulé then obtained it as a decomposition product of the so-called nitrotartaric acid, $C_2H_2(ONO_2)_2(CO_2H)_2$. Its constitution is, therefore, as follows:

$$\begin{array}{c} CO_2H \\ | \\ C\overset{OH}{\underset{OH}{<}} \\ | \\ C\overset{OH}{\underset{OH}{<}} \\ | \\ CO_2H. \end{array}$$

and it is dihydroxytartaric acid or tetrahydroxysuccinic acid.[3] Although its constitution does not exactly contradict the truth of the prismatic formula, its formation from catechol is certainly more easy to explain if we assume Kekulé's formula for benzene.

961 The same is true for the formation of another acid obtained by Carius by acting with chlorous acid on benzene, and afterwards investigated by Kekulé and Strecker,[4] who have found that it is β-trichloracetylacrylic acid, $CCl_3CO.CH{=}CH.CO_2H$, for on

[1] Gruber, *Ber.* **12**, 514.
[2] Barth, *Monatsh.* **1**, 869.
[3] *Annalen*, **221**, 230.
[4] *Ibid.*, **223**, 170

warming with baryta-water it decomposes into chloroform and maleic acid :

$$C_3H_3Cl_3O_3 + H_2O = C_4H_4O_4 + CCl_3H.$$

It combines with bromine, forming trichloracetyldibromopro·pionic acid, $CCl_3CO.CHBr.CHBr.CO_2H$, which on boiling with lime-water is converted into chloroform and inactive tartaric acid.

β-trichloracetylacrylic acid is not obtained from benzene by a simple reaction, as several by-products are formed, amongst others, chlorinated quinones. The acid itself is produced by the action of chlorous acid on quinone, and its formation is most simply explained by the supposition that monochloroquinone yields it, according to the following equation :

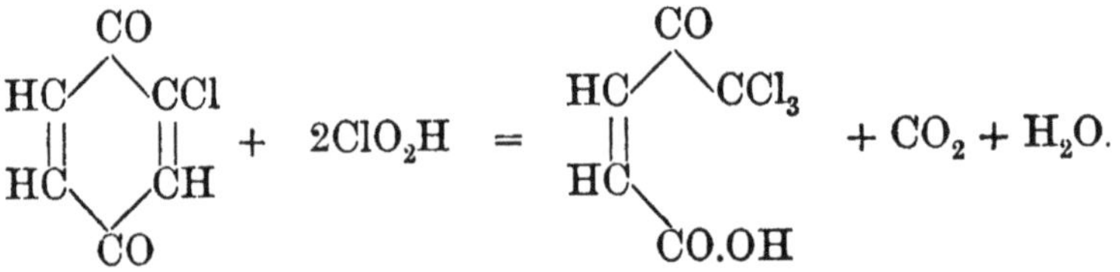

If we attempt to explain this reaction by means of the prism formula, Kekulé and Strecker have shown that if we do not meet with insuperable difficulties we are, at any rate, surrounded by evident improbabilities. Thus we have to assume that in five cases a single linkage of two carbon atoms is broken up. Three of these are resolved by the separation of the carbon atom which is evolved as carbon dioxide, in which of course there is nothing unusual, but the other two are resolved in such a manner as to form the normal chain of carbon atoms which is contained in β-trichloracetylacrylic acid. Hence the improbable supposition must be made that a double linkage of the carbon atoms is brought about from a single linkage under circumstances in which we should rather expect to find the formation of a single from a double linkage. If the formation of the above-named acid does not prove the truth of Kekulé's formula, it renders it more probable than any of those previously mentioned.

962 During the past few years the question of the constitution of benzene has again given rise to a great deal of discussion, chiefly owing to the researches of Baeyer.[1] This investigator has shown that the reduction products of terephthalic acid, $C_6H_4(COOH)_2$, no longer exhibit the characteristic reactions of

[1] *Ber.* **19**, 1797 ; **23**, 1272 ; *Annalen*, **245**, 103 ; **251**, 257 ; **256**, 1 ; **258**, 1, 145.

the aromatic compounds, but behave exactly like fatty compounds, the dihydro- and tetrahydro-addition-products resembling the unsaturated, and the hexhydro-derivatives the saturated compounds. Terephthalic acid itself behaves in most of its reactions as a saturated compound. Thus whilst all unsaturated fatty acids are instantaneously oxidized by alkaline potassium permanganate, terephthalic acid is unaltered by that reagent. Baeyer therefore concluded that in benzene no double bonds corresponding to those in ethylene exist, and that its formula is best represented as follows :

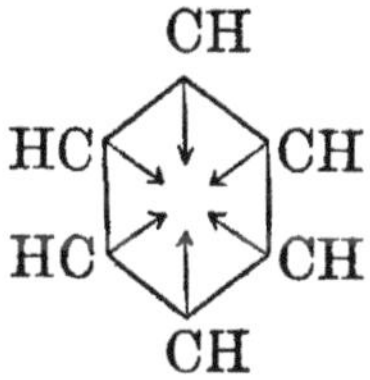

it being supposed that the six combining units neutralize one another in the centre in some manner without being directly combined with one another as in Claus's formula. To this formula which had been previously proposed by Armstrong,[1] the term "centric" is usually applied.

As soon as the equilibrium of these six valencies is destroyed, *e.g.* by the addition of two atoms of hydrogen to terephthalic acid, the remaining valencies are supposed to pass into the ordinary ethylene linkage, as shown in the following formula :

Terephthalic Acid.

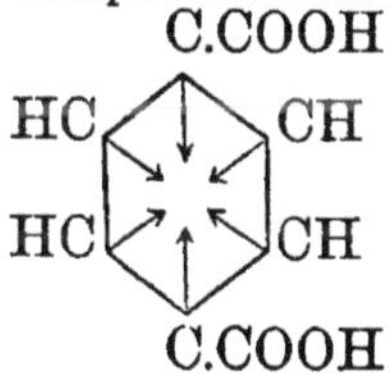

Dihydroterephthalic Acid.

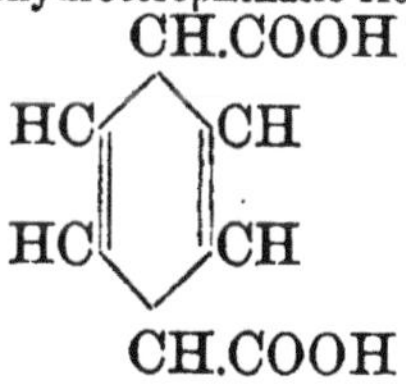

Tetrahydroterephthalic Acid.

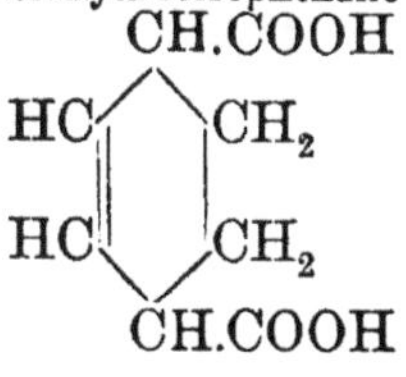

Hexhydroterephthalic Acid.

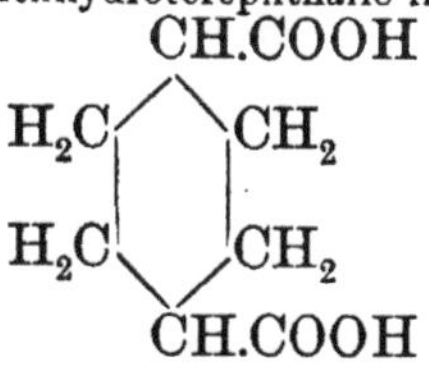

Further investigation brought to light the fact that phenanthrene, which is generally supposed to have the formula

[1] *Journ. Chem. Soc.* 1887, **1,** 264.

$$\begin{array}{c} C_6H_4-CH \\ | \quad\quad || \\ C_6H_4-CH \end{array}$$, is also unacted on by alkaline potassium perman-

ganate, whilst stilbene $$\begin{array}{c} C_6H_5.CH \\ || \\ C_6H_5.CH \end{array}$$ is readily oxidized, whence it

would appear that no conclusion can be drawn from this reaction as to the absence of an ethylene linkage in a *closed*
chain of carbon atoms, and that objection to Kekulé's formula
therefore falls to the ground.

963 Baeyer has further endeavoured to disprove the prism
formula in the following manner :

In the prism formula the carbon atoms occupying the para-
positions are supposed to be directly connected together, as
shown in the formula given below (para-positions = 1·4, 2·5, 3·6).

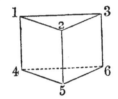

Now if on reduction two of these para-linkages were split up
we should get a derivative of di-trimethylene :

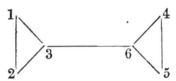

It has been shown that ethyl dihydroxyterephthalate is converted on reduction into ethyl succinylsuccinate, which as has
already been stated is a derivative of hexamethylene (Vol. III.,
Part II., p. 212).

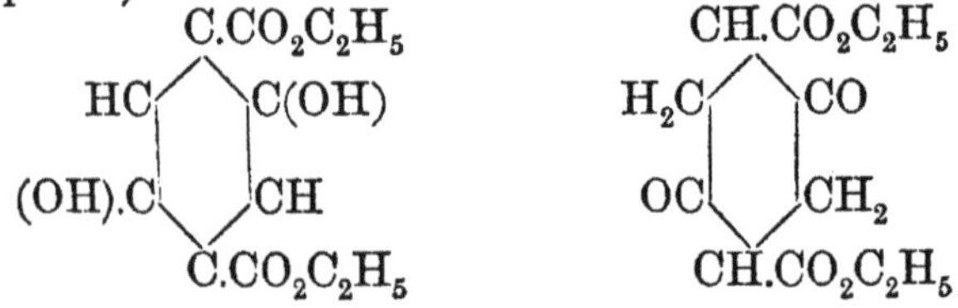

$$\begin{array}{cc} C.CO_2C_2H_5 & CH.CO_2C_2H_5 \\ HC\diagup\quad\diagdown C(OH) & H_2C\diagup\quad\diagdown CO \\ (OH).C\diagdown\quad\diagup CH & OC\diagdown\quad\diagup CH_2 \\ C.CO_2C_2H_5 & CH.CO_2C_2H_5 \end{array}$$

In this instance it will be seen that the benzene derivative
contains two pairs of radicals in the para-position, and that these
retain their position in the hexamethylene compound formed,
and that therefore had the carbon atoms in the para-position

been linked together, these must have been resolved in the reduction. As, however, the compound obtained is a derivative of hexamethylene, and not of di-trimethylene as required by the prism formula, the latter must be regarded as incorrect.

The validity of this proof has been disputed by Miller[1] and Ladenburg,[2] who point out that in the reaction in question not only a reduction but also an intramolecular change take place, and that therefore from such a reaction no reliable deductions as to the constitution of benzene can be made.

964 Claus's formula (Fig. A., p. 59) has also again been under consideration. Its author regards the diagonal or para-linkages as less strong and therefore more easily resolved than the remaining single-linkages, which supposition enables us to explain the existence of three isomeric di-substitution products of benzene. Baeyer pointed out that the first reduction product of terephthalic acid, viz., dihydroterephthalic acid, should, if Claus's formula be correct, have the formula given in Fig. A, whereas experiment has shown that its constitution is as given in Fig. B.

$$CHCOOH$$
$$HC \diamondup CH$$
$$HC \diamonddown CH$$
$$CHCOOH$$

FIG. A.

$$CHCOOH$$
$$HC \diagup CH$$
$$HC \diagdown CH$$
$$CHCOOH$$

FIG. B.

To this Claus replies[3] that here, as in the centric formula, the resolution of one of the para-linkages would destroy the equilibrium, and the remaining valencies would then rearrange themselves in the manner shown in Fig. B.

965 The formation of a dihydroterephthalic of the above constitution appears at first sight to be an argument against Kekulé's formula, from which one would expect to obtain a compound having the constitution:

$$CHCOOH$$
$$H_2C \diagup CH$$
$$HC \diagdown CH$$
$$C.COOH$$

[1] *Journ. Chem. Soc.* 1887, **1**, 208.   [2] *Ber.* **20**, 62.
[3] *J. Pr. Chem.* II. **40**, 69.

It was however found that in the reduction of muconic acid, a similar migration of the double-linkage takes place.[1]

$$\begin{array}{ll}
\text{COOH.HC} & \\
\quad\quad\searrow\!\!\text{CH} & \\
\quad\quad\mid \quad\quad + \text{ H}_2 = \\
\quad\quad\nearrow\!\!\text{CH} & \\
\text{COOH.HC} &
\end{array}
\qquad
\begin{array}{l}
\text{COOH.H}_2\text{C} \\
\quad\quad\searrow\!\!\text{CH} \\
\quad\quad\parallel \\
\quad\quad\diagup\!\!\text{CH} \\
\text{COOH.H}_2\text{C}
\end{array}$$

Other formulæ besides these here mentioned have been proposed for benzene, for which reference must be made to the original papers.[2]

966 The question as to the exact constitution of benzene cannot therefore be yet regarded as settled. Kekulé's formula is most generally accepted, and has also the advantage of readily explaining the constitution of naphthalene, phenanthrene, and similar hydrocarbons. The exact constitution of these latter will be discussed in Part VI. It will be sufficient to point out here that these may be derived from benzene in several ways, viz. :—

1. The hydrogen in benzene can be replaced by phenyl :

Phenylbenzene	$C_6H_5.C_6H_5$
Diphenylbenzene.	$C_6H_4.(C_6H_5)_2$
Triphenylbenzene	$C_6H_3(C_6H_5)_3$

Compounds are also found containing alcohol radicals as side chains:

Phenyltoluene	$C_6H_5.C_6H_4.CH_3$
Methylphenyltoluene . . .	$CH_3.C_6H_4.C_6H_4.CH_3$

2. In others the aromatic nuclei are not combined directly but through a carbon atom :

Diphenylmethane	$CH_2(C_6H_5)_2$
Triphenylmethane	$CH(C_6H_5)_3$
Diphenylethane	$CH_3.CH.(C_6H_5)_2$
Phenyltolylmethane. . . .	$CH_2\begin{cases} C_6H_5 \\ C_6H_4.CH_3 \end{cases}$

[1] Rupe, *Annalen*, **256**, 5.

[2] R. Meyer, *Ber.* **15**, 1823 ; Thomsen, *ibid.* **19**, 2944 ; Marsh, *Phil. Mag.* **162**, 426 ; Herrmann, *Ber.* **21**, 1949, 2338 ; Sworn, *Phil. Mag.* V. **28**, 402, 443.

3. The aromatic nuclei are united by two carbon atoms:

Dibenzyl . . . . $C_5H_6.CH_2—CH_2.C_6H_5$

Toluylene . . . . $C_6H_5.CH=CH.C_6H_5$

Tolane . . . . $C_6H_5.C≡C.C_6H_5$

Tetraphenylethane . . . . $(C_6H_5)_2CH—CH(C_6H_5)_2$

And to this group belongs anthracene :

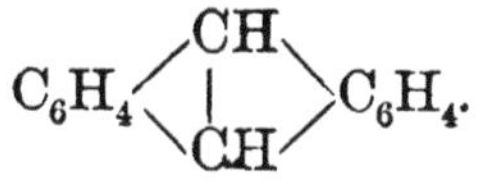

$$C_6H_4 \underset{\diagdown CH \diagup}{\overset{\diagup CH \diagdown}{\big|}} C_6H_4.$$

4. Of the hydrocarbons with condensed nuclei, it is sufficient to mention here naphthalene and phenanthrene, the formulæ of which, according to Kekulé's hypothesis, are:

Naphthalene.

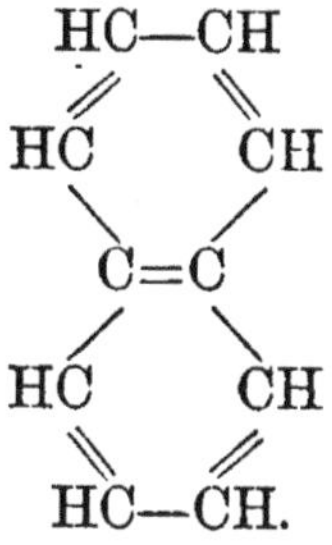

HC—CH
HC   CH
C=C
HC   CH
HC—CH.

Phenanthrene.

HC=CH . HC=CH
HC   C—C   CH
HC—C   C—CH
HC=CH

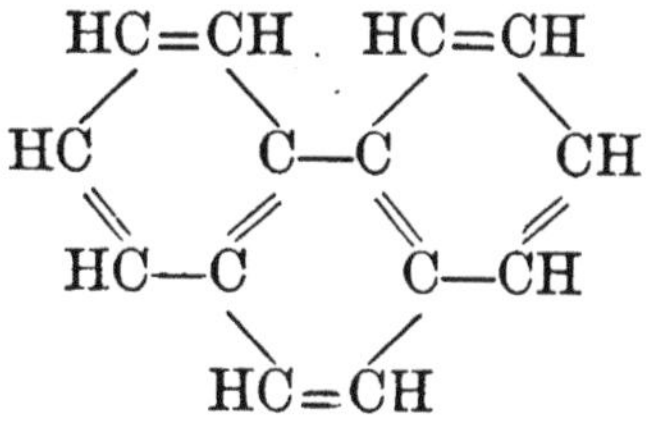

Many others besides these are known, and from each of these a large number of derivatives can be obtained, affording a further proof of the very large number of the aromatic bodies.

In almost all cases the simple hexagonal formula of benzene shown on p. 8, in which the vexed question of the distribution of the six extra valencies is left open, is quite sufficient for all discussions as to the constitution of aromatic compounds, and will therefore usually be employed.

CHARACTERISTIC REACTIONS OF THE DI-SUBSTITUTION PRODUCTS.

967 A characteristic property of the ortho-compounds is their easy transformation with separation of water into inner anhydrides, such as have been mentioned under salicylic and phthalic acids. As another example we may mention coumaric

acid or o-hydroxycinnammic acid, whose anhydride, coumarin, gives to sweet woodruff its pleasant smell:

$$C_6H_4{<}{\overset{\text{OH}}{\underset{\text{CH=CH.CO.OH}}{}}} = C_6H_4{<}{\overset{\text{O——CO}}{\underset{\text{CH=CH}}{|}}} + H_2O.$$

The amido-acids of this group also exhibit a similar reaction; thus o-amidophenylacetic acid passes readily into an anhydro-compound, which has been termed oxindol from the fact that it was first obtained from indigo:

$$C_6H_4{<}{\overset{\text{NH}_2}{\underset{\text{CH}_2.CO.O'H}{}}} = C_6H_4{<}{\overset{\text{NH}}{\underset{\text{CH}_2}{}}}{>}CO + H_2O.$$

The ortho-compounds having side-chains containing carbon are distinguished from those of the two other series, inasmuch as when heated with chromic acid solution, they do not undergo a simple oxidation, but are completely decomposed with formation of carbon dioxide, acetic acid and oxalic acid, &c. Dilute nitric acid, or an alkaline solution of potassium permanganate, on the other hand, easily bring about an oxidation in which the side chains are converted into carboxyl.

Ladenburg found that o-diamido-compounds when heated with monobasic organic acids give rise to peculiar condensation products:

Ethenylphenylenediamine.

$$C_6H_4{<}{\overset{\text{NH}_2}{\underset{\text{NH}_2}{}}} + HO.CO.CH_3 = C_6H_4{<}{\overset{\text{NH}}{\underset{\text{N}}{}}}{>}C.CH_3 + 2H_2O.$$

In the meta- and para-compounds, however, a hydrogen atom of each amido-group is replaced by an acid radical, and thus, $e.g.$, β-diacetyldiamido-benzene, $C_6H_4(NH.CO.CH_3)_2$, is obtained.

968 The o-diamines also form with aldehydes a series of mono-acid bases, the reaction being represented by the following equation:

Benzaldehyde.   Phenylbenzaldehydine.

$$C_6H_4(NH_2)_2 + 2COH.C_6H_5 = C_6H_4(NCH.C_6H_5)_2 + 2H_2O.$$

The diamines of the two other series act similarly; but the compounds thus obtained are so unstable that it is impossible to prepare their salts, since they decompose readily into their constituents with assumption of water.

If the hydrochloride of an *o*-diamine be heated with benz-aldehyde from 100°—120°, half the hydrochloric acid is evolved, whilst the salts of the two other series yield no trace of acid when thus treated.

Ladenburg found that the *o*-diamido-compounds are distinguished from the other isomerides by the action of nitrous acid, inasmuch as only one molecule of the base enters into the reaction,[1] and inner condensation occurs accompanied by elimination of water. *o*-Diamidobenzene yields amidoazophenylene or azimidobenzene :

$$C_6H_4\!\!\big\langle{}^{NH_2}_{NH_2} + NO_2H = C_6H_4\!\!\Big\langle{}^{N}_{N}\!\!\Big\rangle NH + 2H_2O.$$

The above constitution of the azimido-compounds is that given by Griess. Recent investigations[2] show that they have really the constitution $C_6H_4\!\!\big\langle{}^{N}_{NH}\!\!\big\rangle N$, and that a second series of azimides exists corresponding to Griess' formula.

The *o*-diamines further unite with all *a*-diketones, forming a well-marked series of bases, termed *quinoxalines*.[3] Thus, *o*-diamidobenzene unites with diacetyl, forming dimethylquinoxaline.

$$C_6H_4\!\!\big\langle{}^{NH_2}_{NH_2} + {}^{CO.CH_3}_{CO.CH_3} = C_6H_4\!\!\big\langle{}^{N=C.CH_3}_{N=C.CH_3} + 2H_2O.$$

These compounds are discussed in a later volume.

969 In the action of nitrous acid upon meta-compounds, on the other hand, two molecules of the latter take part in the reaction, and yellow or brown azo-colours are formed. From *m*-diamidobenzene, triamidoazobenzene or phenylene-brown is obtained :

$$2C_6H_4(NH_2)_2 + NO_2H = C_6H_4(NH_2)N = NC_6H_3(NH_2)_2 + 2H_2O.$$

According to Griess this reaction is so characteristic and delicate that it may be used to detect the nitrites in potable water ; for

---

[1] *Ber.* **9**, 219, 1524 ; **17**, 147.

[2] Nölting and Abt, *Ber.* **20**, 2999 ; Zincke and Campbell, *Annalen*, **255**, 339 ; *Ber.* **23**, 1315.   [3] *Annalen*, **237**, 327.

this it is only necessary to mix the solution of the base with the solution of the nitrite.[1]

Another characteristic reaction for the m-diamidobenzencs is that they unite with diazo-salts to form diamidoazo-compounds, yellow or yellowish red colours, one of which is well known as chrysoïdine.

970 Nitrous acid acts quite differently on p-diamido-compounds; if potassium nitrite be added to a solution of hydrochloride of p-diamido-benzene, a brown powder separates out after some time, the amount of this being increased on warming with violent evolution of gas and formation of quinone (Ladenburg).

In the para-series many compounds occur which yield quinone, $C_6H_4O_2$, on oxidation with manganese dioxide and dilute sulphuric acid; this substance being recognised by its powerful smell, as well as by the ease with which it sublimes in yellow needles. It is, however, not only obtained from para-substitution products, such as hydroquinone, p-diamidobenzene, &c., but also from monosubstitution products, such as aniline.

The influence of the nitroxyl is remarkable when it occurs in the ortho- or para-positions with respect to the halogens.

When the latter replace hydrogen in the aromatic group, the compounds formed are usually not capable of double decomposition, but they become so when nitroxyl occupies the above position. Both o- and p-bromonitrobenzene when heated with alcoholic potash are converted into the corresponding nitrophenols:

$$C_6H_4(NO_2)Br + KOH = C_6H_4(NO_2)OH + KBr.$$

If the former be heated with alcoholic ammonia, nitranilines are formed, which may be also obtained from the corresponding nitranisols:

$$C_6H_4(NO_2)OCH_3 + NH_3 = C_6H_4(NO_2)NH_2 + HO.CH_3.$$

These reactions do not take place in the meta-series.

The o-dinitro-compounds readily exchange a nitroxyl for other radicals.[2]

If o-dinitrobenzene be boiled with caustic soda o-nitrophenol is formed, and on heating the former with ammonia, o-nitraniline and ammonium nitrite are obtained.

---

[1] *Ber.* **11**, 625.   [2] Laubenheimer, *ibid.* **9**, 761; **11**, 303, 1151; **15**, 597.

If dinitrochlorobenzene, $C_6H_3Cl(NO_2)_2$, which has the following constitution—

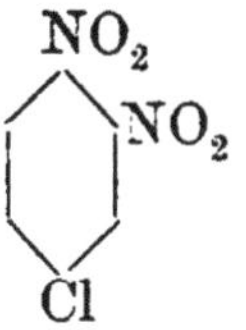

$$\begin{array}{c} NO_2 \\ \langle\,\rangle NO_2 \\ Cl \end{array}$$

be boiled with a solution of sodium sulphite, sodium nitrochloro-benzenesulphonate, $C_6H_3(NO_2)Cl(SO_3Na)$, and sodium nitrite are produced.

BENZENE GROUP.

BENZENE C_6H_6.

971 *Historical.* It has already been stated that Faraday in 1825, in his examination of the oil from "portable gas," dis-covered two new hydrocarbons (Vol. III., Part II., 2nd Ed., p. 180) one of which was afterwards found to be butylene. To the second, Faraday gave the name "bicarburet of hydrogen," as he found its empirical formula to be C_2H. (C=6) By exploding its vapour with oxygen, he observed that one volume contains thirty-six parts by weight of carbon to three parts by weight of hydrogen, and its sp. gr. compared with hydrogen is therefore 39.[1]

Mitscherlich obtained the same hydrocarbon in 1834, by distilling benzoic acid $C_7H_6O_2$ with slaked lime, and termed it benzine. He assumed that it is formed from benzoic acid simply by the removal of carbon dioxide.[2] Liebig denied this, adding the following note to Mitscherlich's memoir : " We have changed the name of the body obtained by Professor Mitscherlich by the dry distillation of benzoic acid and lime, and termed by him benzine, into benzole, because the termination 'ine' appears to denote an analogy bewteen strychnine, quinine &c., substances to which it does not bear the slightest resemblance, whilst the ending in ' ole ' corresponds better with its properties and mode of production. It would have been perhaps better, if the name given to it by the discoverer Faraday had been retained, as its relation to benzoic acid and the benzoyl compounds is not any

[1] *Phil. Trans.* **1825,** 440 ; *Pogg. Ann.* **3,** 306.   [2] *Annalen,* **9,** 43.

closer than it is to that of the tar or coal from which it is obtained."

Almost at the same time Péligot found that the same hydrocarbon occurs, together with benzone, $C_{13}H_{10}O$ (diphenylketone $CO(C_6H_5)_2$), in the products of the dry distillation of calcium benzoate.[1]

The different results obtained by Mitscherlich and Péligot are represented by the following equations.

$$C_7H_6O_2 + CaO = C_6H_6 + CaCO_3.$$
$$(C_7H_5O_2)_2Ca = C_{13}H_{10}O + CaCO_3.$$

Péligot obtained benzene only as a by-product, exactly as in the preparation of acetone (dimethylketone) from calcium acetate, a certain quantity of marsh-gas is always formed.

It is not clear how Liebig became acquainted with the fact that benzene is formed by the dry distillation of coal, as his pupil Hofmann, who obtained it in 1845 from coal-tar, observes : " It is frequently stated in memoirs and text-books that coal-tar oil contains benzene. I am, however, unacquainted with any research in which this question has been investigated." [2] It is, however, worthy of remark that about the year 1834, at the time when Mitscherlich had converted benzene into nitrobenzene, the distillation of coal-tar was carried out on a large scale in the neighbourhood of Manchester ; the naphtha which was obtained was employed for the purpose of dissolving the residual pitch, and thus obtaining black varnish. Attempts were made to supplant the naphtha obtained from wood-tar, which at that time was much used in the hat factories at Gorton near Manchester for the preparation of " lacquer," by coal-tar naphtha. The substitute however did not answer, as the impure naphtha gave off on evaporation so unpleasant a smell, that the workmen were unable to endure it. It was also known about the year 1838, that wood-naphtha contained oxygen, whilst that from coal-tar did not, and hence Mr. John Dale attempted to convert the latter into the former, or into some similar substance. By the action of sulphuric acid and potassium nitrate, he obtained a liquid possessing a smell resembling that of bitter-almond oil, the properties of which he did not further investigate.[3] This was, however, done in 1842 by Mr. John Leigh, who exhibited considerable quantities

---

[1] *Ann. Chim. Phys.* III. **56**, 59. [2] *Annalen*, **55**, 200.
[3] Private communication.

of benzene, nitrobenzene, and dinitrobenzene to the chemical section of the British Association meeting that year in Manchester. His communication is, however, so printed in the Report, that it is not possible from the description to identify the bodies in question.[1]

972 Coal-tar is obtained, as is well known, together with ammonia-liquor in the manufacture of coal-gas. The same products are also produced in the coking process, now usually carried on in ovens connected with vessels in which the aqueous distillate and the tar are collected. It is interesting to learn that these products were collected even before the manufacture of gas became known. Thus De Gensanne describes coke ovens connected with the ironworks at Sulzbach, near Saarbrücken, about 1764, in connection with which was an arrangement for collecting the volatile products " les huiles et le bitume." The oil thus obtained resembled distilled petroleum, had a strong bituminous odour, and burnt with a smoky flame ; the miners and peasants in the neighbourhood used it for their lamps.

These are probably the same ovens which Goethe, in 1771, saw in the neighbourhood of the burning hill near Dutweiler, a village near Sulzbach.[2] Here he met old Stauf, a coal philosopher —"philosophus per ignem," as one used to say—" who complained that the enterprise did not pay," for as Goethe says, they not only wanted to desulphurize the coal for use in iron-works, " but at the same time they wished to turn the oil and resin to account; nay, they would not lose the soot, and thus all failed together, on account of the many ends in view." [3]

Mansfield in 1848 prepared large quantities of benzene under Hofmann's direction, and proved that the naphtha in coal-tar contains homologues of benzenes, which may be separated from it by fractional distillation.[4]

Mansfield foresaw that the hydrocarbon would find further technical applications, but on the 17th of February 1856, whilst he was engaged in the preparation of specimens for the Paris Exhibition, the liquid in the retort boiled over and took fire,

---

[1] Private communication. *Brit. Assoc. Report*, **1842**, 39. See also *Moniteur Scient.* **1865**, 446.

[2] Private communication of Dr. Guilt to Watson Smith. The title of Gensanne's work is : *Traité de la Fonte des Mines par le Feu du Charbon de Terre.* 2 vols. Paris, **1770** et **1776**.

[3] *Aus Meinem Leben,* " Wahrheit und Dichtung." Oxenford's Translation, **1**, 364.

[4] *Quart. Journ. Chem. Soc.* **1**, 244.

burning Mansfield so severely that he died in a few days.[1] The
construction of the still employed is shown in Fig. 1.

Sources of Benzene.—Benzene is also found in the products of
distillation of wood and many organic bodies. It is likewise produced when the vapours of its homologues and other aromatic
compounds are passed through a red-hot tube. Even the simplest
compounds of the fatty group, such as marsh-gas, alcohol and·
acetic acid, yield some benzene when thus treated, together with
a variety of other bodies.[2]

Of special interest is its synthesis from acetylene, C_2H_2, discovered by Berthelot. If this gas be heated for some time at the
temperature at which glass begins to soften, it is converted into

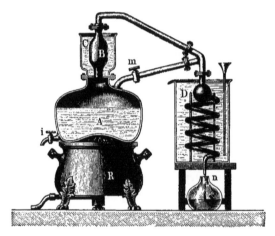

Fig. 1.

a mixture of polymerides, amongst which benzene is present
in considerable quantity.[3] Marignac has shown that it is also
formed by the distillation of phthalic acid, $C_8H_6O_4$, with caustic
lime.[4] The reaction is quite similar to that by which it is formed
from benzoic acid :

$$C_6H_5.CO_2H = C_6H_6 + CO_2$$
$$C_6H_4(CO_2H)_2 = C_6H_6 + 2CO_2.$$

Since that time, it has been observed that all benzenecarboxylic
acids, *i.e.* acids which are derived from benzene by replacement of
hydrogen by carboxyl, are in this way converted into benzene.

[1] *Quart. Journ. Chem. Soc.* **8**, 110.
[2] Berthelot, *Jahresb.* **1851**, 437, 504 ; **1868**, 333.
[3] *Ibid.* **1870**, 1.   [4] *Annalen*, **42**, 217·

Benzene also occurs, together with its homologues, in small quantities in petroleum from Burmah (Rangoon tar),[1] in that from Galizia,[2] and in that from Canadian and Pennsylvanian wells. (Schorlemmer.)

Manufacture.—Benzene is largely employed in the arts, especially in the manufacture of aniline. In preparing it, coal-tar, or tar from certain coking processes, is distilled in large retorts made of malleable iron. The usual English retort is shown in Figs. 4 and 5, whilst those used in France and Germany are represented in Figs. 2 and 3.

FIG. 2.

The vapours are cooled by passing through lead or copper pipes, or by means of a system of iron pipes as shown in Fig. 6.

In this process the following products are obtained, combustible gases being also evolved at the commencement of the distillation.

The *first runnings* amount to about 2 to 4 per cent. of the tar, and consist of carbon disulphide, paraffins, olefines, benzene and other hydrocarbons, acetonitrile, ethyl alcohol, &c. At the same time ammonia-water comes over, and for this reason the dis-

---

[1] Warren de la Rue and H. Müller, *Proc. Roy. Soc.* **8**, 221.
[2] Freund, *Annalen,* **115**, 19.

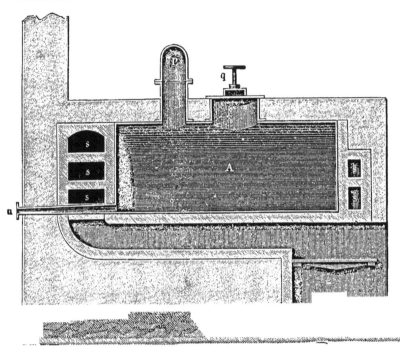

FIG. 3.

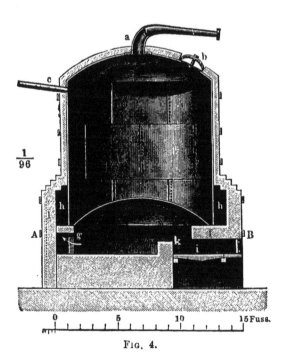

FIG. 4.

tillate separates into two layers. If a thermometer is used in working, the receiver is changed when the temperature rises to about 110°.

The next product is termed *light oil* or *crude naphtha*. As soon as the water is removed, the peculiar hissing noise ceases,

FIG. 5.

and the light oil (which swims on water) now comes over, and amounts to about 7 to 8 per cent. of the tar, the temperature, rising to about 210°. In some works the light oil is collected, below 170°, and that coming over between 170° and 230° is called *middle oil.*

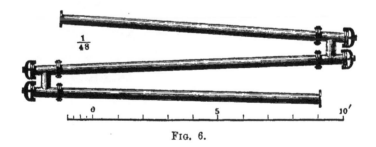

FIG. 6.

Between 210° and 400° *heavy* or *dead oil* comes over, which sinks in water. This is also frequently collected in two portions that which comes over below 270° being known as *creosote oil* whilst the higher boiling portion is termed *anthracene oil.* The further treatment of these oils, as well as that of pitch, will be referred to hereafter.

973 Benzene mixed with other hydrocarbons may be obtained from the first runnings by fractional distillation. The *light oil,*

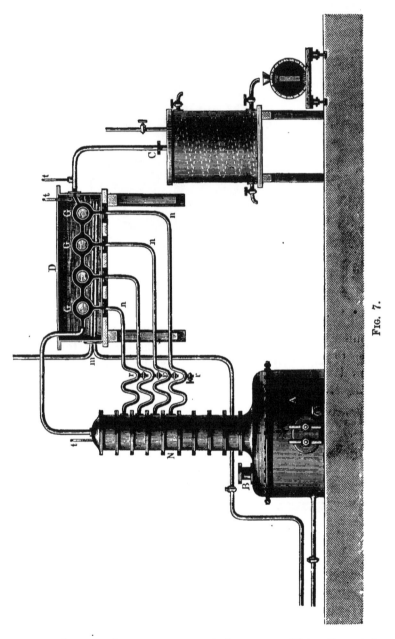

Fig. 7.

however, is the chief source of crude benzene. This oil contains a small quantity of paraffins, olefines, benzene and its homologues, phenol, cresol, naphthalene and bases. It is first rectified once or several times in light oil stills, which are similar in structure to the tar stills, but usually smaller. The portion boiling between 80° and 150°, together with that fraction of the first runnings which boils above 80°, and the portion of the creosote oils which, on rectification, boils below 150°, are then worked up for benzene and its homologues. The first boiling portions of the light oils are frequently collected separately and brought into commerce as crude naphtha, being employed for carbureting gas, and as a solvent.

The rectified light oils are then treated with 5 per cent. of concentrated sulphuric acid, for the purpose of freeing them from bases, olefines and other hydrocarbons; well washed with water and thoroughly agitated with caustic soda, which dissolves phenol and its homologues.

In 1860, E. Kopp suggested the employment of the apparatus used in the rectification of alcohol for the purpose of fractionating the neutral oils thus obtained, and this suggestion was first carried out on the large scale by Coupier at Poissy, near Paris, in 1863.

The apparatus he employed, Fig. 7, consists of a boiler heated by steam and a column that serves as a dephlegmator, in which a partial condensation takes place. The vapours which do not here liquefy pass into a series of condensers surrounded by a solution of calcium chloride, which for the purpose of obtaining benzene is heated to a temperature of 80°.

When this has all distilled over, the bath of calcium chloride is heated to 110°, toluene then passing over. To obtain the mixed xylenes, a paraffin bath is employed heated by high pressure steam to a temperature of 140°. Coupier obtained the following results:

10 litres of light oil boiling between 80° and 150° gave :

			B.P.
6 litres	of	crude benzene	62—80°
44	„	benzene	80—82°
6	„	intermediate product . .	82—110°
17	„	toluene	110—112°
5	„	intermediate	112—137°
9	„	xylenes	137—140°
5	„	intermediate	140—148°
8 to 9	„	trimethylbenzenes . . .	148—150° [1]

[1] These, however, boil from 163° to 166°.

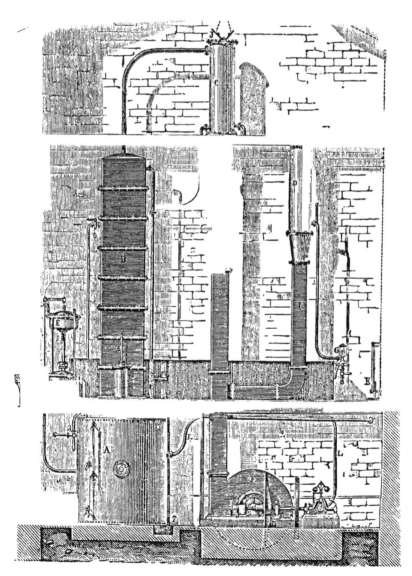

FIG. 8.

In larger works the hydrocarbons are separated by means of large copper column apparatus (similar to those employed in the rectification of spirits). Fig. 8 shows one made by Savalle, in Paris, a vertical section being shown in Fig. 9. The condensation of the vapours is effected by a current of air, the strength of which can be regulated.

The benzene boiling from 80—82° is pure enough for the preparation of pure aniline. It always, however, contains a small

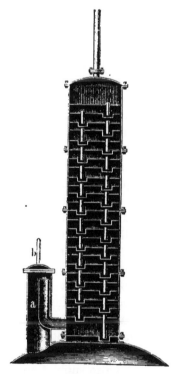

FIG. 9.

quantity of paraffins and non-saturated hydrocarbons. In order to purify it Mansfield's process is employed. This consists in crystallizing the crude benzene in a freezing mixture, when the impurities remain liquid and can be separated from the pure solidified benzene. A simple apparatus for this purification has been described by Hofmann.[1]

Chemically pure benzene, obtained from benzoic acid, does not colour sulphuric acid brown, though that obtained from coal-tar

[1] *Ber.* **4,** 162.

does so even after crystallization. This latter product exhibits a very characteristic reaction. If a granule of isatin, $C_8H_5NO_2$, be dissolved in concentrated sulphuric acid, and benzene obtained from coal-tar added, the liquid after a short time assumes a fine blue colour; if, however, the benzene be previously shaken with sulphuric acid, until it is no longer coloured brown, or if the benzene obtained from benzoic acid be used, the isatin reaction is not given.[1]

The body which produces this reaction is thiophene, C_4H_4S, a liquid possessing a most remarkable similarity to benzene. It will be described, together with its very numerous derivatives, in a later volume.

974 *Properties.*—Benzene is a colourless, mobile, strongly refracting liquid, possessing a peculiar smell, boiling at 80°·39, and solidifying, when cooled, to rhombic crystals which melt at 4°·5. It is easily inflammable, burning with a luminous and smoky flame. Like other hydrocarbons, it is very slightly soluble in water, but readily dissolves in alcohol, ether, chloroform, &c. On the other hand, it serves as an excellent solvent for iodine, sulphur, phosphorus, fats, resins, various alkaloids, and many other organic compounds. Its specific gravity is 0·89408 at 0°, 0·87868 at 15°, and 0·87360 at 20°.[2]

Benzene taken internally is found in the urine as a salt of phenylsulphuric acid. If the vapour of about 10 grams be inhaled, dizziness and sickness, as well as a tendency to sleep, occur. In doses of from 40 to 50 grams it produces a similar action to chloroform, accompanied by profuse perspiration; the same quantity given to cats acts fatally, with epileptic seizures. In works in which the fat is extracted from woollen fibre by benzene, the vapours of this substance produce intoxication amongst the workmen, who also suffer from a peculiar irritation and dryness of the skin, due, according to Perrin, to the solution of its fatty constituents.[3]

When the induction spark is passed through benzene a solid substance is formed, together with carbon, whilst a mixture of 42 to 43 per cent. of acetylene, and 57 to 58 per cent. of hydrogen, is given off.[4]

If benzene be oxidized with manganese dioxide and sulphuric acid various products are formed, amongst which are formic,

---

[1] V. Meyer, *Ber.* **16**, 1465, 2172, 2968.　　　[2] *Ber.* **21**, 2210.
[3] Grandhomme, *Die Farbwerke von Meister, Lucius und Brüning in sanitärer und socialer Beziehung.*　　　[4] Destrem, *Compt. Rend.* **99**, 138.

benzoic, and phthalic acids.[1] If formic acid be added to the above oxidizing mixture a larger proportion of benzoic acid is formed, from which Carius concluded that it is produced by a simultaneous oxidation of benzene and formic acid.

$$C_6H_6 + CHO.OH + O = C_6H_5.CO.OH + H_2O.$$

It appears, however, more probable that a part of the benzene is oxidized to diphenyl, $C_6H_5.C_6H_5$, and that this on further oxidation yields benzoic acid, whilst phthalic acid, $C_6H_4(CO.OH)_2$, is formed from diphenylbenzene, $C_6H_4(C_6H_5)_2$, which is probably formed at the same time as the diphenyl.

Benzene, when brought into contact with a glowing spiral of platinum wire, in presence of air, is oxidized to benzaldehyde and benzoic acid.[2]

By the action of potassium chlorate and sulphuric acid on benzene, Carius obtained a peculiar acid, which he considered to be a chlorine substitution product of an acid homologous with malic acid, and to which he gave the name of trichlorophenomalic acid, $C_6H_7Cl_3O_5$. He believed it to be produced according to the following equation :

$$C_6H_6 + 3ClO_2H = C_6H_7Cl_3O_5 + H_2O.$$

At the same time he obtained chlorobenzene, oxalic acid, chlorinated quinols and quinones, amongst which was dichloroquinone, $C_6H_2Cl_2O_2$, and other bodies.

By heating trichlorophenomalic acid with baryta water, phenaconic acid, $C_6H_6O_6$, a substance which Carius believed to be isomeric with aconitic acid, was produced :

$$C_6H_7Cl_3O_5 + H_2O = C_6H_6O_6 + 3HCl.$$

When, however, this body was heated with hydriodic acid it was completely converted into succinic acid, whilst by the action of bromine two isomeric dibromosuccinic acids were formed. Carius concluded that these, as well as the succinic acid, must be produced from phenaconic acid by " polymeric rearrangements." On warming a mixture of the two dibromosuccinic acids with baryta water Carius obtained racemic acid and bromomaleic acid.

A short time after this Carius, from some experiments made

---

[1] *Annalen*, **148**, 50. [2] Coquill, *Compt. Rend.* **80**, 1089.

on the acid potassium salt of fumaric acid, came to the conclusion that phenaconic acid is identical with fumaric acid. Hence phenaconic acid disappeared from the list of chemical compounds; trichlorophenomalic acid remained, however, as a mysterious body, and its decomposition into hydrochloric acid and fumaric acid was unexplained.

In 1877 Krafft repeated Carius's experiments, and found in the product of the reaction a large quantity of trichloroquinone, $C_6HCl_3O_2$, and trichloroquinol, $C_6H_3Cl_3O_2$, which latter substance he considered to be identical with trichlorophenomalic acid. This compound then disappeared from chemical literature, the fact being overlooked that Krafft had only proved the formation of trichloroquinone and trichloroquinol in the reaction, and had in no way shown the identity of the latter substance with Carius's acid, not having tried the only decisive reaction, viz. the conversion of trichlorophenomalic acid into fumaric acid.

The investigation of trichlorophenomalic acid was next taken up by Kekulé and Strecker. In the introduction to their memoir they say : " To sum up shortly the results of our critical studies, we simply say : that it is nothing but a " comedy of errors." Trichlorophenomalic acid is not formed according to the equation given by Carius, neither has it the formula which Carius ascribed to it ; on the other hand, it is not, as Krafft believed, identical with trichloroquinol. It does not yield hydrochloric acid or fumaric acid by the action of alkalis, but, notwithstanding, Carius's phenaconic acid is identical with fumaric acid."

It has already been stated (p. 66) that trichlorophenomalic acid proved to be β-trichloracetylacrylic acid, $C_5H_3Cl_3O_3$, which is soluble with difficulty in cold, but more readily in hot water, and crystallizes in small glistening scales, which possess a pleasant smell, and melt at 131—132°.

On warming with baryta water it decomposes into chloroform and maleic acid, $C_4H_4O_4$, and it unites with bromine to form trichloracetyldibromopropionic acid, $C_5H_3Cl_3Br_2O_3$, which on heating with lime-water splits up into chloroform and inactive tartaric acid. Carius's statements can now be simply explained. In his first experiments part of the maleic acid was converted into fumaric acid ; by the action of bromine on this mixture he obtained dibromosuccinic acid together with isodibromosuccinic acid, the former of which yielded inactive tartaric acid, then believed to be racemic acid. In his later experiments the conversion of the maleic acid was complete, and his pure phenaconic acid consisted entirely of fumaric acid.[1]

If boiling benzene be saturated with picric acid (trinitrophenol), the compound, $C_6H_6 + C_6H_2(NO_2)_3OH$, separates out on cooling in bright yellow crystals, which give off benzene on exposure to air (Fritzsche).

If antimony chloride be dissolved in hot benzene, transparent monosymmetric tables of the compound, $(C_6H_6)_2(SbCl_3)_3$, separate out on cooling, which deliquesce on exposure to air.[2]

When a current of hydrochloric acid is passed into a mixture of benzene and aluminium chloride, a compound, $3C_6H_6 + AlCl_3$, is obtained as a thick orange-coloured liquid, which crystallizes at $-5°$, and is decomposed by water into its constituents. Aluminium bromide forms a similar compound.[3]

When benzene is heated with potassium at 230—250°, a bluish-black crystalline body is formed without any hydrogen being evolved; on exposure to air this substance ignites spontaneously with explosive violence. If this body be covered with benzene, and water slowly added, diphenylbenzene, $C_6H_4(C_6H_5)_2$, is formed together with a small quantity of diphenyl, $(C_6H_5)_2$ and of a hydrocarbon, $C_{12}H_{16}$, which is a thick liquid smelling like aniseed and boiling at 222°. From this it would appear that the above blue body is a mixture of monopotassium benzene, C_6H_5K, dipotassium benzene, $C_6H_4K_2$, and potassium hydrogenide, K_4H_2 (Vol. II. Part I. p. 61).[4]

ADDITIVE PRODUCTS OF BENZENE.

975 *Benzene hexhydride,* or *Hexhydrobenzene,* C_6H_{12}, was first obtained by Berthelot, and believed by him to be hexane. It is formed when benzene is heated for five hours at 280°, with a large excess of aqueous hydriodic acid, saturated at 0°. It forms a constituent of Caucasian petroleum,[5] and is a liquid boiling at 69°, and having a specific gravity 0·76 at 0°.[6]

---

[1] *Annalen* **223**, 170 (where also the complete literature will be found).
[2] Smith and Davis, *Journ. Chem. Soc.* 1882, **1**, 412.
[3] Gustavson, *Ber.* **11**, 2151.
[4] Abeljanz, *ibid.* **9**, 10.
[5] Beilstein and Kurbatow, *ibid.* **13**, 1818.
[6] Wreden and Znatowicz, *Annalen*, **187**, 163.

Benzene hexchloride, $C_6H_6Cl_6$, was first prepared by Mitscher-lich,[1] Peligot,[2] and Laurent,[3] by exposing benzene in large flasks filled with chlorine to the action of sunlight. It is also obtained when chlorine is passed into boiling benzene.[4]

According to Hugo Müller, benzene dichloride $C_6H_6Cl_2$, and benzene tetrachloride, $C_6H_6Cl_4$, bodies which have not been more closely examined, together with substitution products, are first formed.[5] Additive products are also formed when hydrochloric acid and potassium bichromate act upon benzene.[6]

In order to prepare benzene hexchloride, chlorine is allowed to act in the sunlight on the surface of benzene contained in large flasks.[7] The crystalline mass thus formed is then recrystallized from hot benzene. It forms large monosymmetric prisms melting at 157° and boiling, with liberation of hydrochloric acid, at 288°; alcoholic potash decomposes it completely into hydrochloric acid and asymmetric trichlorobenzene. Fuming nitric acid does not attack it even on warming, whilst zinc reduces it in alcoholic solution to benzene.[8]

On passing chlorine into boiling benzene in the sunlight an isomeric benzene hexchloride is formed, which crystallizes in regular octahedra and tetrahedra, etc. This body melts at about 310° and volatilizes about the same temperature; it is only slowly attacked by boiling alcoholic potash, and is unaltered by a boiling alcoholic solution of potassium cyanide, by which reaction it can be separated from the ordinary hexchloride, as this is at once converted into trichlorobenzene.[9]

According to Schüpphaus, Meunier's compound, which Hübner previously observed, is perhaps diphenyl dodecachloride, $C_{12}H_{10}Cl_{12}$.[10]

Benzene hexbromide, $C_6H_6Br_6$, was obtained by Mitscherlich by the action of bromine on benzene in sunlight,[11] bromo- and tribromobenzene being also formed.[12] It separates from its ethereal solution in microscopic, opaque, rhombic prisms

---

[1] *Pogg. Ann.* **35**, 370.   [2] *Ann Chim. Phys.* II. **56**, 66.

[3] *Annalen,* **23**, 68.

[4] Lesimple, *Annalen,* **137**, 123 ; Heys, *Zeitsch. Chem.* **1871**, 293.

[5] *Jahresb.* **1862**, 414.   [6] Jungfleisch, *ib.* **1868**, 355.

[7] Leeds and Everhart, *Am. Chem. Journ.* **2**, 205.

[8] Zinin, *Zeitsch. Chem.* **1871**, 284.

[9] Meunier, *Compt. Rend.* **98**, 436.

[10] *Ber.* **17**, 2256.   [11] *Annalen,* **16**, 173.

[12] Meunier, *Compt. Rend.* **101**, 578.

(Lassaigne), melts at 212°, and is decomposed by alcoholic potash into tribromobenzene and hydrobromic acid.

Benzene trichlorhydrin, $C_6H_6Cl_3(OH)_3$, is formed when benzene is brought into contact with aqueous hypochlorous acid. It is difficultly soluble in water, but readily in ether, and crystallizes in fine scales which melt at 10°, and are volatile without decomposition. Very dilute alkalis decompose it with formation of various products, among which phenose, $C_6H_{12}O_6$, a body isomeric with the glucoses or " hexoses " is found. This is to be described as an amorphous, hygroscopic mass, which possesses a sweetish taste and reduces an alkaline copper solution. Concentrated hydriodic acid is said to convert it into secondary hexyl iodide on heating.[1]

CHLORINE SUBSTITUTION PRODUCTS OF BENZENE.

976 It has already been stated that by the action of chlorine on benzene a mixture of additive and substitution products is obtained. When iodine is added, a smooth and regular substitution takes place in consequence of an alternate decomposition and reformation of iodine chloride. Still more energetic is the action of antimony chloride, which is therefore used in the preparation of higher chlorinated derivatives.[2]

All the theoretically possible compounds are not formed by this direct substitution. The monochlorobenzene which is first formed is chiefly converted into *p*-dichlorobenzene, some *o*-dichlorobenzene being produced at the same time. On further chlorination both yield asymmetric trichlorobenzene, and this again yields symmetric tetrachlorobenzene, in which the two hydrogen atoms are in the para-position. This is then further converted into pentachlorobenzene and hexchlorobenzene.

Monochlorobenzene is also formed by the action of phosphorus pentachloride on phenol, as well as when the amido-group of aniline is replaced by chlorine, by means of the diazo-reaction. In the same way other chlorobenzenes are obtained from chlorine substitution products of aniline and phenol.

Monochlorobenzene, C_6H_5Cl, was obtained by Laurent and

---

[1] Carius, *Annalen,* **136**, 323 ; **140**, 322.
[2] Hugo Müller, *Journ. Chem. Soc.* **15**, 41.

Gerhardt by the action of phosphorus pentachloride on phenol (phenyl alcohol) and called chloride of phenyl (chlorure de phényle).[1]

The product thus obtained, however, was a mixture of chlorobenzene with phenyl orthophosphate,[2] and probably also the chlorides of monophenyl- and diphenyl-phosphoric acid;[3] this explains the statements of the above chemists that chloride of phenyl when treated with alkalis is again converted into phenyl alcohol.

In order to prepare chlorobenzene, rather less than the theoretical amount of chlorine is passed into a solution of one part of iodine in four parts of benzene, the product washed with caustic soda, and purified by fractional distillation.[4]

It is also formed when benzene is heated with sulphuryl chloride to 150°.[5]

$$C_6H_6 + SO_2Cl_2 = C_6H_5Cl + HCl + SO_2.$$

Monochlorobenzene is a pleasant-smelling liquid, which boils at 132°, and has a specific gravity of 1·12387 at 0°.[6] It solidifies to a crystalline mass, melting at − 40°.

Concentrated nitric acid converts it into a mixture of solid p-chloronitrobenzene and liquid o-chloronitrobenzene.

977 *Monochlorobenzene hexchloride*, $C_6H_5Cl_7$, is formed when diphenyl sulphone, $(C_6H_5)_2SO_2$, is treated with chlorine in the sun-light. It crystallizes from hot alcohol in small quadratic prisms which melt at 255—257°.[7]

p-*Dichlorobenzene*, $C_6H_4Cl_2$, was obtained by H. Müller by passing chlorine into a solution of iodine in benzene until a sample of the product sank rapidly in water; it was then washed with caustic soda, any benzene and monochlorobenzene still present removed by distillation, and the portion boiling above 160° cooled to 0°. The dichloro-compound crystallizes out, and is purified, after removing the mother-liquor, by re-crystallization from alcohol. p-Dichlorobenzene is also readily obtained when the requisite quantity of chlorine is passed into benzene, to which 1 per cent. of molybdenum chloride has been added,[8] and is also formed, with other products, by the action of phosphorus

---

[1] *Annalen*, **75**, 79.　　　　　　[2] Scrugham, *Journ. Chem. Soc.* **7**, 237.
[3] Jacobsen, *Ber.* **8**, 1519.　　　　[4] Hugo Müller, *Zeitsch. Chem.* **1864**, 65.
[5] Dubois, *ibid.* **1866**, 705.　　　　[6] Adrieenz, *Ber.* **6**, 441.
[7] Otto, *Annalen*, **141**, 101.　　　　[8] Aronheim, *Ber.* **8**, 1400.

pentachloride on p-chlorophenol[1] and p-phenolsulphonic acid, $C_6H_4(OH)SO_3H$.[2]

p-Dichlorobenzene crystallizes in monosymmetric plates, which possess a pleasant sweet smell, and sublime in closed vessels, at the ordinary temperature, in large, four-sided tables. It melts at 56·4° and boils at 173·2°.[3] It is converted by fuming nitric acid into p-dichloronitrobenzene, melting at 54·5°.

o-*Dichlorobenzene*, $C_6H_4Cl_2$, was first prepared by Beilstein and Kurbatow by the action of phosphorus pentachloride on o-chlorophenol.[4] They also found it in the mother liquor obtained in the preparation of p-dichlorobenzene by Müller's method. In order to prepare it from this, the liquid is heated for two days with fuming sulphuric acid to 210°, when o-dichloro-benzenesulphonic acid is formed, any p-dichlorobenzene still present not being acted upon. The sulphonic acid, which is obtained in the pure state from the barium salt, yields pure o-dichlorobenzene[5] on dry distillation. It is a liquid boiling at 179° and not solidifying at − 14°. Fuming nitric acid converts it into the dichloronitrobenzene which melts at 43°.

m-*Dichlorobenzene*, $C_6H_4Cl_2$, was obtained by Körner, by converting m-nitraniline into the diazo-chloride and fusing its platinochloride with soda. He thus obtained m-chloronitro-benzene, which he reduced to m-chloraniline, in which he then replaced the amido-group by chlorine. It is also formed when the corresponding dichloraniline is covered with absolute alcohol, nitrogen trioxide passed in, and the solution, after saturation, heated to boiling. It is a liquid boiling at 172°, and having a specific gravity of 1·307 at 0°. On nitration it yields a dichloronitrobenzene melting at 33°, which on reduction is converted into the original dichloraniline.[6]

Dichlorobenzene hexchloride, $C_6H_4Cl_8$, was obtained by Jung-fleisch, by the action of chlorine on chlorobenzene in the sun-light. It crystallizes from chloroform in oblique rhombic prisms which do not melt at 250°.[7]

[1] Beilstein and Kurbatow, *Annalen*, **176**, 40.
[2] Kekulé, *Ber.* **6**, 944.
[3] Körner, *Jahrcsb.* **1875**, 318.
[4] *Annalen*, **176**, 42.
[5] *Ibid.* **182**, 94.
[6] *Ibid.* **182**, 97.
[7] Jungfleisch, *Zeitsch. Chem.* **1868**, 486.

Trichlorobenzenes, $C_6H_3Cl_3$.

		Melting-point.	Boiling-point.
[1] Symmetric	(1.3.5) long needles	63 4°	208·5°
[2] Asymmetric	(1.2.4) rhombic crystals . . .	16·0°	213·0°
[3] Adjacent	(1.2.3) large tables	53-54°	218-219°

Tetrachlorobenzenes, $C_6H_2Cl_4$.

		Melting-point.	Boiling-point.
[4] Symmetric	(1.2.4.5) monosymmetric . crystals	137-138°	243-246°
[5] Asymmetric	(1.3.4.5) needles	50-51°	246°
[6] Adjacent	(1.2.3.4) needles	45-46°	254°

Pentachlorobenzene, C_6HCl_5, crystallizes from hot alcohol in fine needles, melting at 85—86°, and boiling at 275—276°.[7]

978 *Hexchlorobenzene*, C_6Cl_6, was obtained by Hugo Müller as a final product of the action of chlorine on benzene in presence of iodine, or better, of antimony pentachloride. He expressed the opinion that it was identical with the compound already known under the name of Julin's chloride of carbon, which was prepared in a very peculiar way. Julin manufactured nitric acid in Åbo, in Finland, by heating crude saltpetre in cast-iron retorts with calcined green vitriol, obtained from the drainage water of the mines of Fahlun, in Sweden. He thus obtained a body of which he sent a small sample to Richard Phillips, editor of the *Annals of Philosophy*, stating that it was a similar substance to the perchloride of carbon discovered by Faraday.[8]

Faraday and Phillips investigated the compound more accurately, but, owing to the small quantity they possessed they were unable to come to any satisfactory conclusions re-

---

[1] Jungfleisch, *Ann. Chim. Phys.* IV. **15**, 186 ; Beilstein and Kurbatow, *Annalen*, **192**, 236.

[2] Mitscherl ch, *ibid.* **16**, 172 ; Otto, *ibid.* **161**, 105 : Jungfleisch, *loc. cit.* ; Beilstein and Kurbatow.   [3] Beilstein and Kurbatow.

[4] Jungfleisch ; Beilstein and Kurbatow.

[5] Otto ; Jungfleisch ; Beilstein and Kurbatow.

[6] Beilstein and Kurbatow.

[7] Jungfleisch ; Otto and Ostrop, *Annalen*, **141**, 93 ; **154**, 182 ; Beilstein and Kuhlberg, *ibid.* **152**, 247 ; Ladenburg, *ibid.* **172**, 344.

[8] *Ann. Phil.* **17**, 216.

specting its composition. They noticed, however, that it possessed certain peculiar properties, and came to the conclusion that it probably contained a new modification of carbon, or perhaps some analogous element.[1] Julin gave them a larger quantity after his return from the Continent, and they then found that the body was a chloride of carbon, whose vapour on being passed through a red-hot tube, filled with small pieces of rock crystal, is decomposed into its elements. Analysis showed that the formula was $C_4Cl_2(C=6)$, but they gave the compound no name.[2] Gmelin described it as dichloride of carbon : according to him the carbon needed for its formation was probably derived from the cast iron, and the chlorine from the crude saltpetre.[3]

Regnault obtained a compound of chlorine and carbon by passing the vapour of chloroform and of tetrachlorethylene through a red-hot tube filled with pieces of porcelain.[4] Bassett[5] showed that this compound is hexchlorobenzene, and Ramsay and Young[6] have also shown that if chloroform be employed, hexchlorethane and tetrachlorethylene are also formed.

It is also formed when acetylene tetrachloride, $C_2H_2Cl_4$, is heated for 100 hours at 360°,[7] as well as by the prolonged chlorination of methylbenzene and of dimethylbenzene in presence of antimony chloride.[8] It is likewise formed when several benzene derivatives are heated with iodine chloride to 350°. Several aromatic hydrocarbons, and oil of turpentine, $C_{10}H_{16}$, also yield it by exhaustive chlorination, together with tetrachloromethane, or hexchlorethane.[9]

If secondary hexyl iodide be heated with excess of iodine chloride for a long time to 240°, hexchlorobenzene, together with tetrachloromethane, is formed.[10]

Hexchlorobenzene is insoluble in cold, slightly soluble in hot alcohol, but readily in benzene. From a mixture of the two it crystallizes in long thin prisms, and from carbon disulphide in rhombic prisms. It melts at 226°, boils at 326°, and is not attacked either by boiling concentrated acids or by alkalis.

[1] *Ann. Phil.* p. 217.
[2] *Phil. Trans.* **1821**, 392.
[3] Gmelin, *Handbook*, **8**, 160.
[4] *Annalen*, **30**, 350.
[5] *Journ. Chem. Soc.* II. **5**, 443.
[6] *Journ. Soc. Chem. Ind.* **5**, 232.
[7] Berthelot and Jungfleisch, *Annalen*, Suppl. **7**, 256.
[8] Beilstein and Kuhlberg, *ibid.* **150**, 309.
[9] Ruoff, *Ber.* **9**, 1483.
[10] Krafft, *ibid.* **9**, 1085.

BROMINE SUBSTITUTION PRODUCTS OF BENZENE.

979 Bromine acts on benzene in a similar manner to chlorine but more slowly; the presence of iodine accelerates the reaction.

Monobromobenzene, C_6H_5Br, was first obtained by Couper by the prolonged action of bromine on benzene in diffused daylight.[1]

Riche obtained it from phenol and phosphorus pentabromide,[2] and Griess by the decomposition of diazobenzene perbromide.[3]

It is produced when a mixture of equal molecular weights of bromine and benzene are allowed to act upon one another for a week; the product is then washed with caustic potash, and the unaltered benzene, and a small quantity of dibromobenzene removed by distillation.[4]

According to Michaelis it is best prepared by heating 2,500 grams of benzene and 50 grams of iodine in a bolthead connected with an inverted condenser, and gradually adding 2,500 grams of bromine. As soon as no further evolution of hydrobromic acid takes place on continued boiling, the excess of benzene is distilled off, the residue washed with caustic soda, and distilled in a current of steam. The dry distillate is finally purified by fractionation.[5]

Monobromobenzene is a liquid boiling from 154·8—155·5° and having at 0° a specific gravity 1·51768.[6]

On nitration it yields chiefly *p*-nitrobromobenzene, together with a small quantity of *o*-nitrobromobenzene. If from three to four grams of bromobenzene be given to a large dog daily, several different bodies make their appearance in the urine; amongst these are *p*-bromophenylsulphuric acid, $C_6H_4BrSO_4H$, and bromophenylmercapturic acid, $C_{11}H_{10}BrNSO_3$. This latter substance crystallizes from hot water in long needles, decomposing on boiling with caustic soda into *p*-bromothiophenol, C_6H_4BrSH ammonia, acetic acid, and other bodies as yet unexamined;[7] according to Jaffé chlorobenzene acts on the animal economy in a similar manner.

---

[1] *Annalen*, **104**, 225.　　[2] *Ibid.* **121**, 359.
[3] *Ibid.* **137**, 86.　　[4] Fittig, *ibid.* **132**, 201.
[5] *Ibid.* **181**, 289.　　[6] Adrieenz, *Ber.* **6**, 443.
[7] Baumann and Preusse, *ibid.* **12**, 806 ; Jaffé, *ibid.* 1092.

p-Dibromobenzene, $C_6H_4Br_2$, was discovered by Couper.[1] It is produced when one part of benzene and eight parts of bromine are boiled for some days in an apparatus connected with a reflux condenser; the excess of bromine is removed by heating, the residue washed with caustic soda, and cooled down until *p*-dibromobenzene crystallizes out. This is purified by pressing out the mother liquor, and recrystallized from alcohol.[2] It is also formed by heating *p*-bromophenol with phosphorus pentabromide,[3] and from *p*-dibromaniline by the diazo-reaction.[4] It crystallizes in monosymmetric tables or prisms melting at 89·3° and boiling at 219°.

o-Dibromobenzene, $C_6H_4Br_2$, is formed in small quantities during the preparation of the para-compound,[5] but more readily from *o*-bromaniline.[6] It is a light mobile liquid, having at 0° a specific gravity 2·003, boiling at 224°, and on cooling forms crystals melting at −1°.

m-Dibromobenzene, $C_6H_4Br_2$, was obtained by V. Meyer and Stüber, by heating an alcoholic solution of *m*-dibromaniline with ethyl nitrite.[7]

$$C_6H_3Br_2.NH_2 + C_2H_5NO_2 =$$
$$C_6H_4Br_2 + C_2H_4O + N_2 + H_2O.$$

It has also been prepared from *m*-dibromaniline.[8] It is a liquid boiling at 219·4°, not solidifying at −26°, and having a specific gravity of 1·955 at 18·6°.

Tribromobenzenes, $C_6H_3Br_3$.

			Melting-point.	Boiling-point.
[9] Symmetric (1.3.5)	needles		119·6°	278°
[10] Asymmetric (1.2.4)	needles		44·0°	275—276°
[11] Adjacent (1.2.3)	rhombic tables	87·4°		—

The symmetric tribromobenzene is obtained from *s*-tribromaniline by removal of the amido-group, and is also formed by the polymerization of bromacetylene under the influence of light (Part II. p. 524).

---

[1] *Annalen,* **104**, 225. [2] Riche and Bérard, *ibid.* **133**, 51.
[3] Mayer, *ibid.* **137**, 221.
[4] Griess, *Jahresb.* **1866**, 454; Körner, *Journ. Chem. Soc.* 1876, **1**, 212.
[5] Riese, *Annalen,* **164**, 176. [6] Körner, *Journ. Chem. Soc.* 1876, **1**, 212.
[7] *Annalen,* **165**, 161. [8] Körner, Wurster, *ibid.* **176**, 170.
[9] Körner; Meyer and Stüber, *loc. cit.*
[10] Mitscherlich; Meyer; Körner; Griess; Wurster, *Ber.* **6**, 1490; Wroblewsky, *ibid.* **7**, 1060. [11] Körner.

Tetrabromobenzenes, $C_6H_2Br_4$.

[1] Symmetric (1.2.4.5) long needles   137—140°   —
[2] Asymmetric (1.3.4.5) slender needles 98·5°   329°
[3] Adjacent   (1.2.3.4) small needles   160°   —

Pentabromobenzene, C_6H_5Br, was obtained by Kekulé, together with s-tetrabromobenzene, by heating nitrobenzene with bromine to 250°. It is also formed together with tribromobenzenesulphonic acid, when s-tribromobenzene is heated, for from eight to fourteen days, with fuming sulphuric acid to 100°.[4] It crystallizes in needles, melting at 260°.

Hexbromobenzene, C_6Br_6, is formed by the action of bromine chloride on benzene, phenol, azobenzene, and toluene at a temperature of from 350—400°.[5] It is also formed by the action of phosphorus pentabromide on bromanil.[6] It is, however, most readily prepared by dropping benzene into an excess of dry bromine containing a few grains of ferric chloride.[7] It crystallizes from toluene in long needles, which are almost insoluble in boiling alcohol, and melt at 315°.

IODINE SUBSTITUTION PRODUCTS OF BENZENE.

980. *Moniodobenzene,* C_6H_5I.—Scrugham obtained this substance by the action of iodine and phosphorus on phenol,[8] and Kekulé, by heating benzene with iodine, iodic acid and water at 200—240°. The following reaction takes place :

$$5C_6H_6 + HIO_3 + 2I_2 = 5C_6H_5I + 3H_2O.$$

At the same time a part of the iodic acid acts as an oxidizing agent, converting the benzene into water and carbon dioxide.[9] Iodobenzene is also formed when sodium benzoate is treated with iodine chloride.[10]

$$C_6H_5.CO_2Na + ICl = C_6H_5I + NaCl + CO_2.$$

[1] Riche and Bérard ; Kekulé, *Annalen,* **137**, 172.
[2] Körner, *ibid.* p. 218 ; Mayer, *ibid.* p. 227 ; Wurster and Nölting, *Ber.* **7**, 1564.
[3] Halberstadt, *ibid.* **14**, 911.   [4] Bässman, *Annalen,* **191**, 208.
[5] Gessner, *Ber.* **9**, 1505.   [6] Ruoff, *ibid.* **10**, 403.
[7] Schengelen, *Annalen,* **231**, 189.   [8] *Ibid.* **92**, 318.
[9] *Ibid.* **137**, 162.   [10] Schützenberger, *Jahresb.* **1862**, 251.

It is also formed, together with higher substitution products, by gradually adding iodine chloride to a mixture of aluminium chloride and benzene,[1] and by the action of an excess of iodine in potassium iodide solution on phenylhydrazine.[2] It is, however, best obtained by the action of concentrated hydriodic acid on acid diazobenzene sulphate.[3]

It is a liquid which turns red on exposure to light, and boils at 190—190·5°. Sodium amalgam converts it, in an alcoholic solution, into benzene, while it remains unchanged on heating with caustic potash.

p-Di-iodobenzene, $C_6H_4I_2$, is formed, with moniodobenzene, in the preparation of the latter according to the methods of Kekulé and Schützenberger.

It is also easily obtained from p-iodaniline by the diazo-reaction. It crystallizes in plates or hexagonal tables, melting at 129·4°, and boiling at 280—281°.[4]

o-Di-iodobenzene, $C_6H_4I_2$, was obtained by Körner from o-iodaniline[5]; it forms thick prisms or tables, melts at 27°, and boils at 286·5°.

m-Di-iodobenzene, $C_6H_4I_2$, can be obtained by means of the diazo-reaction from m-iodaniline[6] and m-di-iodaniline.[7] It crystallizes from a mixture of alcohol and ether in rhombic tables, melting at 40·4°, and boiling at 282°.

Tri-iodobenzene, $C_6H_3I_3$, probably the asymmetric modification was obtained by Kekulé by heating benzene with iodine and iodic acid. It crystallizes in small needles which melt at 76°, and sublime without decomposition.

s-Tri-iodobenzene $C_6H_3I_3$, is obtained by the polymerization of iodacetylene. It separates from ether in almost odourless crystals, which melt at 171°, and readily sublime.[8]

Hexiodobenzene, C_6I_6. By the action of light on di-iodacetylene a compound is formed which is sparingly soluble in alcohol, and melts at 184°. It is probably hexiodobenzene.[9]

[1] Greene, Compt. Rend. **90**, 40.   [2] E. v. Meyer, J. Pr. Chem. II. **36**, 115.
[3] Griess, Annalen, **137**, 76.   [4] Körner and Wender, Gazzetta, **17**, 486.
[5] Ibid.   [6] Korner; Rudolph, Ber. **11**, 81.
[7] Paternò and Oliveri, ibid. **17**, 109 R.
[8] Baeyer, ibid. **18**, 2275.   [9] Ibid.

FLUORINE SUBSTITUTION PRODUCTS OF BENZENE.

981 *Monofluobenzene*, C_6H_5F, was first obtained by the action of hydrochloric acid on the potassium salt of fluobenzenesulphonic acid.[1] It is more readily prepared by the action of hydrofluoric acid on diazoamido-compounds containing fatty residues, *e.g.* benzenediazopiperidide.[2] It is a colourless liquid which smells like benzene, boils at 85°, and has a sp. gr. of 1·0236 at 20°.

p-Difluobenzene, $C_6H_4F_2$, is obtained from *p*-fluoraniline by converting it into fluobenzene-*p*-diazopiperidide, and treating this with hydrofluoric acid.[3] It is a liquid which becomes crystalline at a low temperature, boils at 87—89°, and has a sp. gr. of about 1·11.

NITROSO-SUBSTITUTION PRODUCTS OF BENZENE.

982 *Nitrosobenzene*, $C_6H_5.NO$ (?), has hitherto only been prepared in solution. It is obtained when a solution of nitrosyl bromide, NOBr, or nitrosyl chloride, NOCl, in benzene, is added to a solution of mercury phenyl $(C_6H_5)_2Hg$. The compound $SnCl_4+NOCl$, which is easily obtained in large yellow crystals by passing the vapours of aqua regia over stannic chloride, acts still better. On distillation with steam a beautiful green liquid is obtained smelling like mustard oil, which on treatment with tin and hydrochloric acid yields aniline.[4]

---

[1] Paternò and Oliveri, *Ber.* **11**, 109 R.
[2] Wallach and Heusler, *Annalen*, **243**, 219.
[3] *Ibid.* p. 225   [4] Baeyer, *Ber.* **7**, 1638.

NITRO-SUBSTITUTION PRODUCTS OF BENZENE.

983 *Nitrobenzene*, $C_6H_5NO_2$, was first obtained by Mitscherlich in 1834, by the action of fuming nitric acid on benzene, and termed by him nitrobenzide.[1] In order to prepare it in small quantities, equal parts of fuming nitric acid and benzene are gradually mixed, the vessel being kept cool; the mixture is then poured into water, and the heavy oil which separates, washed first with water and then with caustic soda, and afterwards distilled with steam. It is now prepared on the large scale, by a process which will be subsequently described.

Nitrobenzene is an almost colourless, strongly refractive liquid, having a specific gravity of 1·200 at 0°. It has a peculiar smell, similar to that of oil of bitter almonds, at the same time reminding one of oil of cinnamon, and possesses a sweet and burning taste. It boils at 210°, and at a low temperature solidifies in large needles, melting at 3°. In water it is scarcely soluble, but it dissolves readily in alcohol, ether, benzene, and concentrated nitric acid, and is itself an excellent solvent for many organic substances, which are sparingly or not at all soluble in the ordinary solvents.

Nitrobenzene is poisonous, especially when the vapour is inhaled; it produces a burning sensation in the mouth, nausea and giddiness, also cyanosis of the lips and face, and in serious cases, which frequently end fatally, symptoms of a general depression, such as fright, coma, humming in the ears, convulsions, and pallor, are observed, the breath and vomited matter smelling of nitrobenzene.[2] When introduced into the animal organism, nitrobenzene is transformed into aniline, whilst its homologues, such as *p*-nitrotoluene, are oxidized to acids, and are not poisonous.[3]

Nitrobenzene is almost always obtained as a yellow oil, but if prepared from benzene free from thiophene, may be obtained almost colourless, and does not then become coloured yellow on exposure to the light.[4]

---

[1] *Pogg. Ann.* **31**, 625.

[2] Grandhomme, *Die Theerfabriken des Herren Meister, Lucius, und Brüning in sanitärer und socialer Bezichung.*

[3] Jaffé, *Ber.* **7**, 1673. [4] Bidet, *Compt. Rend.* **108**, 520.

Nitrobenzene was first introduced into commerce by Collas under the name of essence of mirban, or artificial bitter-almond oil; and Mansfield, in 1847, patented a process for its preparation from coal-tar. It is now prepared on a very large scale and employed for a variety of purposes.

Nitrobenzene is manufactured by allowing a well-cooled mixture of fuming nitric acid free from chlorine, and concentrated sulphuric acid to flow into benzene, contained in cast-iron

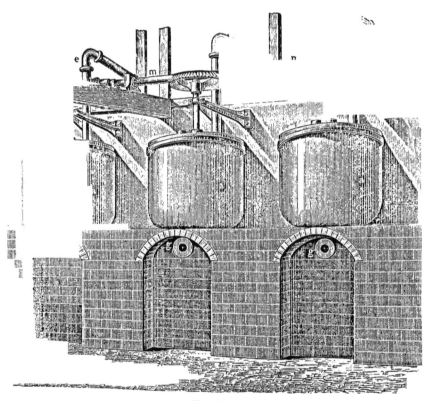

FIG. 10.

vessels provided with agitators (Fig. 10). The mixture must at first be kept cool, but towards the end of the reaction the temperature may rise to 80—90°. When the reaction is over the product is run into tanks; the acid mixture separates as a layer at the bottom, whilst nitrobenzene, being insoluble in the acid goes to the top.

The acid layer is drawn off, and the nitric acid recovered.

Crude nitrobenzene contains more or less benzene which has escaped the reaction. To remove the latter, the crude product is treated with steam, when the benzene distils over with a small quantity of nitrobenzene, and this mixture is used again for the preparation of nitrobenzene. The residual nitrobenzene is washed with caustic soda and water, and if necessary purified by distillation in high pressure steam.

Nitrobenzene is employed in perfumery, especially for scenting toilet soaps, but by far the greater quantity is used in the manufacture of pure aniline for the colours known as aniline blue and aniline black, also in the manufacture of magenta by the modern process, and for various other purposes to be hereafter mentioned.

The nitrobenzene prepared from commercially pure benzene is used in commerce under the name of *light nitrobenzene*, or *nitrobenzene for blue or black*. The *heavy nitrobenzene*, or *nitrobenzene for red*, is obtained in a similar way from a mixture of toluene and benzene, containing about 40 per cent. of the latter, and is used for the preparation of magenta by the older processes. Finally, *very heavy nitrobenzene* consists principally of *p*- and *o-nitrotoluene* from which the pure compounds can be prepared by fractional distillation. This is likewise used in the manufacture of colours.

We may here remark that the above technical terms are not chosen with respect to the specific gravity of the different products, since that of the pure nitrobenzene is greater than that of its homologues.[1]

984 *Dinitrobenzene*, $C_6H_4(NO_2)_2$, was obtained by Deville by boiling benzene with fuming nitric acid for a long time, and was called by him binitrobenzide or "nitrobenzinèse."[2] It was then more accurately investigated by Hofmann and Muspratt, who prepared it by dissolving benzene in a mixture of equal parts of fuming nitric and sulphuric acids, and then heating for a few minutes.[3] The chief portion of the product is *m*-dinitrobenzene, but it also contains smaller quantities of the ortho- and para-compounds.[4]

In order to prepare the crude product, benzene is run without

---

[1] Schultz, *Chemie des Steinkohlentheers*, 2nd ed. I. 246.

[2] *Ann. Chim. Phys.* III. **3**, 187.

[3] *Annalen*, **57**, 214.

[4] Rinne and Zincke, *Ber.* **7**, 869, 1372 ; Körner, *Gazzetta*, **4**, 305 ; *Journ. Chem. Soc.* 1876, **1**, 204.

cooling into a mixture of concentrated nitric and sulphuric acids
and then boiled for a short time. The product is purified by
washing with water, pressing and crystallizing from alcohol, when
the meta-compound separates out first. After standing for
some time p-dinitrobenzene crystallizes out from the mother
liquor, and may be purified by recrystallization from absolute
alcohol. The first mother liquor of the para-compound is freed
from alcohol by distillation, when some of the meta-compound
separates out; the o-dinitrobenzene is then deposited and may
be purified by recrystallization from 25 per cent. acetic acid
(Rinne and Zincke).

o-Dinitrobenzene crystallizes from hot water or acetic acid in
needles, and from alcohol or chloroform in monosymmetric
tables,[1] melting at 117·9°. On boiling with caustic soda it is
converted into o-nitrophenol, $C_6H_4(NO_2)OH$, and on heating with
alcoholic ammonia into o-nitraniline,[2] $C_6H_4(NO_2)NH_2$, whilst
its isomerides are not affected by this reagent.

m-Dinitrobenzene is best obtained, according to Beilstein and
Kurbatow, by dissolving one volume of benzene in two volumes
of nitric acid of sp. gr. 1·52, and finishing the reaction by heating.
After cooling, 3·3 volumes of sulphuric acid are added. The
whole is then boiled up, and after again cooling, precipitated
with water. The washed precipitate is purified by recrystallization from alcohol. It crystallizes in long needles or thin rhombic
tables (Bodewig), melting at 89·9° (Körner). It is somewhat
more readily soluble in alcohol than its isomerides; nevertheless
when these are present only in small quantity it separates out
first. It is used in the colour industry, and is prepared on the
large scale in the apparatus used for the preparation of nitrobenzene; the usual method is to run a mixture of 100 kilos of
nitric acid of sp. gr. 1·38, and 156 kilos of concentrated sulphuric
acid into 100 kilos of benzene. After the completion of the
reaction the acids are separated, and a fresh mixture of acids
prepared as above stated, run in. The whole is then gently
warmed for some time, and the product separated while it is still
liquid. It is then repeatedly washed with hot and cold water,
and is not generally further purified.

It dissolves in alcohol, and on addition of a few drops of caustic
potash to this solution, gives a magenta colouration, caused by

[1] Bodewig, *Pogg. Ann.* **158**, 239.
[2] Laubenheimer, *Ber.* **9**, 1828 ; **11**, 1155.

some adhering dinitrothiophene. This reaction is given by all nitrobenzene prepared from benzene containing thiophene.[1]

p-Dinitrobenzene forms monosymmetric needles melting at 171—172° and subliming readily.

985 *s-Trinitrobenzene*, $C_6H_3(NO_2)_3$(1 : 3 : 5), is obtained by heating two parts of *m*-dinitrobenzene with six parts of the most concentrated nitric acid, and fifteen parts of pyro-sulphuric acid to 80—120°. It crystallizes from hot alcohol in white silky plates, or fern-like needles ; when the cold saturated solution is allowed to evaporate slowly, small rhombic tables are obtained, melting at 121—122.°[2]

When oxidized with potassium ferricyanide in weak alkaline solution, it is converted into picric acid or *s*-trinitrophenol, $C_6H_2(NO_2)_3OH$, which, as will be shown later, has the following constitution :

$$
\begin{array}{c}
\text{OH} \\
\text{NO}_2 \diagup\!\!\!\diagdown \text{NO}_2 \\
\diagdown\!\!\!\diagup \\
\text{NO}_2
\end{array}
$$

Trinitrobenzene forms with benzene the compound $C_6H_3(NO_2)_3+C_6H_6$, crystallizing in hard shining prisms, and quickly decomposing on exposure to the air. With other hydrocarbons it forms analogous compounds.[3]

as-Trinitrobenzene, $C_6H_3(NO_2)_3$ (1 : 2 : 4), is prepared by boiling *p*-dinitrobenzene with a mixture of fuming nitric and sulphuric acids.[4] It has not yet been obtained free from dinitrobenzene.

Chloronitrobenzenes, $C_6H_4ClNO_2$.

	Melting-point.	Boiling-point.
Ortho ([5]) needles	32·5°	243°
Meta ([6]) rhombic crystals . .	44·4°	235·6°
Para ([7]) rhombic plates . . .	83°	242°

[1] V. Meyer and Stadler, *Ber.* **17**, 2778.

[2] Hepp, *Annalen*, **215**, 344.   [3] *Ibid.* p. 375.   [4] *Ibid.* p. 361.

[5] Engelhardt and Latschinow, *Zeitsch. Chem.* **1870**, 229 ; Beilstein and Kurbatow, *Annalen*, **182**, 107.

[6] *Ibid.* p. 102 ; Griess, *Jahresb.* **1866**, 457 ; Laubenheimer, *Ber.* **7**, 1765 ; **8**, 1622 ; **9**, 764.

[7] Riche, *Annalen*, **121**, 357 ; Sokolow, *Zeitsch. Chem.* **1866**, 621 ; Beilstein and Kurbatow, *loc. cit.* p. 105.

Bromonitrobenzenes, $C_6H_4Br.NO_2$.

	Melting-point.	Boiling-point.
Ortho ([1]) lance-like crystals. .	43·1°	261°
Meta ([2]) light yellow plates .	56·4°	256·5°
Para ([3]) needles	126—127°	255—256°

Iodonitrobenzenes, $C_6H_4INO_2$.

	Melting-point.	Boiling-point.
Ortho ([4]) yellow needles	49·4°	—
Meta ([5]) monosymmetric plates	36°	—
Para ([6]) needles	171·5°	—

Fluonitrobenzenes, $C_6H_4F.NO_2$

	Melting-point.	Boiling-point.
Ortho ([7]) crystals	26·5°	205°

These disubstitution products, which are of great theoretical interest, can be easily converted into the corresponding amido-compounds, and these are transformed by means of the diazo-reaction into many other bodies whose constitution has been thus ascertained.

With regard to the formation of these compounds it may be remarked that by the action of concentrated nitric acid on chlorobenzene and bromobenzene, the para-compound is obtained, together with small quantities of the ortho-compound. *m*-Chloronitrobenzene, on the other hand, is formed when nitrobenzene, containing iodine or antimony chloride, is treated with chlorine. They can also be obtained from the nitrophenols by replacing the hydroxyl by chlorine or bromine, and, in a similar way, from the nitranilines by the diazo-reaction.

By heating the ortho-compounds with caustic soda, the halogen

---

[1] Hübner and Alsberg, *Annalen,* **156**, 316 ; Walker and Zincke, *Ber.* **5**, 114 ; Fittig and Mayer, *ibid.* **7**, 1179 ; Körner, *Jahresb.* **1875**, 302.

[2] Griess, *ibid.* **1863**, 423 ; Wurster and Grubenmann, *Ber.* **7**, 416 ; Körner, *loc. cit.* ; Fittig and Mayer, *Ber.* **8**, 364.

[3] Couper, *Annalen,* **104**, 225 ; Griess, *loc. cit.* ; Hübner and Alsberg, *loc. cit.* Körner, *loc. cit.*   [4] Körner, *Jahresb.* **1875**, 302.

[5] Griess, *Zeitsch. Chem.* **1866**, 218 ; Körner.

[6] Kekulé, *Annalen,* **137**, 168 ; Griess.

[7] Wallach and Heusler, *ibid.* **235**, 264 ; **243**, 222.

is replaced by hydroxyl, and o-nitrophenol is obtained, whereas by heating with alcoholic ammonia, o-nitraniline is formed. The para-compound is acted upon by the above reagents in an exactly similar manner, but with greater difficulty, whilst the meta-compounds are not attacked.

Many halogen derivatives of the higher nitrobenzenes have also been obtained, which resemble the foregoing in their general properties.

BENZENE SULPHONIC ACIDS.

986 *Benzenemonosulphonic acid*, $C_6H_5.SO_2.OH$. Mitscherlich obtained this compound by the action of fuming sulphuric acid on benzene, and called it benzinsulphuric acid. He states that, like Faraday, he did not succeed when ordinary strong sulphuric acid [1] was used; but Freund has shown that on long standing in the cold it may thus be obtained.[2] It is more quickly prepared by heating benzene with sulphuric acid.[3] For this purpose a mixture of equal volumes of sulphuric acid and benzene are allowed to boil gently for from twenty to thirty hours in a flask connected with an inverted condenser, when about 80 per cent. of the benzene passes into solution. Further heating does not increase the yield.[4] From the solution thus obtained the barium, lead or calcium salt is prepared, and these then converted into other benzenesulphonates or the free acid.

Benzenesulphonic acid crystallizes in small four-sided tables, which deliquesce in damp air, and when dried over sulphuric acid contain two molecules of water of crystallization.[5] When its aqueous solution is distilled water passes over first, and then decomposition sets in with formation of benzene, diphenyl sulphone, $(C_6H_5)_2SO_2$, sulphur dioxide, and carbonization of the residue (Freund).

Barium benzenesulphonate $(C_6H_5.SO_3)_2Ba + H_2O$, crystallizes in pearly tablets which are slightly soluble in alcohol.

Copper benzenesulphonate, $(C_6H_5.SO_3)_2Cu + 6H_2O$, forms large light blue tables.

Ethyl benzenesulphonate, $C_6H_5.SO_2.C_2H_5$, is formed by the

---

[1] *Pogg. Ann.* **31**, 283 and 634.   [2] *Annalen*, **120**, 76.
[3] Stenhouse, *Proc. Roy. Soc.* **14**, 351.   [4] Michael and Adair, *Ber.* **10**, 585.
[5] Hübner, *Annalen*, **223**, 235.

action of benzenesulphonic chloride on sodium ethylate, and is a peculiarly smelling, oily liquid, which on boiling with water decomposes into alcohol and benzenesulphonic acid.[1]

Benzenesulphonic chloride, $C_6H_5.SO_2Cl$, is formed by the action of phosphorus pentachloride on a benzenesulphonate,[2] and is an oily liquid, which, on standing for a long time at 0°, solidifies to large rhombic crystals.[3] It is insoluble in water and is scarcely affected by it.

Under diminished pressure it can be distilled without decomposition; at the ordinary pressure it boils, undergoing considerable decomposition, at 246—247° (Otto).

Benzenesulphonamide, $C_6H_5.SO_2.NH_2$, was obtained by Chancel and Gerhardt by trituration of the chloride with ammonium carbonate.[4] It is also easily obtained by treating the chloride with alcoholic ammonia. It is readily soluble in alcohol, slightly in water, and crystallizes from the latter in thin rhombic prisms, melting at 149° (Otto); with an ammoniacal solution of silver it forms the compound $C_6H_5.SO_2.NHAg$—a crystalline precipitate (Gerhardt and Chiozza).

987 *Benzenedisulphonic acids,* $C_6H_4(SO_3H)_2$. By heating benzonitril with fuming sulphuric acid, Buckton and Hofmann obtained an acid of this composition[5] consisting chiefly of the para-compound.[6] It is also formed as chief product when the vapour of benzene, heated to 240°, is passed into sulphuric acid.[7] If benzene, on the other hand, be heated with fuming sulphuric acid, the meta-acid is chiefly formed, and this on heating more strongly is partly converted into the para-compound. The isomerides can be separated by converting the mixture into the potassium salts, separating the crystals mechanically and purifying by recrystallization.[8]

Benzene-p-disulphonic acid is a very hygroscopic crystalline mass.

		Melting point.
$C_6H_4(SO_3K)_2 + H_2O$	small thin plates	—
$C_6H_4(SO_2Cl)_2$	long needles	131°
$C_6H_4(SO_2NH_2)_2$	—	288°

[1] Schiller and Otto, *Ber.* **9,** 1638 ; Hübner, *loc. cit.*
[2] Gerhardt and Chiozza, *Annalen,* **87,** 299.   [3] Otto, *ibid.* **145,** 321.
[4] *Ibid.* **87,** 296 ; *Jahresb.* **1852,** 434.   [5] *Annalen,* **100,** 157.
[6] Garrick, *Zeitschr. Chem.* **1869,** 550 ; Fittig, *Annalen,* **174,** 123.
[7] Egli, *Ber.* **8,** 817.
[8] Barth and Senhofer, *ibid.* **8,** 1478 ; Körner and Monsclise, *ibid.* **9,** 583.

By heating the potassium salt with potassium cyanide, the nitril of terephthalic acid, $C_6H_4(CO_2H)_2$, which is a para-compound, is formed; whilst by fusing the salt with caustic potash an intramolecular change takes place and m-dihydroxybenzene or resorcinol, $C_6H_4(OH_2)$, is formed.

Benzene-m-disulphonic acid is formed almost exclusively when benzene is dissolved in an equal volume of fuming sulphuric acid, another volume of the latter added, and the liquid strongly heated in a retort with an upright neck for two to three hours, till the vessel is filled with white vapours.[1] The free acid is a very deliquescent crystalline mass containing $2\frac{1}{2}$ molecules of water.

		Melting-point.
[2] $C_6H_4(SO_3K)_2 + 1\frac{1}{2} H_2O$	needle-shaped oblique prisms	—
$C_6H_4(SO_2Cl)_2$	large prisms	63°
$C_6H_4(SO_2.NH_2)_2$	needles	229°

Benzene-o-disulphonic acid is obtained from the corresponding amidobenzenedisulphonic acid by the diazo-reaction. Its potassium salt is readily soluble in water and crystallizes tolerably well.[3]

		Melting-point.
$C_6H_4(SO_2Cl)_2$	large 4-sided tables	105°
$C_6H_4(SO_2.NH_2)_2$	needles	233°

Benzenetrisulphonic acid, $C_6H_3(SO_3H)_3 + 3H_2O$, is formed by heating benzene with sulphuric acid and phosphorus pentoxide for some hours to 280—290°, or by heating the potassium salt of benzene-m-disulphonic acid with concentrated sulphuric acid to 300°. The free acid crystallizes in flat, very hygroscopic needles. The potassium salt, $C_6H_3(SO_3K)_3 + 3H_2O$, forms oblique prisms.[4]

Benzenesulphinic acid, $C_6H_5SO_2H$. By the action of zinc ethyl on benzenesulphonyl chloride, Kalle expected to form the compound, $C_6H_5.SO_2C_2H_5$, but instead of this he obtained the zinc salt of the above acid, which he called benzylsulphurous acid.[5]

$$2C_6H_5.SO_2Cl + (C_2H_5)_2Zn = (C_6H_5.SO_2)_2Zn + 2C_2H_5Cl.$$

It is also formed by the direct combination of sulphur dioxide and benzene in the presence of aluminium chloride,[6] but is best

[1] Heinzelmann, *Annalen*, **188**, 157. [2] Reiche, *ibid.* **203**, 69.

[3] Drebes, *Ber.* **9**, 552. [4] Senhofer, *Annalen*, **174**, 243.

[5] *Ibid.* **119**, 153.

[6] Friedel and Crafts, *Ann. Chim. Phys.* VI. **14**, 442.

prepared by gradually adding zinc-dust to a well-cooled alcoholic solution of benzenesulphonic chloride, and then washing out the alcohol and zinc chloride from the thick mass with cold water. The zinc salt is almost insoluble in cold water, and is converted into the sodium salt by the addition of carbonate of soda, the benzenesulphinic acid being precipitated from the concentrated solution by the addition of hydrochloric acid.[1]

It is soluble with difficulty in cold water and crystallizes from hot water in stellated prisms, which melt at 83—84° and are readily soluble in alcohol and ether. Oxidizing agents easily convert it into benzenesulphonic acid, even the oxygen of the air gradually transforming it. Phosphorus pentachloride converts it into benzenesulphonic chloride.

$$C_6H_5.SO_2H + PCl_5 = C_6H_5.SO_2Cl + HCl + PCl_3.$$

By fusing with caustic potash it is decomposed, forming benzene and potassium sulphite, and on heating with water to 130° it is converted into benzenesulphonic acid, and phenyl benzenethiosulphonate. Phenylhydrazine converts it into *benzene disulphoxide*, $C_6H_5.S.SO_2.C_6H_5$, melting at 44—45°, and *phenylbenzenesulphazide*,[2] $C_6H_5NH.NH.SO_2.C_6H_5$, melting at 146—147°.

Its *ethyl* salt is obtained by the action of ethyl chloroformate on the potassium salt. It is an oily liquid insoluble in water which cannot be distilled.

Nitrobenzenesulphonic Acids, $C_6H_4.(NO_2).SO_3H.$

988 An acid of this composition was first obtained by Laurent by treating benzenesulphonic acid with nitric acid,[3] and Schmitt showed that it is also obtained [4] by warming nitrobenzene with fuming sulphuric acid. This was supposed to be a definite compound until Limpricht pointed out that, while the chief product in both cases is the meta-acid, the other two isomerides are formed at the same time. These may be roughly separated by means of their barium salts, but the separation is only rendered complete by conversion into the amides, which are decomposed by heating with hydrochloric acid to 150°.[5]

[1] Otto, *Ber.* **9**, 1584.   [2] Escales, *ibid.* **18**, 893.
[3] *Jahresb.* **1850**, 418.   [4] *Annalen,* **120**, 163.
[5] *Ibid.* **177**, 60.

According to Limpricht these compounds are prepared by mixing 200 grms. of benzene with 300 grms. of fuming sulphuric acid, removing the undissolved benzene after two to three hours, and adding to the solution nitric acid, of about sp. gr. 1·5, drop by drop till the action ceases; water is then added and the liquid poured off from the dinitrobenzene formed at the same time. The solution is neutralized with milk of lime, and concentrated, when the calcium salt of the meta-acid first crystallizes out. This is decomposed with potassium carbonate and the dry potassium salt ground up with phosphorus pentachloride; the product is washed with water, which scarcely attacks the nitrobenzenesulphonic chloride, and then dissolved in ether and dried over calcium chloride. On evaporation the chloride crystallizes out in large, shining prisms. By treating these with strong ammonia the amide is obtained and this is purified by recrystallization.

The last crystallizations of the calcium salt are worked up in the same way, but the mixed chlorides are at once converted into the amido-compounds, these being separated by frequent recrystallizations from hot water—the ortho-acid being the least soluble, the para-acid the most, and the solubility of the meta-acid lying between that of the other two.

Of the free acids, *m*-nitrobenzenesulphonic acid is the only one which has been prepared pure; it crystallizes in large, flat, deliquescent tables, and it is not attacked when boiled with fuming nitric acid.

The properties of the chlorides and amides are given in the following table :

Nitrobenzenesulphonic chlorides, $C_6H_4(NO_2)SO_2Cl$.

		Melting-point.
Ortho.	white prisms	67°
Meta.	4-sided prisms	60·5°
Para.	red oil	—

Nitrobenzenesulphonamides, $C_6H_4(NO_2)SO_2.NH_2$.

Ortho.	slender needles	186°
Meta.	needles or prisms	161°
Para.	slender needles . . .	131°

HYDROXYBENZENES AND ALLIED BODIES.

989 *Phenol*, C_6H_5OH.—This compound was discovered by Runge in 1834 in coal-tar, and called by him "Kohlenölsäure" or carbolic acid.[1] Laurent obtained it pure in 1841 and determined its composition ; he gave it the name of phenylhydrate (hydrate de phényle) or phenic acid (acide phénique),[2] from φαίνειν, to emit light, probably because it was found as a by-product in the oils from the manufacture of illuminating gas. Gerhardt named it phenol in order to mark the fact that this body is a kind of alcohol, whence it has also been termed phenyl alcohol. Phenol was for a long time confounded with creosote, discovered by Reichenbach in 1832, although Runge had already pointed out that phenol is distinguished from wood creosote by having an acid reaction and by the fact that, if used instead of creosote for the preservation of meat, it imparts to the latter a strong and very unpleasant taste. Later investigations have proved that wood creosote is a mixture of different compounds and only contains very small quantities of phenol. This latter is a product of the dry distillation of many organic bodies, being found in largest quantities in the tar obtained by the dry distillation of coal, lignite, and peat.

Wöhler found it in small quantities in *castoreum*,[3] a substance secreted by the præputial glands of the beaver, and Städler has indicated its presence in the urine of man, the horse, and the cow.[4] It occurs in urine as the potassium salt of the unstable phenyl-sulphuric acid,[5] and for this reason it can only be detected by distillation with hydrochloric acid.[6] Healthy urine from a mixed diet contains 0·004 grm. of phenol per litre,[7] but under certain pathological conditions it may rise as high as 1·5575 grm. (Salkowski). Phenol is also formed when albuminoid bodies are allowed to putrefy in presence of some water and pancreas,[8] and hence it is found in small quantities in excrements.[9]

[1] *Pogg. Ann.* **31**, 65 ; **32**, 308.　　[2] *Ann. Chim. Phys.* III. **3**, 195.
[3] *Annalen*, **67**, 360.　　[4] *Ibid.* **77**, 18.
[5] Baumann, *Ber.* **9**, 55.　　[6] Salkowski, *ibid.* **9**, 1595.
[7] Munk, *ibid.* p. 1596.
[8] Baumann, *ibid.* **10**, 685 ; Odermatt, *J. Pr. Chem.* II. **18**, 249.
[9] Prieger, *ibid.* II. **17**, 133.

According to Griffiths it also occurs in the trunk, leaves, and sap of the Scotch fir (*Pinus sylvestris*).[1]

When benzene and water are shaken up in the air with palladium hydride (**Vol.** II. Part II. p. 425), a little phenol, together with other products, is formed.[2] It is obtained in larger quantities, together with oxalic acid, when water is poured on to phosphorus, some benzene added, and the whole exposed to the action of sunlight. In both cases the oxidation is effected by hydrogen dioxide, which also oxidizes benzene directly.[3]

$$C_6H_6 + HO.OH = C_6H_5.OH + HOH.$$

Phenol is also readily formed when dry oxygen or air is passed into a mixture of benzene and aluminium chloride. The reaction takes place very slowly at the ordinary temperature, but the oxygen is readily absorbed when the mixture is kept at a temperature near its boiling-point. After sufficient gas has been absorbed the mixture is poured into water, when the unaltered benzene and some coloured products formed in the reaction, separate as a light layer. The aqueous solution is diluted with water, and extracted with ether; on evaporating the ethereal solution pure phenol remains.[4]

Hofmann[5] and Hunt[6] obtained it from aniline by treating a hydrochloric acid solution with silver nitrite. Diazobenzene chloride, which was afterwards obtained by Griess, is formed as an intermediate product.

Wurtz[7] and Kekulé[8] found that phenol may be easily obtained by fusing benzenesulphonic acid with caustic potash; caustic soda gives a smaller yield than potash, and the best proportions are six molecules of the latter to one molecule of potassium benzenesulphonate. The excess of alkali assists the reaction and by dilution prevents the overheating of the melt.[9]

Berthelot obtained phenol from acetylene by dissolving it in fuming sulphuric acid and fusing the product with caustic potash.[10] It is also formed in smaller quantities, together with

[1] *Chem. News,* **49**, 59. [2] Hoppe-Seyler, *Ber.* **12**, 1551.
[3] Leeds, *ibid.* **14**, 975.
[4] Friedel and Craffts, *Ann. Chim. Phys.* VI. **14**, 435.
[5] *Annalen,* **75**, 356. [6] Silliman, *Americ. Journ.* **1849**.
[7] *Annalen,* **144**, 121. [8] *Lehrb. Org. Chem.* **3**, 13.
[9] Degener, *J. Pr. Chem.* II. **17**, 394. [10] *Compt. Rend.* **68**, 539.

the so-called glycerine ether (**Vol. III. Part II. p.** 402), and other products, when glycerol is distilled with calcium chloride.[1]

Phenol was first prepared on the large scale from coal-tar in 1845 by Sell[2] in Offenbach, and Brönner in Frankfort.[3] Its manufacture in large quantities was commenced in England in 1861, by Messrs. Crace Calvert and Charles Lowe, at Bradford, near Manchester. The latter discovered in 1862 that phenol forms a hydrate with water which crystallizes at a low temperature, a property which is not possessed by its homologues. This hydrate splits up on distillation into water and pure phenol, and the latter compound was thus first brought into commerce by Messrs. Charles Lowe and Co., Lowe's carbolic acid melting at a higher temperature and boiling at a lower temperature than the ordinary commercial product.

It was formerly obtained by distilling the portions of tar boiling between 150—250°, but it is now prepared from the portion boiling between 150—200°, or the so-called *middle-oil*, which after the larger quantity of the naphthalene it contains has crystallized out, is treated with caustic soda of sp. gr. 1·34, and the mixture well worked in a cylinder by means of an agitator. After standing for some time, the upper layer, which consists of hydrocarbons, is removed, and the remainder diluted with water and allowed to stand in contact with air, when naphthalene and tarry products separate. The aqueous solution is fractionally precipitated by dilute sulphuric acid, tarry matters separating out first, then the homologues of phenol, and, lastly, phenol itself.

In other works the decomposition is completed in one operation and the tar-acids separated by distillation, the portion boiling between 175° to 200° being collected separately as the crude carbolic acid of commerce. By repeated fractional distillation, and by cooling the portions boiling at 180—190°, crystalline phenol is obtained, which, however, still contains more or less *p*-cresol. In order to prepare chemically pure phenol, either Coupier's apparatus is employed, or the abovementioned hydrate is decomposed by distillation.

At the present time pure phenol is also manufactured in Germany from pure benzene, the latter being converted into the sulphonic acid, and this fused with alkali.

---

[1] *Annalen*, Suppl. **8**, 254.   [2] Hofmann, *Report London Exhibition*, **1862**.
[3] C. Brönner prepared "creosote as clear as water," which was frequently returned, because it crystallized in winter. Lunge, *Dist. Coal-Tar*, &c.

Phenol crystallizes in long rhombic needles melting at 42° and boiling at 182° (Lowe). Its melting-point is lowered by the admixture of even insignificant quantities of its homologues, as well as of naphthalene and water. A few drops of the latter suffice to liquefy a large quantity of phenol. The above-mentioned hydrate, $2C_6H_6O + H_2O$, solidifies on cooling to crystals melting at 17° (Lowe). The specific gravity of liquid phenol at 46° is 1·0560, and at 56° is 1·0469.[1]

990 Phenol combines with liquid carbon dioxide to form a crystalline compound,[2] which is also obtained when salicylic acid, or the isomeric p-hydroxybenzoic acid, is heated for two hours to 250—260° in a closed tube, both these acids decomposing easily into phenol and carbon dioxide. On cooling, the compound separates out as a crystalline mass, similar to the pyramidal form of common salt, and melting at 37°. The compound is, of course only stable under pressure.[3]

Phenol also forms with sulphur dioxide a compound which crystallizes in yellow, rhombic tables melting at 25—30° and distilling at 140° in a stream of sulphur dioxide. On exposure to air it quickly gives off sulphur dioxide. Its composition is probably $SO_2 + 4C_6H_6O$.[4]

Phenol possesses a peculiar smell and has a burning, caustic taste : it is soluble at the ordinary temperature in fifteen parts of water ; its solubility increases with increase of temperature, so that at 84° both liquids may be mixed in all proportions.[5] It is readily soluble in the fatty oils and glycerol, and dissolves in alcohol and ether in every proportion.

Pure phenol remains colourless on exposure to light and air, but if it contains a trace of impurity it is quickly coloured red or brown, and deliquesces in moist air.

It acts on the skin as a powerful caustic ; it coagulates albumen and precipitates solutions of gelatine. Taken internally it is a violent poison ; a few drops kill a dog, and plants expire in a dilute aqueous solution (Frerichs and Wöhler). In cases of poisoning by carbolic acid, which not unfrequently occur, Crace Calvert recommends olive or almond oil as an antidote ; Husemann advises the use of saccharosate of lime prepared by adding five parts of slaked lime to sixteen parts of sugar in forty parts of water ; after digesting for three

---

[1] Ladenburg, *Ber.* **7**, 1687. [2] Barth, *Wien. Akad. Ber.* **58**, 2, 16.
[3] Klepl, *J. Pr. Chem.* II. **25**, 464. [4] Holzer, *ibid.* p. 463.
[5] Alexejew, *Ber.* **9**, 1810.

days it is filtered, the solution evaporated, and the residue dried at 100°.[1]

When the vapour of phenol is passed over red hot zinc dust it is reduced to benzene.[2] In this way all phenols and various other oxidized aromatic compounds can be converted into the corresponding hydrocarbons. This reaction, discovered by Baeyer, is of great interest, as it led to the discovery of artificial alizarin, the colouring matter of madder. When phenol is merely passed through a red-hot tube it forms benzene, toluene, C_7H_8, xylene, C_8H_{10}, naphthalene, $C_{10}H_8$, anthracene, $C_{14}H_{10}$, and a small quantity of phenanthrene, $C_{14}H_{10}$.[3]

Reactions of Phenol.—It has been remarked in the introduction that most of the phenols give characteristic colours with a solution of ferric chloride, this property having been first noticed in the phenol *par excellence* by Runge, who obtained a violet colouration with ferric salts. This reaction is, however, not a delicate one, and does not occur in presence of alcohol.[4] By adding some ammonia, and then a solution of bleaching powder to an aqueous solution of phenol a blue colouration is obtained.[5] Millon's reagent (a solution of mercuric nitrate containing nitrous acid) gives on boiling with a phenol solution a yellow precipitate, which dissolves in nitric acid, yielding a deep red-coloured solution. By this reaction, $\frac{1}{2000000}$ of phenol can be detected. Salicylic acid likewise gives this reaction and the presence of phenol must therefore be confirmed by adding to the solution ammonia and sodium hypochlorite. When the solution contains not less than $\frac{1}{50000}$ of phenol, the blue colour is obtained after standing for twenty-four hours.[6]

By dissolving 6 per cent. of potassium nitrite in strong sulphuric acid and adding phenol, the liquid is coloured brown, then green, and afterwards blue.[7] The colouring matters here formed will be subsequently described. This reaction is given by all nitroso-compounds, and is frequently employed for their detection. On adding bromine water to an aqueous solution of phenol a yellowish white precipitate of tribromophenol is obtained, which disappears until an excess of bromine has been added. If the solution contains 1 part of phenol to 50,000 to 60,000

[1] Schultz, *Steinkohlentheer*, 446.   [2] *Annalen*, **140**, 295.
[3] Kramers, *ibid.* **189**, 129.   [4] Hesse, *ibid*, **182**, 161.
[5] Berthelot, *Rép. Chim. Appl.* **1**, 284 ; Lex, *Ber.* **3**, 458 ; also Salkowski, *Fresenius' Zeitschr.* **11**, 316.
[6] Almén, *Pharm. Journ. Trans.* III. **7**, 812.   [7] Liebermann, *Ber.* **7**, 247.

parts of water, a crystalline precipitate separates out in a few hours. Aniline, and some other bodies, give similar precipitates, but these may be easily distinguished from tribromophenol by bringing some of the washed precipitate into a test-tube with some water and sodium amalgam, shaking and warming the mixture. The liquid is then poured into a basin and dilute sulphuric acid added, when the characteristic smell of free phenol is noticed, and, if the quantity is not too small, it separates out in oily drops.[1]

The easy transformation of phenol into tribromophenol can be used for the quantitative estimation of phenol by determining the weight of the tribromophenol (Landolt). It is, however, simpler to add to the phenol solution a normal solution of bromine in caustic soda, and hydrochloric acid, and titrate the excess of bromine with potassium iodide and sodium thiosulphate.[2]

Uses of Phenol.—This body, commonly known as carbolic acid is used in the pure state, mixed with water, glycerol, or olive oil, as a powerful antiseptic in medicine and surgery, its dilute aqueous solution being employed as a spray in surgical operations (Lister's dressing). It is also used for destroying parasitic and other organisms infesting both animals and plants. Its aqueous solution is also employed in a more or less pure state for impregnating or creosoting wood, &c. For the latter purpose the crude carbolic acid which contains cresol can be used in many cases, or cresol alone may be employed. For certain purposes it is desirable to have the antiseptic in a solid form, and phenol is then mixed with such substances as marl, clay, chalk, infusorial earth, saw-dust, &c. McDougall's disinfecting powder is a mixture of calcium phenate and magnesium sulphite.

A considerable quantity of phenol is now used for the manufacture of salicylic acid, and of various colouring matters, such as picric acid, aurin, azo-colours, &c.

991 *Phenates.*—These bodies, which are also called carbolates, were first prepared by Runge and Laurent.

Potassium phenate, C_6H_5OK, is obtained by warming equal equivalents of phenol and caustic potash, or by dissolving potassium in phenol. In order to prepare the pure compound, phenol is gradually heated with the requisite quantity

[1] Landolt, *Ber.* **4**, 770.
[2] Koppeschar, *Fresenius' Zeitschr.* **15**, 233 ; see also Degener, *J. Pr. Chem.* II. **17**. 390 ; Chandelon, *Bull. Soc. Chim.* **38**, 69.

of the metal in an atmosphere of hydrogen. As soon as this is completely dissolved, the solution is cooled, when a radiating crystalline mass is obtained, containing cavities in which well-developed needles have been formed.[1] The salt is extremely hygroscopic. It was formerly supposed that phenol does not decompose potassium carbonate, inasmuch as an alkaline phenate rapidly absorbs carbon dioxide (Runge). Baumann has however shown that the crystalline phenate can be obtained by boiling phenol with a solution of potassium carbonate.[2]

Sodium phenate, C_6H_5ONa, is prepared in a similar way to the potassium compound. It is obtained in large quantities by dissolving the calculated quantity of phenol in strong caustic soda, an excess of phenol being avoided, as, otherwise, a dark-coloured product is obtained. On evaporating the solution to dryness, and continually stirring the heated residue, the whole falls to a dry powder. Sodium phenate is very hygroscopic, and must, therefore, be placed while hot in an air-tight vessel.[3] It is employed in the manufacture of salicylic acid.

The phenates of calcium, barium, lead (Runge, Laurent) thallium,[4] and aluminium,[5] have likewise been prepared.

PHENYL ETHERS.

992 *Phenyl oxide*, or *diphenyl ether*, $(C_6H_5)_2O$, was first obtained by Limpricht and List, in small quantities, together with other products, by the dry distillation of copper benzoate. It is prepared by warming a solution of diazobenzene sulphate with phenol :[6]

$$C_6H_5.N_2.SO_4H + \left.{C_6H_5 \atop H}\right\}O = \left.{C_6H_5 \atop C_6H_5}\right\}O + SO_4H_2 + N_2.$$

Phenyl oxide is also formed, together with other products, by warming phenol with aluminium chloride.[7] It separates out from its solutions as an oil, which solidifies at a low temperature to needles or four-sided prisms melting at 28°. It smells like geranium, and boils at 252—253°.

Phenyl oxide is an extremely stable body, which is neither attacked by phosphorus pentachloride nor by a solution of

[1] Hartmann, *J. Pr. Chem.* II. **16**, 36.   [2] *Ber.* **10**, 686.
[3] H. Kolbe, *J. Pr. Chem.* II. **10**, 89.   [4] Kuhlmann, *Jahresb.* **1864**, 254.
[5] Gladstone and Tribe, *Journ. Chem. Soc.* 1881, **1**, 9.
[6] Hofmeister, *Annalen*, **159**, 191.   [7] Merz and Weith, *Ber.* **14**, 189.

chromium trioxide in acetic acid. Even hydriodic acid does not act on it at 250°, and it remains unchanged when ignited with zinc-dust.

Phenyl methyl ether or *Anisoïl* was first obtained by Cahours by distilling anisic acid (*p*-methoxybenzoic acid) with baryta, and was therefore called anisol, which name has now been altered to anisoïl, to avoid confusion with the phenols and alcohols.[1]

$$C_6H_4 \left\{ \begin{matrix} OCH_3 \\ CO_2H \end{matrix} \right. = C_6H_5OCH_3 + CO_2.$$

In the same way he obtained it from wintergreen oil, which chiefly consists of methyl salicylate, this being metameric with anisic acid.[2] He then found that anisoïl is also formed by using Williamson's method of preparing mixed ethers, viz., by heating potassium phenate with methyl iodide, or by distilling it with potassium methyl sulphate.[3] It is obtained in larger quantities by heating sodium phenate to 190—200°, and passing methyl chloride through the mass by a tube reaching to the bottom of the vessel.[4]

Anisoïl is a pleasantly-smelling liquid, boiling at 152°, and having a sp. gr. of 0·991 at 15°. It is not attacked by concentrated hydrochloric acid at 120—130°, but hydriodic acid decomposes it at 130—140° into phenol and methyl iodide.[5] Heated zinc-dust does not decompose it.[6]

Phenyl ethyl ether, or *Phenetoïl*, $C_6H_5OC_2H_5$.—Baly first obtained this compound by the action of ethyl salicylate on anhydrous baryta, and by the distillation of the product; the new compound he termed salithol.[7] Soon afterwards Cabours obtained it by the same process and called it phenetol (*phénétol*[8]), and then prepared it from ethyl iodide and potassium phenate.[9] It is also easily obtained by heating sodium phenate with sodium ethyl sulphate for several hours to 150°, or by the action of alcohol on diazobenzene nitrate.[10] It is a liquid possessing similar properties to anisoïl, and boiling at 172°.

---

[1] *Annalen*, **41**, 69.
[2] *Ibid.* **48**, 65.
[3] *Ibid.* **78**, 226.
[4] Vincent, *Bull. Soc. Chim.* **40**, 106.
[5] Gräbe, *Annalen*, **139**, 149.
[6] *Ibid.* **152**, 66.
[7] *Journ. Chem. Soc.* **2**, 28 ; *Annalen*, **70**, 269.
[8] *Ibid.* **74**, 314.
[9] *Ibid.* **78**, 226.
[10] Kolbe, *J. Pr. Chem.* II. **27**, 425. Remsen and Orndorff, *Am. Chem. Journ.* **9**, 317 ; **10**, 368.

We are also acquainted with the following mixed phenyl ethers containing monatomic alcohol radicals:

		Boiling-point.
Phenyl isopropyl ether,[1]	$\left.\begin{array}{l}(CH_3)_2CH \\ C_6H_5\end{array}\right\}O$	176°
Phenyl propyl ether,[2]	$\left.\begin{array}{l}C_3H_7 \\ C_6H_5\end{array}\right\}O$	185°
Phenyl isobutyl ether,[3]	$\left.\begin{array}{l}(CH_3)_2C_2H_3 \\ C_6H_5\end{array}\right\}O$	198°
Phenyl isoamyl ether,[4]	$\left.\begin{array}{l}C_5H_{11} \\ C_6H_5\end{array}\right\}O$	224—225°
Phenyl allyl ether,[5]	$\left.\begin{array}{l}C_3H_5 \\ C_6H_5\end{array}\right\}O$	192—195°

Diphenyl methylene ether or *Diphenoxymethane*, $(C_6H_5O)_2CH_2$, is obtained by heating methylene bromide with potassium phenate. It is a liquid with a faint odour resembling that of phenol, and boiling at 293—295°.[6]

Diphenyl ethylene ether or *Diphenoxyethane*, $(C_6H_5O)_2C_2H_4$, is formed when potassium phenate is heated with ethylene bromide to 140°. It forms crystals melting at 98·5°, sparingly soluble in cold, but readily in hot alcohol, and ether.[7]

By allowing equal molecules of sodium phenate and ethylene bromide to act upon one another in alcoholic solution, phenyl bromethyl ether, $C_6H_5OC_2H_4Br$, is formed. It crystallizes in colourless needles melting at 39°, boils with slight decomposition at 240—250°, and easily loses its bromine. By heating the ether with alcoholic ammonia to 100—120°, the hydrobromide of imidodiethylene phenyl ether, $(C_6H_5OC_2H_4)_2HN$, is formed, and by acting on this with soda the free base is obtained as a thick, strongly basic oil.[8]

Phenolglucoside, $C_6H_{11}O_5(OC_6H_5)$, is formed by the action of alcoholic potassium phenate on acetochlorohydrose (Vol. III. Part II. p. 581) :

$$C_6H_7(C_2H_3O)_4ClO_5 + C_6H_5OK + 4C_2H_5OH$$
$$= C_6H_{11}O_5(OC_6H_5) + 4C_2H_5.OC_2H_3O + KCl.$$

It is soluble in cold, but more readily in hot water, and crystallizes from its aqueous solution in concentrically grouped

---

[1] Silva, *Zeitsch. Chem.* **1870**, 249.   [2] Cahours, *Bull. Soc. Chim.* **21**, 78.

[3] Riess, *Ber.* **3**, 780.   [4] Cahours, *Annalen*, **78**, 227

[5] Henry, *Ber.* **5**, 455.

[6] Henry, *Ann. Chim. Phys.* V. **30**, 266.

[7] Burr, *Zeitsch. Chem.* **1869**, 165 ; Lippmann, *ibid.* 447.

[8] Weddige, *J. Pr. Chem.* II. **24**, 241.

needles which melt at 171—172°.[1] By heating it with acetic anhydride and sodium acetate, the tetracetyl-compound $C_6H_7(C_2H_3O)_4O_5(OC_6H_5)$, which crystallizes from hot alcohol in long shining needles, is formed.[2]

It has been mentioned in the previous volume that glucose frequently occurs as a product of decomposition of bodies belonging to the vegetable kingdom, termed glucosides, which are split up by dilute acids, or by certain ferments, with assumption of water, yielding sugar-like substances and other bodies. (Vol. III. Part II. p. 575.) Phenolglucoside is not only the simplest compound of this group, but also the first one which was artificially prepared. When boiled with dilute acids, or when its solution, warmed to 40°, is brought into contact with emulsin, a ferment contained in almonds, it splits up into phenol and glucose.

$$C_6H_{11}(OC_6H_5)O_5 + H_2O = HO.C_6H_5 + C_6H_{12}O_6.$$

ETHEREAL SALTS OF PHENYL WITH INORGANIC ACIDS.

993 *Phenyl sulphates.*—According to theory, two sulphates should exist, viz., the normal and the acid ethereal salt ; these bodies, however, cannot be prepared in a similar way to the corresponding ethyl compounds, for phenyl iodide (iodobenzene) does not act on silver sulphate, and sulphuric acid converts phenol into the phenolsulphonic acid. It appeared therefore very doubtful whether it was possible to obtain sulphate of phenyl, till Baumann found in the urine of herbivora, and in smaller quantities in that of man and the dog, the potassium salt of an acid which he first believed to be a phenolsulphonic acid, but afterwards found to be *phenylsulphuric acid*, $C_6H_5HSO_4$. He observed moreover that the quantity of this compound is greatly increased if phenol be taken internally, and that if sufficient be taken the sulphates disappear from the urine. He then prepared phenylsulphuric acid artificially.[3]

Its potassium salt is obtained by mixing 100 parts of phenol and 60 parts of caustic potash with from 80—90 parts of water, and gradually adding 125 parts of finely powdered potassium

---

[1] A. Michael, *Jahresb.* **1879**, 58.   [2] *Ber.* **16**, 2510.
[3] *Ibid.* **9**, 54, 1715 ; **11**, 1907.

disulphate as soon as the solution has cooled to 60—70°. The mass is kept at this temperature for from eight to ten hours, and frequently agitated, after which it is extracted with boiling 95 p. c. alcohol, and the salt which separates out on cooling recrystallized from alcohol. Its formation is shown by the following equation :

$$C_6H_5.OK + O{\Large<}{\,^{SO_2.OK}_{SO_2.OK}} = SO_2(OK)_2 + C_6H_5O.SO_2.OK.$$

It crystallizes in small shining tablets which feel greasy to the touch ; it is obtained in transparent rhombic tablets from alcohol at 60°. At 15° it dissolves in seven parts of water; in cold absolute alcohol it is scarcely soluble, but dissolves in boiling alcohol somewhat more readily.

It decomposes on exposure to moist air, sometimes in a few minutes, into phenol and acid potassium sulphate. This change is also effected by heating the salt with water for some hours to 100°, or by warming it with dilute hydrochloric acid for a few minutes. Towards alkalis it is, on the contrary, very stable, and is only gradually attacked on treatment with caustic potash at 150°. By heating it to 150—160°, in absence of moisture, it is converted into the isomeric potassium p-phenolsulphonate, $C_6H_4(OH)SO_3K$.

Free phenylsulphuric acid is so unstable that its aqueous or alcoholic solution decomposes almost immediately.

Phenyl phosphates.—By the action of phosphorus pentoxide on phenol, mono- and di-phenylphosphoric acids are formed; these can be separated by means of their copper salts, since that of the former acid is more readily soluble than that of the latter.[1]

Monophenylphosphoric acid, $PO(OC_6H_5)(OH)_2$, crystallizes in thick needles, readily soluble in water and alcohol, and melting at 97—98°. On distillation it decomposes into phenol and metaphosphoric acid.

Diphenylphosphoric acid, $PO(OC_6H_5)_2OH$, forms crystals melting at 50°, soluble with difficulty in water, but readily in alcohol. It has a great tendency to remain in the fused condition, and hence it has been described as an oily liquid.[2] The chlorides of this acid are formed by the action of phosphorus oxychloride on phenol, and can be separated by fractional distillation.

[1] Rembold, *Zeitsch. Chem.* **1866**, 651.   [2] Rapp, *Annalen,* **224,** 156.

Phenylphosphoric chloride, $PO(OC_6H_5)Cl_2$, is a heavy, strongly refractive, not unpleasant smelling liquid, boiling at 241—243°, and is easily decomposed by water.

Diphenylphosphoric chloride, $PO(OC_6H_5)_2Cl$, is a thick liquid boiling at 314—316°, and having a similar smell to the preceding compound. It is only slowly attacked by water and cold dilute alkalis.

Normal phenyl phosphate, $PO(OC_6H_5)_3$, is formed, together with chlorobenzene, by the action of phosphorus pentachloride on phenol.[2]

$$4C_6H_5OH + PCl_5 = C_6H_5Cl + PO(OC_6H_5)_3 + 4HCl.$$

It crystallizes in small needles melting at 45° (Jacobsen), is readily soluble in alcohol and ether, and can be recrystallized from strong sulphuric acid without alteration.

Phenyl carbonate, $CO(OC_6H_5)_2$, is formed when phenol is heated with carbonyl chloride to 140—150°. It crystallizes from alcohol in silky needles melting at 78°.[3]

Phenyl ethylcarbonate, $CO \begin{cases} OC_2H_5 \\ OC_6H_5 \end{cases}$, is formed by the action of ethyl chlorocarbonate on potassium phenate, as well as by that of aluminium chloride on a mixture of phenol and ethyl carbonate.[4] It is an oily liquid boiling at 234°.[5]

Phenyl carbamate, $CO(NH_2)OC_6H_5$.—In the preparation of the carbonates phenyl chlorocarbonate is also formed, and this comes over first on distilling the product, though it is not pure. On passing ammonia through its ethereal solution, sal-ammoniac separates out, and on evaporating the ether, phenyl carbamate crystallizes out in tablets which melt at 141°, and are readily soluble in hot, but with difficulty in cold water. By heating it with ammonia to 140° phenol and carbamide are formed (Kempf).

Phenyl orthosilicate, $(C_6H_5)_4SiO_4$, is obtained by the action of silicon tetrachloride on phenol. It forms long prisms, which melt at 47—48°, and boil at 417—420°. It is readily soluble in alcohol, ether, and benzene, and is quickly decomposed by water.[6]

---

[1] Jacobsen, *Ber.* **8**, 1519.
[2] Scrugham, *Annalen*, **92**, 317.
[3] Kempf, *J. Pr. Chem.* II. **1**, 404.
[4] Pawlewsky, *Ber.* **17**, 1205.
[5] Fatianow, *Jahresb. Chem.* **1864**, 477.
[6] Hertkorn, *Ber.* **18**, 1679.

ETHEREAL SALTS OF PHENYL WITH ORGANIC ACIDS.

994 *Phenyl orthoformate*, $CH(OC_6H_5)_3$, is formed, together with salicylaldehyde, benzaldehyde and other products, when an alkaline solution of phenol is heated with chloroform. It crystallizes from alcohol in long white needles melting at $71·5°$, and distils without decomposition under diminished pressure.

It is not saponified by continued boiling with alkalis, but acids transform it very readily into phenol and formic acid.[1]

Phenyl acetate, $C_2H_3O_2.C_6H_5$.—Cabours obtained this substance by the action of acetyl chloride on phenol,[2] and Scrugham by heating phenyl phosphate with potassium acetate and alcohol.[3]

According to Kreysler anhydrous sodium acetate may be substituted for potassium acetate and alcohol.[4] It may also be prepared[5] by heating diazobenzene nitrate with glacial acetic acid to $70°$, by boiling phenol with acetamide,[6] and by heating phenol with lead acetate and carbon disulphide to $170°$.[7]

$$4C_6H_5OH + 2Pb(C_2H_3O_2)_2 + CS_2$$
$$= 4C_6H_5.C_2H_3O_2 + 2PbS + CO_2 + 2H_2O.$$

Phenyl acetate is a peculiarly smelling liquid boiling at $193°$ (Perkin and Hodgkinson), $195°$ at 733 mm. (Remsen and Orndorff), and possessing the same refractive index as ordinary soda-lime glass, so that a tube of this glass dipped in the liquid is invisible (Broughton). Sodium acts violently upon it, with formation of acetic acid, ethyl acetate, phenol, salicylic acid, and two crystalline bodies, $C_{15}H_{12}O_3$, and $C_{18}H_{14}O_4$, which have not yet been examined.[8]

Phenyl chloracetate, $C_2H_2ClO_2.C_6H_5$.—Provost obtained this compound by the action of chloracetyl chloride on phenol. It is readily soluble in alcohol, and crystallizes in needles melting at $40·2°$ and distilling without decomposition at $230—235°$. By heating it with alcoholic ammonia it forms phenyl amid-

[1] Tiemann, *Ber.* **15**, 2685. [2] *Annalen*, **92**, 316.
[3] *Ibid.* **92**, 317. [4] *Ber.* **18**, 1716.
[5] Remsen and Orndorff, *Am. Chem. Journ.* **10**, 368.
[6] Guareschi, *Annalen*, **171**, 142.
[7] Broughton, *ibid.* Suppl. **4**, 121.
[8] Perkin and Hodgkinson, *Journ. Chem. Soc.* 1880, **1**, 487.

acetate, $C_2H_2(NH_2)O_2.C_6H_5$, which crystallizes in needles and dissolves in weak acids and water, but is scarcely soluble in alcohol.[1]

Phenoxyacetic acid, $C_6H_5OCH_2.CO_2H$, is formed by the action of sodium phenate on chloracetic acid.[2] It is best obtained by mixing an aqueous solution of sodium chloracetate with sodium phenate, and evaporating until the whole becomes thick; it is difficultly soluble in cold water, readily in alcohol and ether, and crystallizes from hot water in long lustrous needles which are acid and bitter to the taste and have a peculiar smell. It melts at 96°, and boils with slight decomposition at 285°. Phenoxyacetic acid is not poisonous, but acts as a strong antiseptic.[3] It forms salts which crystallize well.

Ethyl phenoxyacetate, $C_6H_5.OCH_2.CO_2.C_2H_5$. may be obtained, by passing hydrochloric acid through a hot alcoholic solution of the above acid. It is a thick oily liquid, having a peculiar, not unpleasant, but persistent smell, and boiling at 251°. The methyl ether is a similar compound, boiling at 245° (Fritzsche).

Phenoxypropionic acid or *Phenyl-lactic acid*, $CH_3.CH(OC_6H_5)$ CO_2H, is obtained in a similar way to its analogue, the preceding acid, from a-chloropropionic acid. It crystallizes from hot water in long needles melting at 112—113°. Its ethyl ether is a body smelling like chloroform and boiling at 243—244°.

Phenyl oxalate, $C_2O_2(OC_6H_5)_2$, is formed when phenol is heated with anhydrous oxalic acid and phosphorus oxychloride:

$$2C_6H_5.OH + C_2O_4H_2 + POCl_3$$
$$= C_2O_4(C_6H_5)_2 + PO_3H + 3HCl.$$

It crystallizes from absolute alcohol in fine prisms, is insoluble in water, but decomposes on long boiling with it, or more rapidly in presence of alkalis or acids.[4]

By distilling a mixture of one molecule of anhydrous oxalic acid and two molecules of phenol, thin lustrous plates are obtained melting at 126—127°. This body is also formed when the two compounds are dissolved in glacial acetic acid. It distils between 150—180°, partly decomposing into phenol and formic acid. Water and alcohol decompose it into its components. Its

[1] *J. Pr. Chem.* II. **4**, 579.
[2] Heintz, *Jahresb.* **1859**, 361 ; Giacosa, *J. Pr. Chem.* II. **19**, 369.
[3] Fritzsche, *ibid.* II. **20**, 269.   [4] Nencki, *ibid.* II. **25**, 282.

empirical formula is $C_2H_2O_4 + 2C_6H_6O$, and it may be regarded as the phenyl ether of orthoxalic acid, $C_2(OH)_6$: [1]

$$(HO)_2COC_6H_5$$
$$|$$
$$(HO)_2COC_6H_5.$$

Phenyl succinate, $C_4H_4O_2(OC_6H_5)_2$, is formed by warming phenol with succinyl chloride. It is insoluble in water, and crystallizes from boiling alcohol in pearly plates; these melt at 118°, and the liquid boils without decomposition at 330°.[3]

CHLORINE SUBSTITUTION PRODUCTS OF PHENOL.

995 *Monochlorophenols*, $C_6H_4Cl(OH)$.—By the action of chlorine on phenol, *p*-chlorophenol, together with a small quantity of the ortho-compound, is formed.[3] These bodies, as well as *m*-chlorophenol, are also obtained by means of the diazo-reaction from the corresponding chloranilines [4] as well as from the amidophenols.[5]

o-Chlorophenol is formed, together with *a*- and *γ*-dichlorophenol and trichlorophenol, by the action of sodium hypochlorite on an aqueous solution of phenol.[6] It is a liquid having an unpleasant, persistent smell, boiling at 175—176°, and solidifying at a low temperature in needles, melting at 7°. By the action of phosphorous pentachloride it is converted into *o*-dichlorobenzene.

m-Chlorophenol crystallizes in white needles, melting at 28·5° [7] and boiling at 214°.

p-Chlorophenol is prepared, in addition to the above-mentioned methods, by the action of sulphuryl chloride on phenol.[8]

$$C_6H_5.OH + SO_2Cl_2 = C_6H_4Cl.OH + SO_2 + HCl.$$

It forms crystals which melt at 37° and boil at 217°; it has a weak but unpleasant and very persistent smell.

---

[1] Claparède and Smith, *Journ. Chem. Soc.* 1883, i. 358 ; Staub and Smith, *Ber.* **17**, 1740.

[2] Weselsky, *ibid.* **2**, 518.   [3] Faust and Müller, *Annalen*, **173**, 303.

[4] Beilstein and Kurbatow, *ibid.* p. 176.

[5] Schmitt, *Ber.* **1**, 67.   [6] Chandelon, *ibid.* **16**, 1479.

[7] Uhlemann, *ibid.* **11**, 1161.

[8] Dubois, *Zeitschr. Chem.* **1866**, 705 ; **1867**, 205.

a-Dichlorophenol, $C_6H_3Cl_2OH$. Laurent obtained this compound by the action of chlorine on phenol, and named it *"Acide chlorophénisique."*[1] It was afterwards prepared by F. Fischer, by passing chlorine through phenol for several days and fractionating the product.[2] It is insoluble in water and crystallizes from benzene in long six-sided needles which have an unpleasant persistent smell, melt at 43° and boil at 209—210°. Its solution in alcohol is acid, and it decomposes carbonates on boiling with water, though its salts are almost completely decomposed by carbon dioxide in the cold. Phosphorus pentachloride converts it into the asymmetric trichlorobenzene.

Two other dichlorophenols are known, obtained from the dichloramidophenols.

		Melting-point.	Boiling-point.
β-Dichlorophenol,[3]	thin needles.	54—55°	—
γ-Dichlorophenol,[4]	fine needles.	65°	218—220°

Trichlorophenol, $C_6H_2Cl_3(OH)$, was likewise discovered by Laurent and named *Acide chlorphénisique ;*[5] it is obtained by the action of chlorine, not only on phenol, but also on aniline,[6] indigo,[7] and other aromatic compounds. It is prepared by passing chlorine into phenol, warming gradually until the melting point rises to about 67°, and isolating the pure compound by fractional distillation.[8] It is easily soluble in alcohol and ether, and crystallizes in needles melting at 67—68°, boiling at 243·5—244·5°, and having an acid reaction. Its salts are only slightly soluble in water.

An isomeric trichlorophenol is obtained from trichloramidophenol; it melts at 54·1—54·5° and boils at 248·5—249·5°.[9]

Pentachlorophenol, $C_6Cl_5(OH)$. Erdmann obtained this compound by the continued action of chlorine on isatin and on trichlorophenol (chlorindoptic acid), and named it chlorinated indoptic acid.[10] Laurent who likewise prepared it from isatin named it *Acide chlorphénusique.*[11] It is also prepared by the action of iodine chloride on phenol,[12] or, better, by passing chlorine continuously through a mixture of three parts of

---

[1] *Ann. Chim. Phys.* **63**, 27.   [2] *Annalen*, Suppl. **7**, 180.
[3] Hirsch, *Ber.* **11**, 1981.   [4] Seifart, *Annalen*, Suppl. **7**, 203.
[5] *Ann. Chim. Phys.* III. **3**, 206.   [6] *Annalen*, **53**, 8.
[7] Erdmann, *J. Pr. Chem.* **19**, 332; **22**, 276; **25**, 472.
[8] Faust, *Annalen*, **149**, 149.   [9] Hirsch, *Ber.* **13**, 1908.
[10] *J. Pr. Chem.* **22**, 272.   [11] *Ann. Chim. Phys.* III. **3**, 497.
[12] Schützenberger, *Bull. Soc. Chim.* **4**, 102.

phenol and one part of antimony trichloride at 100—110° till the action ceases. The antimony chloride is then dissolved in strong hydrochloric acid, the residue treated with boiling soda solution, precipitated by hydrochloric acid, and the separated pentachlorophenol purified by distillation in superheated steam and recrystallization from light petroleum.[1] It crystallizes in rhombic prisms which melt at 187°, and are readily soluble in alcohol, the solution having an acid reaction. When heated it first has a pungent smell and then produces coughing ; its powder excites violent sneezing. By carefully heating it may be sublimed in white needles; it does not boil until a higher temperature is reached, when it decomposes with separation of hydrochloric acid and pentachlorophenylene oxide, which is described below. Phosphorus pentachloride converts it into perchlorobenzene. Its salts are for the most part only slightly soluble in water.

Potassium pentachlorophenate, C_6Cl_5OK, crystallizes from concentrated caustic potash in prisms with a diamond lustre. At a high temperature it decomposes into potassium chloride and *pentachlorophenylene oxide*, $(C_6Cl_4)_2O_2$, which is scarcely soluble in alcohol and ether, and crystallizes from hot nitrobenzene in broad needles resembling benzoic acid. It melts at about 320°, and boils above the boiling point of mercury. Its constitution is probably the following :

$$O \diamondsuit O \quad \begin{matrix} C_6Cl_4 \\ C_6Cl_4 \end{matrix}$$

Pentachlorophenol chloride, $C_6Cl_5(OH)Cl_2$, is obtained by the continued action of chlorine on aceto-*m*-chloranilide, $C_6H_4Cl.N(C_2H_3O)H$, and crystallizes from light petroleum in large, thick prisms melting at 78·5—80°. By heating it with absolute alcohol to 230° it is converted into pentachlorophenol.[2]

a-Hexchlorophenol, C_6Cl_6O, is obtained when chlorine is led for several days into pentachlorophenol suspended in hydrochloric acid. It forms yellow crystals, melting at 46°, and gives off chlorine on heating.[3]

β-Hexchlorophenol is obtained by passing chlorine into pentachloraniline suspended in glacial acetic acid.[4] It forms yellow

[1] Merz and Weith, *Ber.* **5**, 458.   [2] Beilstein, *ibid.* **11**, 2182.
[3] Benedikt and Schmidt, *Monatsh.* **4**, 607.   [4] Langer, *Annalen*, **215**, 122.

prisms which melt at 106°. A compound probably identical with this has been obtained by Hugounenq [1] by·acting on a mixture of anisoïl and antimony pentachloride with chlorine.

BROMINE SUBSTITUTION PRODUCTS OF PHENOL.

996 *Monobromophenols,* $C_6H_4Br(OH)$, are obtained in a similar way to the corresponding chlorophenols.

o-Bromophenol is an oily liquid having an unpleasant, strong and persistent smell, and boiling at 194—195°.[2]

m-Bromophenol crystallizes in scales melting at 33° and boiling at 236·5°; it does not smell so unpleasantly as the ortho-compound.[3]

p-Bromophenol crystallizes from chloroform in large octo-hedra resembling those of alum, melting at 63—64°. It is easily soluble in alcohol and boils at 238°.[4]

	Melting-point.
[5] Dibromophenol, $C_6H_3Br_2(OH)(1.3.4)$, snow-white crystals	40°
[6] Dibromophenol, $C_6H_3Br_2OH.(1.3.5)$, crystals	76·5°
[7] Tribromophenol, $C_6H_2Br_3(OH)$, very long hair-like needles	95°
[8] Tetrabromophenol, $C_6H.Br_4(OH)$, needles	120°
[9] Pentabromophenol, $C_6Br_5(OH)$, needles . , ...	225°

These bodies are all obtained by the direct bromination of phenol. It is singular that the last three compounds are further acted upon by bromine, forming derivatives in which the hydrogen of the hydroxyl group appears to be replaced by hydrogen. It is, however, not improbable that the constitution of these compounds is similar to the alternative formula given for penta-chloro- and pentabromoresorcinol (p. 162). The compounds formed are as follows :

Tribromophenol bromide, $C_6H_2Br_3(OBr)$ (?), is obtained by treating an aqueous solution of phenol, or better salicylic acid,

[1] *Compt. Rend.* **109**, 309.   [2] Fittig and Mager, *Ber.* **8**, 362.
[3] Wurster and Nölting, *ibid.* **7**, 905 ; Fittig and Mager, *loc. cit.*
[4] Hubner and Brenken, *ibid.* **6**, 171 ; Fittig and Mager, *ibid.* **7**, 1176.
[5] Körner, *Annalen*, **137**, 205.   [6] Blau, *Monatsh.* **7**, 630.
[7] Laurent, *ibid.* **43**, 212 ; Körner, *loc. cit.*
[8] Körner, *loc. cit.* p. 209.   [9] Körner, *loc. cit.* p. 210.

$C_6H_4(OH)CO_2H$, with strong bromine water. The precipitate thus obtained crystallizes from carbon disulphide in lemon-coloured scales which are not attacked by the boiling aqueous solution of an alkali. By dissolving it in benzene and adding caustic potash or ammonia, tribromophenol is formed. On heating with sulphuric acid it is converted into the isomeric tetrabromophenol.[1]

Tetrabromophenol bromide, $C_6HBr_4(OBr)$ (?), is obtained by dissolving tetrabromophenol in caustic potash, adding hydrochloric acid and then quickly an excess of bromine water. It crystallizes from chloroform in yellow monosymmetric tables; boiling alcohol converts it into tetrabromophenol, and on heating with sulphuric acid it is converted into pentabromophenol.

Hexbromophenol or Perbromophenol bromide, $C_6Br_5(OBr)$ (?), is obtained from pentabromophenol in a similar way to the preceding compound, and forms granular yellow crystals, which are insoluble in cold alcohol, but are converted by boiling alcohol into pentabromophenol (Benedikt).

IODINE SUBSTITUTION PRODUCTS OF PHENOL.

997 *Moniodophenols.* The literature on this subject contained many contradictions, and therefore Nölting and Stricker[2] have reinvestigated these compounds, with the following results:

o-Iodophenol, $C_6H_4I.OH$, is obtained by heating *o*-diazophenol with a solution of potassium iodide, and is also formed together with small quantities of the para-compound, and di- and tri-iodophenols, by the action of iodine on an aqueous solution of phenol and alcoholic ammonia.[3] It crystallizes from light petroleum in flat white needles, melting at 43°, which are slightly soluble in water, readily in other solvents, and is converted on fusion with potash into catechol. It becomes brown on exposure to air and light.

m-Iodophenol, $C_6H_4I.OH$, may be prepared from *m*-iodaniline by diazotizing and boiling, or from *m*-diazophenol in a similar manner to the ortho-compound. It crystallizes from light

[1] Benedikt, *Annalen*, **199**, 128 ; *Monatsh.* **1**, 360.
[2] *Ber.* **20**, 3018.
[3] Willgerodt, *J. Pr. Chem.* II. **37**, 446.

petroleum in snow-white needles, which melt at 39—40°, and on fusion with potash yields resorcinol.

p-Iodophenol, $C_6H_4I.OH$, is obtained by boiling *p*-diazophenol with a solution of potassium iodide, and is also formed together with tri-iodophenol by action of iodide of nitrogen on a solution of potassium phenate in aqueous alcohol.[1] It forms long needles, which melt at 93—94°, and on fusion with potash forms quinol. At a higher temperature however this is converted into resorcinol.

Melting-point.

[2] *a*-Di-iodophenol, $C_6H_3I_2(OH)$, crystals . . 150°

[3] *β*-Di-iodophenol, $C_6H_3I_2(OH)$, crystals . . 68°

γ-Di-iodophenol, (OH.I.I = 1.2.4), crystals . 72°

Di-iodophenol-p-sulphonic acid, (OH.I.SO$_3$H.I = 1.2.4.6), is obtained by treating potassium *p*-phenolsulphonate with iodine dissolved in a solution of potassium iodide and iodate. It crystallizes from hot water with two molecules of water, and is known in commerce as "Sozoiodol," and is used in place of iodoform. That it has the constitution given above is shown by the fact that it is readily converted into picric acid.

Tri-iodophenol, $C_6H_2.I_3.OH$, is obtained by the action of iodine and potash,[4] or iodine and iodic acid[5] on salicylic acid, or by treating phenol with chloride of iodine.[6] It crystallizes from dilute alcohol in needles, which melt at 156°.

Another compound having the same composition, has been obtained by Messinger and Vortmann[7] by the action of an excess of iodine and alkali on phenol. It has a violet colour and is decomposed by hot water to some extent, iodine and di-iodophenol being formed, and is converted into the above tri-iodophenol by alkalis. The authors assign to it the constitution $C_6H_3I_2.OI$.

[1] Willgerodt, *J. Pr. Chem.* II. **37**, 446.

[2] Hlasiwetz and Weselsky, *Ber.* **2**, 524.

[3] Schall, *ibid.* **16**, 1899, 1902.

[4] Lautemann, *Annalen*, **120**, 307.

[5] Kekulé, *ibid.* **131**, 231.

[6] Schutzenberger, *Jahresb.* **1865**, 524.

[7] *Ber.* **22**, 2313.

FLUORINE SUBSTITUTION PRODUCTS OF PHENOL.

998 *p-Fluophenol*, $C_6H_4F.OH$, is prepared by diazotizing *p*-fluoraniline, and boiling with water.[1] It is solid at the ordinary temperature, and boils at 186—188°.

NITRO-SUBSTITUTION-PRODUCTS OF PHENOL.

999 *Mononitrophenols*, $C_6H_4(NO_2)OH$.—In 1839 Fritzsche, by acting with nitric acid on indigo, obtained a peculiar smelling volatile, yellow, crystalline body, which he subsequently found to be identical with Hofmann's nitrophenol, obtained by the action of nitrous acid on aniline, and of nitric acid on phenol.[2] Fritzsche afterwards found that in the latter reaction two isomeric bodies are formed, of which nitrophenic acid can be volatilized in a current of steam, whilst isonitrophenic acid cannot.[3] These bodies were afterwards distinguished as volatile and non-volatile nitrophenol, until it was found that the first is the ortho- and the second the para-compound.

They may be prepared by gradually adding one part of pure phenol to a well-cooled mixture of two parts of nitric acid of sp. gr. 1·34 and four parts of water, stirring, and after some time separating the heavy oil from the aqueous solution, washing it with water, and distilling in a current of steam, till the distillate is no longer coloured yellow.

p-Nitrophenol may be obtained from the residue by extracting with boiling water. The yield amounts to 30 per cent. of *o*-nitrophenol and 32 per cent. of *p*-nitrophenol on the phenol taken.[4]

The higher the temperature rises during the operation, the larger is the yield of *o*-nitrophenol, whilst if the temperature be kept low, more of the para-compound is formed.[5]

The two nitrophenols are also obtained, together with the diazobenzene nitrate, when nitrogen trioxide is passed into an ethereal

[1] Wallach and Heusler, *Annalen*, **243**, 228.
[2] Körner, *ibid.* **75**, 359 ; **103**, 347.   [3] *Annalen*, **110**, 150.
[4] Schmitt and Cooke ; Körner, Kekulé's *Lehrb. Org. Chem.* **3**, 40.
[5] Goldstein, *Journ. Russ. Chem. Ges.* **10**, 353.

solution of phenol,[1] as well as when sodium phenate is mixed with a solution of nitrogen peroxide in carbon disulphide (Schall).

They may also be obtained[2] from aniline by diazotizing, allowing to stand, and afterwards boiling with nitric acid of sp. gr. 1·335.

By warming a mixture of 5 parts of phenol and an equal weight of ethyl nitrate with 16 parts of water, and 32 parts of concentrated sulphuric acid for a long time, 22 p. c. of o-nitrophenol and only 0·5 p. c. of p-nitrophenol is obtained.[3] According to Fittica a fourth isomeric nitrophenol is formed at the same time.

o-Nitrophenol is also obtained in addition to the manner already stated, when o-chloronitrobenzene, o-bromonitrobenzene, or o-dinitrobenzene is heated with dilute alkalis. It is soluble with difficulty in cold, readily in hot water, as also in alcohol and ether, and crystallizes in sulphur-yellow needles or prisms, having a peculiar aromatic smell and melting at 45°. It boils without decomposition at 214°, is readily volatile with steam, and is converted by phosphorus pentachloride into o-nitrophenyl phosphate, only a little o-chloronitrobenzene being formed.

Its salts, which for the most part crystallize well, have a scarlet-red to an orange-yellow colour.

Methyl o-nitrophenate or *Nitranisoïl*, $C_6H_4(NO_2)OCH_3$, is obtained, together with the para-compound, by the nitration of anisoïl, as well as by heating o-nitrophenol with caustic potash and methyl iodide.[4] It is a yellowish oil, boiling at 276·5°; it solidifies on cooling, and melts at 9°. When heated with ammonia to 200° it forms o-nitraniline.[5]

m-Nitrophenol is obtained by the diazo-reaction from m-nitraniline;[6] it separates from an ethereal solution in thick sulphur-yellow crystals, melting at 96°. It can only be distilled without decomposition under diminished pressure, and is not volatile in a current of steam. Its salts are of an orange-yellow colour.

Methyl m-nitrophenate, $C_6H_4(NO_2)OCH_3$, crystallizes from alcohol in flat needles, melting at 38° and boiling at 258°. It is easily volatilized by steam, and when heated with ammonia yields a small quantity of m-nitrophenol together with humus-like products, but no m-nitraniline.[7]

p-Nitrophenol is obtained, like o-nitrophenol, from the cor-

---

[1] Weselsky, *Ber.* **8**, 89.
[2] Nölting and Wild, *ibid.* **18**, 1338.
[3] Natanson, *ibid.* **13**, 415.
[4] Mühlhäuser, *Annalen*, **207**, 237.
[5] Salkowski, *ibid.* **174**, 278.   Bantlin, *Ber.* **11**, 2100.
[7] Salkowski, *ibid.* **12**, 155.

responding haloid compound by heating with alkalis. It is however best prepared from the mixture of the ortho- and para-compounds obtained by acting on phenol with nitric acid (p. 132). After the o-nitrophenol has been removed by distillation with steam, the residue is boiled with dilute caustic soda solution, a very strong solution of caustic soda then added, and the precipitated sodium compound filtered and purified by dissolving in a small quantity of water, reprecipitating with concentrated caustic soda solution, and then decomposed by acids.[1]

p-Nitrophenol crystallizes from boiling water in colourless needles or in monosymmetric prisms, which melt at 114° and on again cooling separate out as monosymmetric crystals, which, however, are dimorphous with the ordinary form.[2] It boils almost without decomposition but is not volatile in a current of steam. On boiling it with water and barium carbonate it decomposes the last-named compound more rapidly than the ortho-compound, and this again acts more rapidly than m-nitrophenol.[3] Its salts are not so deeply coloured as those of its isomerides.

Methyl p-nitrophenate, $C_6H_4(NO_2)OCH_3$, crystallizes in large rhombic prisms melting at 51° and boiling at 258—260°, and forms p-nitraniline on heating with ammonia.[4]

HALOGEN SUBSTITUTION PRODUCTS OF THE MONONITROPHENOLS.

Of these only such as contain one halogen will be mentioned. The first number gives the position of the halogen, the second that of the nitroxyl, the hydroxyl occupying position 1.

		Melting-point.
	2 : 6 yellow needles[5] . . .	70°
	2 : 4 white needles[6] . . .	110—111°
Chloronitrophenols.	3 : 6 lemon-yellow needles[7] .	39°
	4 : 2 monosymmetric prisms[8]	86—87°

[1] E. Fischer, Org. Präparate, p. 33.   [2] Lehmann, Jahresb. 1877, 549.
[3] Post and Mehrtens, Ber. 8, 1549.
[4] Brunck, Zeitschr. Chem. 1867, 205 ; Willgerodt, Ber. 14, 2632.
[5] Faust and Muller, Annalen, 173, 309.
[6] Faust, Zeitschr. Chem. 1871, 591 ; Armstrong, ibid. 596 ; Faust and Müller.
[7] Uhlemann, Ber. 11, 1161.
[8] Faust and Saanse, Annalen Suppl. 7, 190 ; Laubenheimer, Ber. 7, 1601 ;
Bodewig, Jahresb. Chem. 1879, 511.

Melting point.

Bromonitrophenols. $\begin{cases} 2:4 & \text{long white needles}^1 \quad \ldots \quad 102° \\ 5:2 & \text{yellow prisms}^2 \quad \ldots \ldots \quad 44° \\ 4:6 & \text{yellow monosymmetric prisms}^3 \ 88° \end{cases}$

IODONITROPHENOLS.

The first two of these are obtained by the action of iodine and mercuric oxide on a solution of o-nitrophenol in glacial acetic acid, and the third in the same way from p-nitrophenol.[4]

Melting-point.

a-Iodo-o-nitrophenol,   long, yellow needles . . . .   90—91°

β-Iodo-o-nitrophenol,   short, yellow needles   . . .   66—67°

Iodo-p-nitrophenol,   thick, light yellow crystals . 154—155°

DINITROPHENOLS, $C_6H_3(NO_2)_2OH$.

Of the six theoretically possible compounds the following are known. The subjoined figures give the position of the nitroxyls when the hydroxyl occupies position 1.

Ordinary or a-*Dinitrophenol* (2:4) is obtained by nitrating phenol[5] as well as from both o- and p-nitrophenol, by which its constitution is shown.[6] It is best obtained by warming p-nitrophenol with an equal weight of nitric acid of sp. gr. 1·37.[7] It crystallizes from hot water in yellowish-white rectangular tables which melt at 113—114°.

β-*Dinitrophenol* (2:6) is obtained, together with the a-compound, by the nitration of o-nitrophenol, and crystallizes from hot water in fine, bright yellow needles melting at 63—64°.

γ-*Dinitrophenol* (3:6) is obtained together with the two following, when m-nitrophenol is heated with nitric acid.[8] They can be separated by means of their barium salts, and the γ-compound can also be separated from the other two by distilling

---

[1] Brunck and Körner, *Zeitschr. Chem.* **1868**, 323.

[2] Laubenheimer.

[3] Brunck ; Körner ; Laubenheimer ; Hüfner and Brenken, *Ber.* **6**, 170 ; Azruni, *Jahresb. Chem.* **1877**, 547.   [4] Busch, *Ber.* **7**, 462.

[5] Laurent, *Annalen*, **43**, 213.   [6] Körner, *Zeitschr. Chem.* **1868**, 322.

[7] Körner, Kekulé's *Lehrb.* **3**, 42.

[8] Bantlin, *Ber.* **8**, 21 ; **11**, 2102 ; see also Henriques, *Annalen*, **215**, 321.

in a current of steam. It crystallizes from hot water in light yellow needles melting at 104°.

δ-*Dinitrophenol* (3 : 4) crystallizes in long, colourless, silky needles melting at 134°.

ε-*Dinitrophenol* (2 : 3) separates from hot water in small yellow needles, and from alcohol in thick crystals which melt at 144°.

TRINITROPHENOLS, $C_6H_2(NO_2)_3OH$.

1000 We are now acquainted with three of these bodies. Of these the following has been known for a long time.

Symmetric or *a-Trinitrophenol* (2 : 4 : 6). Woulfe found, in 1771, that by the action of nitric acid on indigo a liquid is obtained which dyes silk yellow.[1] Similar observations were made by other chemists ; thus Quatremère Disjonval in 1780 laid before the Paris Academy the results of an investigation on indigo, in which he remarks that this body forms with nitric acid a substance which stains the skin a saffron-yellow colour. Hausmann then observed that in this reaction a bitter acid compound is formed. Indigo bitter was then further investigated by Fourcroy and Vauquélin[2] as well as by Chevreul.[3] The latter chemist considered it to be a compound of nitric acid with a peculiar organic substance. Welter, by treating silk with nitric acid, had already obtained a yellow, crystallized acid, whose potassium salt exploded when heated, like gunpowder.[4] Liebig found that Welter's bitter body is identical with indigo-bitter and aloe-bitter, which Braconnot had obtained by heating aloes with nitric acid, and called it " Kohlenstickstoffsäure," (carbazotic acid),[5] while Berzelius named it "Pikrinsalpetersäure," and Dumas gave it the name by which it is now universally known, picric acid (πικρός, bitter).[6] Laurent first obtained it by the action of nitric acid on phenol, and showed that picric acid (*Acide nitrophénisique*) is trinitrophenol.[7] This is the final product of the action of nitric acid on a large number of substances containing the benzene ring, just as oxalic acid is the result of the oxidation of many fatty bodies. Amongst those which give a good yield of the acid is the acaroïd resin (from *Xanthorrhoea hastilis*),[8] and before phenol was manufactured on a large

[1] *Phil. Trans.* 1771.   [2] Gehlen, *N. Journ.* **2**, 231.
[3] *Ann. Chim. Phys.* **72**, 113.   [4] *Ibid.* **29**, 301.
[5] *Pogg. Ann.* **13**, 191.   [6] *Annalen,* **39**, 350.
[7] *Ann. Chim. Phys.* III. **3**, 221.   [8] Stenhouse, *Mem. Chem. Soc.* **3**, 12.

scale, this resin was employed for the preparation of picric acid. This body is also obtained by the action of nitric acid on o-nitrophenol, p-nitrophenol and the dinitrophenols (2:4 and 2:6), but not on m-nitrophenol, and this fact indicates its constitution.

Picric acid was formerly obtained by the action of nitric acid on phenol; phenolsulphonic acid is now used, since this substance as Laurent had already pointed out, is easily converted into picric acid. In its manufacture the apparatus is employed which is used for making nitrobenzene. A mixture of equal parts of phenol and concentrated sulphuric acid is placed in the vessel, the contents then heated to 100°, and nitric acid of sp. gr. 1·3 allowed to flow in. After cooling, the product solidifies to a crystalline mass, which is filtered and drained, and then washed with cold water. It is further purified by recrystallizing from water containing 0·1 p. c. of sulphuric acid. Another useful mode of purification consists in exactly neutralizing the crude acid with sodium carbonate and adding to the hot filtered solution a few crystals of sodium carbonate, when almost the whole of the sodium picrate separates out and can be decomposed with hydrochloric acid.[1]

Picric acid is difficultly soluble in cold, easily in hot water, and crystallizes from aqueous solution in pale yellow, shining scales, whilst it crystallizes from an ethereal solution in rhombic prisms melting at 122·5°. When carefully heated it sublimes, but on strongly heating it decomposes with detonation, and on warming with calcium hypochlorite and water it yields chloropicrin, $CCl_3(NO_2)$, and chloranil, $C_6Cl_4O_2$. It has an intensely bitter taste, an acid reaction, and colours animal fibre, the skin, etc. pure yellow. As its tinctorial power is very great it is largely used as a dyeing material for wool, silk, leather, etc., either alone, or in combination with red and blue for dyeing oranges, greys, etc. It is not fixed by itself to vegetable fibre, and is therefore employed to distinguish cotton from wool and silk, and to detect the presence of cotton in mixed fabrics, by steeping these for some time in a hot solution of the acid, washing with water and then examining under the microscope.

Both its intensely bitter taste and its power of dyeing wool yellow serve to detect picric acid, but the best test is to add ammoniacal copper sulphate to the solution under examination; this gives a greenish precipitate, which can be seen even when

[1] Carey Lea, *Sill. Amer. Journ.* II. **32**, 180.

only 1 part of picric acid is contained in 5,000 of water (C. Lea). By warming a solution of picric acid with ammonia and potassium cyanide, a deep red colour is produced, isopurpuric acid being formed (p. 140). Picric acid unites with other aromatic hydrocarbons, as it does with benzene, to form compounds, many of which are distinguished by their crystalline form or colour. This property indeed is frequently employed as a test for, and as a means of separating, some of these hydrocarbons. All such compounds are decomposed by ammonia, and some of them by alcohol, whilst others again may be recrystallized without decomposition. Of these latter one of the most characteristic is the naphthalene compound, $C_{10}H_8 + C_6H_3(NO_2)_3O$, crystallizing in golden needles, often united in stellate groups, and melting at 149°. This peculiar compound may be employed to distinguish picric acid from similar nitro-compounds.

Picric acid is poisonous, rabbits and dogs being killed by doses of from 0·06 gram to 0·6 gram. It is said that it is sometimes added to beer to give it a bitter taste, 0·012 gram sufficing to give to a litre of beer an insupportable bitterness. In order to detect the picric acid, the beer is warmed and a skein of white woollen yarn steeped in it; if this becomes coloured yellow it is treated with ammonia, washed with water, the solution concentrated on the water-bath and some potassium cyanide added, when, if picric acid be present, the red colour of isopurpuric acid will be seen. Picric acid can also be detected by shaking a few cubic centimetres of beer with half its volume of amyl alcohol, which takes up the picric acid; after evaporation of the alcohol the residue is tested as above.[1]

Wöhler in 1828 made the remarkable discovery that on boiling picric acid with water and baryta, hydrocyanic acid is formed.[2] This observation has been almost entirely forgotten, probably because it seemed to be highly unlikely, considering the very stable character which the benzene derivatives possess in other respects, and it was not until Hübner and Post carefully repeated and confirmed it that Wöhler's observation obtained the recognition of chemists. Hübner and Post also found that even dinitrobenzene yields prussic acid on boiling with caustic potash, and that it is also obtained by fusing nitrobenzene for a short time with caustic potash.[3]

---

[1] Vitali, *Ber.* **10**, 83. [2] *Pogg. Ann.* **13**, 488.
[3] *Ber.* **5**, 408.

1001 Picric acid forms salts having a yellow colour, which, for the most part crystallize well and are explosive.

Potassium Picrate, $C_6H_2(NO_2)_3OK$, crystallizes in long, yellow, four-sided, lustrous needles, which, obtained from dilute solutions, appear sometimes red and sometimes green (Liebig). It dissolves in 260 parts of water at 15°, and in 4 parts at the boiling point. It is not soluble in alcohol. On account of its sparing solubility it was formerly employed as a reagent for potassium salts. On heating, it becomes of a red colour and then explodes like gunpowder; on being struck with a hammer it detonates powerfully. It is sometimes used mixed with saltpetre for blasting purposes.

Sodium Picrate, $C_6H_2(NO_2)_3ONa$, forms yellow needles, which dissolve at the ordinary temperature in about twelve parts of water, but as already stated the addition of soda precipitates the greater portion of it.

Ammonium Picrate, $C_6H_2(NO_2)_3ONH_4$, is sparingly soluble in water and still less so in alcohol, and crystallizes in rhombic scales. It is employed mixed with the nitrates of barium, strontium, &c., for bengal fire. Brugère's picrate-powder consists of 54 parts of this salt to 46 parts of saltpetre; 2·6 grains of this is said to equal 5·5 grains of gunpowder. According to Abel this mixture is valuable for the blasting charge in shells.

Silver Picrate, $C_6H_2(NO_2)_3OAg + H_2O$, crystallizes from hot water in yellow shining needles, which dissolve at 15° in 113·09 parts of water.[1]

Lead Picrate, $[C_6H_2(NO_2)_3O]_2Pb + H_2O$ crystallizes in brown needles which dissolve at 15° in 113·17 parts of water. It is employed mixed with saltpetre as a blasting agent in the place of potassium picrate.

Methyl picrate or *Trinitranisoïl,* $C_6H_2(NO_2)_3OCH_3$, was obtained by Cahours by the action of concentrated nitric acid on anisoïl; it is also obtained by the action of methyl iodide on silver picrate, and crystallizes in yellow monosymmetric tables, melting at 64°.

Ethyl picrate, $C_6H_2(NO_2)_3OC_2H_5$, is formed in a similar way to

[1] Post and Mehrtens, *Ber.* **8**, 1549.

the methyl ether,[1] and also by adding caustic potash to a solution of chlorotrinitrobenzene (picryl chloride) in absolute alcohol.[2] It crystallizes in long, almost colourless needles which melt at 78·5°; by heating it with alcoholic ammonia it is converted, like the methyl ether, into trinitraniline.[3]

Phenyl picrate, $C_6H_2(NO_2)_3OC_6H_5$, is obtained in a similar way to the preceding compound, using potassium phenate, and crystallizes in colourless needles (Willgerodt).

Trinitrophenyl acetate, $C_6H_2(NO_2)_3OC_2H_3O$, is formed by boiling picric acid with acetic anhydride, and separates from ether in dark yellow crystals, melting at 75—76°.[4]

Picryl chloride or *Chlorotrinitrobenzene*, $C_6H_2(NO_2)_3Cl$, was first prepared by Pisani by the action of phosphorus pentachloride on picric acid and called by him "*Chlorure de picryle*," because he found that, like other acid chlorides, it is converted by water into hydrochloric acid and picric acid.[5]

According to Clemm, the latter observation is not accurate. Pisani's product contained, without doubt, some free picric acid, which naturally enters into solution when water is added, while the pure compound is not attacked even by boiling water.[6]

Chlorotrinitrobenzene crystallizes from alcohol in almost colourless needles, and from ether in amber-coloured monosymmetric tables, melting at 83°.[7]

By boiling it with soda solution it gives picric acid, and by heating it with an aqueous solution of ammonia it is converted into trinitraniline or picramide, $C_6H_2(NO_2)_3NH_2$.

Picryl chloride, like picric acid, forms compounds with aromatic hydrocarbons, which crystallize well.[8] The benzene compound $C_6H_2(NO_2)_3Cl,C_6H_6$, crystallizes in large, pale yellow prisms, which very quickly lose benzene in the air.[9]

1002 *Isopurpuric acid* or *Picrocyamic acid*, $C_8H_5N_5O_6$, is as little known in the free state as its isomeride purpuric acid (Vol. III. Part II. pp. 357), but many of its salts, which are very similar to the purpurates, have been prepared.

[1] H. Müller and Stenhouse, *Annalen*, **141**, 80.
[2] Willgerodt, *Ber.* **12**, 1277.
[3] Salkowski, *Annalen*, **174**, 259.
[4] Tomassi and David, *ibid.* **169**, 167.
[5] *Compt. Rend.* **39**, 852 ; *Annalen*, **92**, 326.
[6] *J. Pr. Chem.* II. **1**, 145.
[7] Bodewig, *Jahresb.* **1879**, 394.
[8] Liebermann and Palm, *Ber.* **8**, 377.
[9] Mehrtens, *ibid.* **11**, 844.

Potassium isopurpurate, $C_8H_4KN_5O_6$, is obtained when picric acid is warmed with water and potassium cyanide : [1]

$$C_6H_3N_3O_7 + 3KCN + 2H_2O = C_8H_4KN_5O_6 + K_2CO_3 + NH_3.$$

It crystallizes in brownish-red scales, having a greenish metallic lustre ; is slightly soluble in cold, more readily in hot water, forming a fine red solution ; on heating, or in contact with concentrated sulphuric acid, it explodes. On adding an acid to the aqueous solution it is coloured brownish-yellow, evolving a pungent odour, and then deposits a brown flocculent precipitate. On evaporating the solution, a brownish-yellow amorphous mass remains behind.

Ammonium Isopurpurate, $C_8H_4(NH_4)N_5O_6$, is obtained by the addition of sal-ammoniac to a concentrated solution of the potassium salt. It forms small brownish-red crystals, with a green metallic lustre, which are analogous in optical properties and crystalline form to murexide (ammonium purpurate). It is employed as a dye for wool and silk, and called in commerce *Grénat soluble.*

1003 *β-Trinitrophenol,* $C_6H_2(NO_2)_3OH(3:4:6)$, is obtained by nitrating γ- and δ-dinitrophenol ; it crystallizes from hot water in white, satin-like needles or scales, melting at 96°. The potassium salt forms light red, brilliant crystals, which are soluble with difficulty in water, forming a light-yellow solution.

γ-Trinitrophenol, $C_6H_2(NO_2)_3OH(2:3:6)$, is obtained from γ-and ε-dinitrophenol, and forms white needles, melting at 117—118°. The potassium salt, easily soluble in water, forms brilliant red needles ; its aqueous solution colours wool and silk a fine orange shade.

Both trinitrophenols taste bitter and detonate on heating ; there salts are, like the picrates, very explosive.[2] By the continued action of nitric acid they are converted into styphnic acid or trinitroresorcinol, $C_6H(NO_2)_3(OH)_2$. By nitrating γ-dinitrobenzene, besides trinitroresorcinol and the two trinitrophenols, a tetranitrodihydroxybenzene, $C_6(NO_2)_4(OH)_2$, probably tetranitroresorcinol, is formed.

The production of styphnic acid by the nitration of dinitrophenol had already been observed by Bantlin, and the compound described under the name of isopicric acid.

---

[1] Baeyer, *Jahresb.* **1859**, 458 ; Hlasiwetz, *Annalen,* **110**, 289 ; Kopp. *Ber.* **5**, 644.

[2] Henriques, *Annalen,* **215**, 321.

The following graphic formulæ show the connection between
the different nitro-substitution products of phenol :

o-Nitrophenol.   p-Nitrophenol.   m-Nitrophenol.

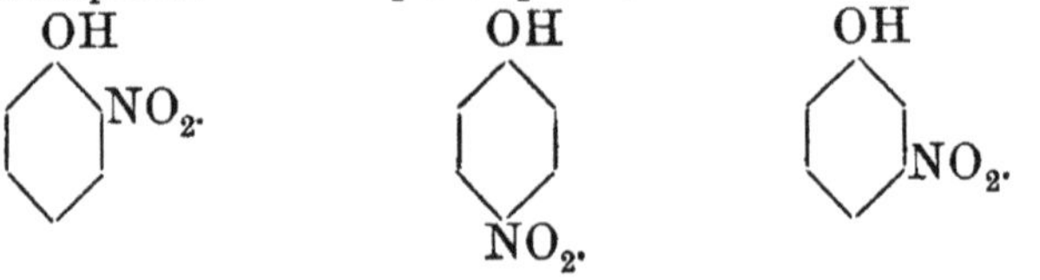

Dinitrophenols.

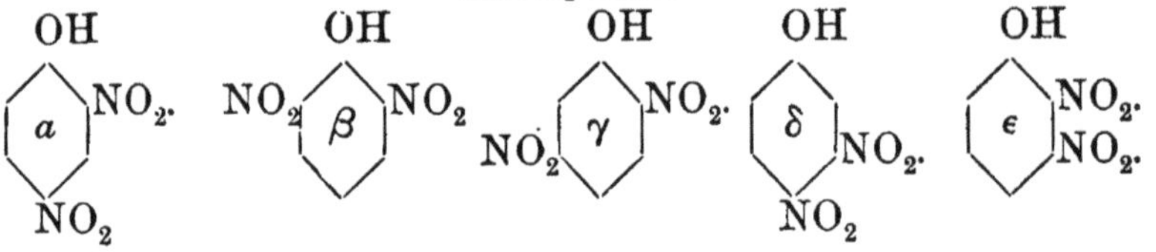

Trinitrophenols.

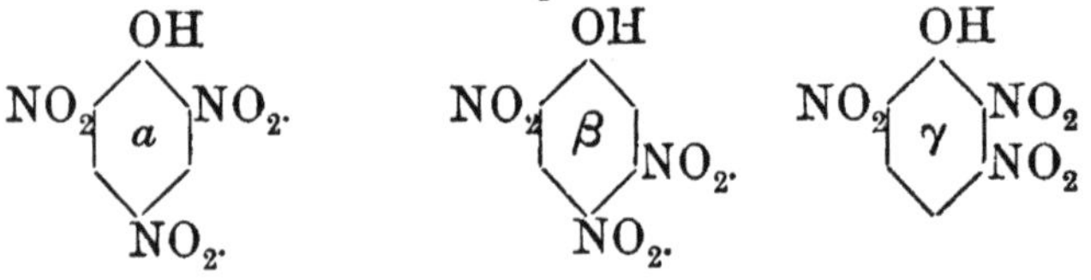

PHENOLMONOSULPHONIC ACIDS, $C_6H_4(OH)SO_3H$.

1004 By dissolving phenol in sulphuric acid Laurent obtained
his " Acide sulphophénique." [1] Kekulé then found that a mixture of o- and p-sulphonic acids [2] is thus formed.

o-Phenolsulphonic acid is principally formed in the cold ; it
is not known in the free state, because by evaporating its
solution on the water-bath it is partly, and by strong heating,
completely, converted into the para-compound.[3] In order to
prepare the potassium salt, equal volumes of phenol and
sulphuric acid are mixed, diluted with water and, after some
days, the excess of sulphuric acid removed by lead oxide, and
the filtrate neutralized with potassium carbonate. On concentrating the solution the para-salt first separates out in anhydrous plates, and then the ortho-salt crystallizes in rhombic
prisms, which contain two molecules of water of crystallization

---

[1] *Ann. Chim. Phys.* III. **3**, 293.   [2] *Zeitschr. Chem.* **1867**, 199.
[3] Kekulé, *Ber.* **2**, 330.

and effloresce in the air; the two compounds can thus be readily separated.[1]

m-Phenolsulphonic acid is formed by heating the potassium salts of the two phenoldisulphonic acids with some water and two or three times their weight of caustic potash to 178—180°, till a uniform thick mass is produced.[2] The acid, which crystallizes in fine needles containing two molecules of water, produces a violet colour with ferric chloride, and on heating with caustic potash to 250° is converted into resorcinol.

p-Phenolsulphonic acid.—The preparation of this body has already been described. It is also obtained by the action of chlorosulphonic acid on phenol,[3] as well as by warming *p*-diazo-benzenesulphonic acid with water.[4] In the free state it is a syrup. By oxidizing its sodium salt with manganese dioxide and sulphuric acid quinone is obtained.[5]

PHENOLDISULPHONIC ACIDS, $C_6H_3(OH)(SO_3H)_2$.

1005 *a-Phenoldisulphonic acid* is formed by the action of concentrated sulphuric acid on diazobenzene sulphate.[6]

In order to prepare it, one part of phenol is heated with four parts of a mixture of ordinary and fuming sulphuric acids on the water-bath, until sulphur dioxide is evolved; it is then diluted with water and saturated with baryta.[7] From the barium salt, purified by recrystallization, the free acid is easily prepared; it crystallizes in deliquescent nodular aggregates of needles and is coloured a ruby red by ferric chloride.[8]

β-Phenoldisulphonic acid is obtained by heating phenoltrisul-phonic acid with caustic potash and water. It forms a syrup and decomposes on heating.

Phenoltrisulphonic acid, $C_6H_2(OH)(SO_3H)_3$, is formed when two parts of phenol are heated with ten parts of sulphuric acid and five parts of phosphorus pentoxide to 180°.[9] It crystallizes *in vacuô* in needles containing water or in short prisms, which give an intense blood-red colour with ferric chloride.

[1] Post, *Annalen,* **205**, 64.   [2] Barth and Senhofer, *Ber.* **9**, 969.
[3] Engelhardt and Latschinow, *Zeitschr. Chem.* **1869**, 298.
[4] Schmitt, *Annalen,* **120**, 148.   [5] Schrader, *Ber.* **8**, 760.
[6] Griess, *Annalen,* **137**, 69.   [7] Kekulé, *Lehrb.* **3**, 263.
[8] Senhofer, *Jahresb.* **1879**, 749.   [9] Senhofer, *Annalen,* **170**, 110.

Substituted Phenolsulphonic acids.—The phenolsulphonic acids being phenols and powerful acids at the same time, form various series of salts. By the action of chlorine, bromine and nitric acid, substitution products are obtained, which may also be prepared from the substituted phenols by the action of sulphuric acid. Just as with the phenols themselves, it is found that in these bodies the larger the number of negative elements or hydroxyls in the compound, the more easily is the hydrogen of the phenol hydroxyl replaceable by metals; the stability of the compound is, however, correspondingly decreased.

SULPHUR COMPOUNDS OF PHENOL.

1006 *Phenyl hydrosulphide,* or *Thiophenol,* $C_6H_5.SH.$—Vogt first obtained this compound, which is also called phenyl mercaptan, by the action of zinc and dilute sulphuric acid on benzenesulphonyl chloride, and named it benzyl mercaptan.[1] In order to obtain a good yield, the chloride is first reduced to benzenesulphinic acid, and the crude zinc salt so obtained added to a well-cooled mixture of zinc and hydrochloric acid. It is thus obtained mixed with phenyl disulphide. To the liquid, which in addition contains a little free hydrochloric acid, zinc-dust is added to convert the disulphide into zinc thiophenate :

$$(C_6H_5)_2S_2 + Zn = (C_6H_5S)_2Zn.$$

It is then treated with hydrochloric acid and the thiophenol distilled off.[2] Stenhouse obtained it, together with phenyl sulphide and diphenylene sulphide, $C_{12}H_8S$, by the dry distillation of sodium benzenesulphonate in an iron retort, but when he used one of copper he only obtained a trace of thiophenol.[3] It is also formed by the action of phosphorus pentasulphide on phenol.[4] The reaction, according to Geuther, is as follows :

$$8C_6H_5.OH + P_2S_5 = 2C_6H_5.SH + 2PO_4(C_6H_5)_3 + 3H_2S.$$

At the same time, small quantities of phenyl sulphide and phenyl disulphide, which are decomposition products of thiophenol, and also some benzene, are produced by the action of

---

[1] *Annalen,* **119**, 142.　　[2] Otto, *Ber.* **10**, 939.　　[3] *Proc. Roy. Soc.* **17**, 62.
[4] Kekulé and Szuch, *Zeitschr. Chem.* **1867**, 193.

the intermixed phosphorus trisulphide; when phenol is heated with the latter, the following reaction takes place:

$$8C_6H_5OH + P_2S_3 = 2C_6H_6 + 2PO_4(C_6H_5)_3 + 3H_2S.$$

A little thiophenol and phenyl sulphide are formed at the same time.[1] Thiophenol is also formed, together with the sulphide, when benzene is warmed with sulphur and aluminium chloride,[2] and may further be obtained by acting on phenyl-thiocyanate with alcoholic potassium hydrosulphide.[3]

Thiophenol is a colourless, strongly refractive liquid, boiling at 172·5° and having a specific gravity of 1·078 at 24°. In the pure state it has an aromatic and somewhat alliaceous odour; it produces on the skin a burning pain, whilst its vapour attacks the eyes and causes temporary dizziness. It is readily converted by oxidation into phenyl disulphide; even the oxygen of the air effects this conversion in presence of ammonia.

As with other mercaptans, the hydrogen combined with the sulphur is easily replaceable by metals. It acts quickly on mercuric oxide, with formation of mercury thiophenate, $(C_6H_5S)_2Hg$, which crystallizes from boiling alcohol in white, silky needles.

Lead thiophenate $(C_6H_5S)_2Pb$ is obtained by the addition of thiophenol to an alcoholic solution of lead acetate, as a yellow, crystalline precipitate, which decomposes on dry distillation, forming lead sulphide and phenyl sulphide.

Ethyl thiophenate, $C_6H_5.SC_2H_5.$, is formed when sodium thio-phenate is heated with ethyl iodide to 120°; it is an unpleasantly smelling liquid, boiling at 204°.[4]

Phenyl orthothioformate, $CH(SC_6H_5)_3$, is prepared by heating an aqueous solution of sodium thiophenate with chloroform. It crystallizes from a mixture of alcohol and a small quantity of benzene in short, thick prisms, melting at 39·5°. Like the corresponding oxy-compound, it is not decomposed at 120° by caustic soda, but fuming hydrochloric acid decomposes it at 120°, under pressure, into formic acid and thiophenol.[5]

Phenyl thiacetate, $C_2H_3O.SC_6H_5$, is formed by the action of acetyl chloride on thiophenol; it is an unpleasant smelling

[1] *Annalen,* **221**, 55.
[2] Friedel and Crafts, *Bull. Soc. Chim.* **31**, 464.
[3] Gattermann and Haussknecht, *Ber.* **23**, 738.
[4] Beckmann, *J. Pr. Chem.* II. **17**, 457.
[5] Gabriel, *Ber.* **10**, 185.

oily liquid, boiling at 228—230° and dissolving in alcohol and ether; phenyl disulphide separates out from the solution when it is exposed to the air. On boiling the ethereal salt with concentrated caustic potash it is decomposed into thiophenol and acetic acid.[1]

1007 *Phenyl sulphide*, $(C_6H_5)_2S$, was first obtained by Stenhouse by the dry distillation of sodium benzenesulphonate.[2] Other methods of formation have already been described under thiophenol. To prepare it, sodium benzenesulphonate is treated with phosphorus pentasulphide, the crude product rectified over powdered copper and then purified by fractional distillation.[3]

Phenyl sulphide is a liquid possessing an alliaceous odour, boiling at 272·5° and having a sp. gr. 1·119. Nitric acid oxidizes it to *sulphobenzide* or *diphenylsulphone*, $(C_6H_5)_2.SO_2$.

Phenyl disulphide, $(C_6H_5)_2S_2$, is obtained, as already stated, by the oxidation of thiophenol; it is best to employ nitric acid of sp. gr. 1·11 to 1·12.[4] It is also obtained by the action of iodine on sodium thiophenate:[5]

$$\begin{matrix} C_6H_5.SNa \\ C_6H_5.SNa \end{matrix} + I_2 = \begin{matrix} C_6H_5.S \\ | \\ C_6H_5.S \end{matrix} + 2NaI.$$

It is further formed when benzenesulphinic acid is heated with thiophenol to 110°.[6]

$$C_6H_5.SO_2H + 3C_6H_5.SH = 2(C_6H_5)_2S_2 + 2H_2O.$$

This reaction explains the fact that, by the action of nascent hydrogen on hot benzenesulphinic acid, the disulphide is formed, but that when the solution is kept cool, thiophenol is obtained. Phenyl disulphide crystallizes from alcohol in brilliant needles, melting at 60—61°. It boils at 310°[7] and decomposes on continued boiling into sulphur and phenyl sulphide. Reducing agents quickly convert it into thiophenol, and concentrated nitric acid oxidizes it to benzenesulphonic acid.

On heating it with alcoholic potash it forms thiophenol and benzenesulphinic acid. Its solution in concentrated sulphuric acid is coloured cherry-red and then blue on warming. Thio-

[1] *Annalen*, **176**, 177.   [2] *Ibid.* **140**, 287.
[3] Spring and Krafft, *Ber.* **7**, 384.   [4] Otto, *Annalen*, **163**, 213.
[5] Hüber and Alsberg, *ibid.* **156**, 330.   [6] Schiller and Otto, *Ber.* **9**, 1589.
[7] Gräbe, *Annalen*, **174**, 189.

phenol gives the same reaction,[1] as it is oxidized by the acid to phenyl disulphide (Stenhouse).

Phenyl tetrasulphide, $(C_6H_5)_2S_4$, is obtained when sulphuretted hydrogen is passed into an alcoholic solution of benzenesulphinic acid, and by the action of sulphur monochloride on thiophenol. It is a thick yellow oil, which smells like mercaptan, and has a sp. gr. of 1·297 at 145°. It is readily converted into the disulphide.[2]

Thiopicric acid, $C_6H_2(NO_2)_3SH$.—The potassium salt of this body separates in reddish-brown needles when a hot alcoholic solution of picryl chloride is gradually added to an alcoholic solution of potassium sulphide, the liquid being well cooled between each addition, and the whole allowed to stand; it detonates violently at 140° and on percussion. The acid obtained from it, or *a-trinitrothiophenol*, is easily soluble in water, alcohol, and ether, has a bitter taste, and crystallizes in yellowish needles, melting at 114° and exploding at 115°.

Picryl sulphide, $[C_6H_2(NO_2)_3]_2S$, is formed by mixing 10 cc. of an alcoholic solution of 10 grms. potassium sulphide with 4·4 grms. of picryl chloride also dissolved in alcohol. It crystallizes from glacial acetic acid in golden-coloured leaves or whitish-yellow prisms, melting at 266°.[3]

SULPHONES AND ALLIED COMPOUNDS.

1008 *Thionyldiphenyl*, $SO(C_6H_5)_2$, is obtained by passing sulphur dioxide into a mixture of 100 grms. benzene and 35 grms. aluminium chloride till no more is absorbed, and subsequently heating till no more hydrochloric acid is evolved. It may also be prepared by the action of thionyl chloride on benzene in presence of aluminium chloride. It crystallizes from light petroleum in transparent asymmetric prisms, which melt at 70·5°. On oxidation it is converted into diphenylsulphone, whilst on reduction with sodium it yields phenyl sulphide.[4]

Methylphenylsulphone, $CH_3.SO_2.C_6H_5$, is obtained by acting on

---

[1] Baumann and Preusse, *Zeit. Physiol. Chem.* **5**, 321.
[2] Milch and Otto, *J. Pr. Chem.* II. **37**, 207.
[3] Willgerodt, *Ber.* **17**, 353 R.
[4] Colby and McLaughlin, *ibid.* **20**, 195.

benzenesulphinic acid with sodium ethylate and methyl iodide,[1] and by heating the salts of phenylsulphonacetic acid.[2] It forms large tablets, melting at 88°. *Iodomethylphenylsulphone* is prepared in a similar manner from benzenesulphinic acid, sodium ethylate, and methylene iodide.

Ethylphenylsulphone, $C_2H_5.SO_2.C_6H_5$, is formed by the oxidation of ethylphenyl sulphide, and by the action of ethyl iodide on sodium benzenesulphinate. It crystallizes from ether in monosymmetric tables, which melt at 42°, and boil without decomposition above 300°.

1009 *Diphenylsulphone*, or *Sulphobenzide*, $(C_6H_5)_2SO_2$, is the most important member of the group of sulphones. It may be prepared in a number of ways, one of which, the oxidation of phenyl sulphide, has already been mentioned. It was first obtained by Mitscherlich by the action of sulphur trioxide [3] on benzene; the sulphur trioxide may be replaced by chlorosulphonic acid,[4] and it is also formed when benzenesulphonic chloride, benzene and aluminium are heated together.[5] It crystallizes from benzene in monosymmetric prisms, and from alcohol in small plates, which melt at 128—129°, may be readily sublimed, and are insoluble in cold water, but readily soluble in hot alcohol and in benzene. It is an extremely stable substance, and is converted on warming with fuming sulphuric acid into benzenesulphonic acid.

A number of substitution products of diphenylsulphone are also known. Among these may be mentioned *Chlorodiphenylsulphone*, $C_6H_5.SO_2.C_6H_4Cl$, obtained by heating benzenesulphonic chloride, chlorobenzene, and aluminium chloride. It crystallizes in plates which are difficultly soluble in cold alcohol. *Nitrodiphenylsulphone*, $C_6H_5.SO_2.C_6H_4.NO_2$, is prepared by treating diphenylsulphone with fuming nitric acid. It forms microscopic crystals, which melt at 92°.

Dihydroxyphenylsulphone, or *Dihydroxysulphobenzide*, $SO_2.$ $(C_6H_4.OH)_2$, is a by-product of the action of sulphuric acid on phenol.[6] In order to prepare it, fuming sulphuric acid is heated with double its weight of phenol for 3—5 hours at 180—190°, and the mass poured into water whilst still warm. It is almost insoluble in cold water, but crystallizes from the

[1] Michael and Palmer, *Am. Chem. Journ.* **6**, 254

[2] Otto, *Ber.* **18**, 156.   [3] Mitscherlich, *Annalen*, **12**, 208.

[4] Knapp, *Zeitsch. Chem.* **1869**, 41.   [5] Beckurts and Otto, *Ber.* **11**, 2066.

[6] Glutz, *Annalen*, **147**, 52.

boiling liquid in long prismatic needles, and from glacial acetic acid in rhombic prisms melting at 239°. On further heating with fuming sulphuric acid it is converted into phenoltrisulphonic acid.[1] Like phenol it forms salts, which have been examined by Glutz.

1010 Sulphones are also known which contain alcoholic or acid residues, such as—$CH_2.OH$, and—$CH_2.COOH$. Of these the following may be mentioned :—

Phenylsulphone-ethyl alcohol, $C_6H_5.SO_2.CH_2.CH_2.OH$. This compound is prepared by the action of ethylene chlorhydrin on sodium benzenesulphinate, or of caustic potash on ethylene-diphenyl-disulphone. It forms a syrup which mixes with alcohol and benzene, but is less soluble in water and ether.[2] Aqueous ammonia converts it at 120° into *diphenylsulphone ethylamine*, and it yields on oxidation phenylsulphonacetic acid.

Phenylsulphoneformic acid, $C_6H_5.SO_2.COOH$. The ethyl salt of this acid is obtained in small quantities by the action of ethyl chloroformate on sodium benzenesulphinate. The chief product is, however, ethylphenylsulphone. The mixture is completely saponified by heating with water to 110°, but no carbon dioxide is evolved, and it would therefore appear that, like pyroracemic acid, this compound does not lose carbon dioxide very readily.[3]

Phenylsulphonacetic acid, $C_6H_5.SO_2.CH_2.COOH$, is formed, as already mentioned, by the oxidation of phenylsulphone ethyl alcohol, and may also be prepared by the oxidation of phenyl-thioglycollic acid,[4] and by the action of caustic soda solution on a solution of equivalent quantities of benzenesulphinic and chloracetic acids.[5] It crystallizes from alcohol or chloroform in monosymmetric tables which melt at 111·5—112·5°, and decomposes on distillation into methylphenylsulphone and carbon dioxide. It is converted by sodium amalgam into acetic and benzenesulphinic acids.

Its *ethyl salt* is prepared by the action of ethyl chloracetate on an alcoholic solution of sodium benzenesulphinate.[6] It crystallizes from alcohol in long prisms which melt at 41—42°. It yields a sodium compound analogous to ethyl sodacetoacetate when treated with sodium ethylate.

[1] Annaheim, *Annalen*, **172**, 36 ; *Ber.* **9**, 1148.
[2] Otto, *J. Pr. Chem.* II. **30**, 189.   [3] Otto, *Ber.* **21**, 91.
[4] *Ibid.* **19**, 3138.   [5] Gabriel, *ibid.* **14**, 833.
[6] Michael and Comey, *Am. Chem. Journ.* **5**, 116.

β-Phenylsulphonepropionic acid, $C_6H_5.SO_2.CH_2.CH_2.COOH$, is obtained by neutralizing a solution of equivalent quantities of β-iodopropionic acid and benzenesulphinic acid, warming the solution on the water-bath, and finally over the naked flame, then adding water and hydrochloric acid, and recrystallizing the product from hot water.[1] It forms monosymmetric or asymmetric tablets, which melt at 123—124°. It is very stable towards caustic potash, and does not lose carbon dioxide when heated with it to 180°.

An examination of the foregoing compounds, especially those containing fatty residues, brings to light a distinct analogy between the sulphones and the ketones, and it would therefore appear that the groups SO_2 and CO, give to the compounds in which they occur, similar properties.[2]

Compounds have also been obtained which contain two sulphone groups, and are therefore termed disulphones. These are divided into two classes, according as the two sulphone groups are attached to the same or to different carbon atoms.[3] Trisulphones have also been prepared.[4]

SELENIUM COMPOUNDS OF PHENOL.

1011 In presence of aluminium chloride, selenium tetrachloride acts on benzene with formation of *selenophenol,* C_6H_5. SeH, and *phenyl selenide,* $(C_6H_5)_2Se$. The former is a solid substance, insoluble in water but soluble in alcohol, which melts at 60°. Phenyl selenide is an amber-yellow oil boiling at 227—228°. A red oil, having the composition, $Se(C_6H_5)_3$. C_6H_4Cl, is also obtained at the same time.[5]

---

[1] Otto and Rössing, *Ber.* **21**, 95.   [2] *Ibid.* **21**, 992.
[3] See Otto and Damköhler, *J. Pr. Chem.* II. **30**, 171, 321 ; R. and W. Otto, *ibid.* II. **36**, 401 ; Otto and Casanova, *ibid.* p. 433 ; Baumann and Escales, *Ber.* **19**, 2806, 2815 ; Otto, *ibid.* **21**, 652 ; **22**, 1965.
[4] Fromm, *Annalen,* **253**, 160.   [5] *Compt. Rend.* **109**, 182.

DIHYDROXYBENZENES, $C_6H_4(OH)_2$, AND RELATED COMPOUNDS.

o-Dihydroxybenzene, Pyrocatechin or Catechol.

1012 Reinsch first obtained this body by the dry distillation of catechin,[1] and it was further examined by Zwenger, who named it "Brenzcatechin."[2] In his investigations on morintannic acid contained in fustic, R. Wagner observed that it is converted on heating into pyromorintannic acid,[3] which, he afterwards found, was identical with pyrocatechin, which he had prepared by heating catechu, and named, according to Erdmann's proposal, oxyphenic acid.[4] He obtained it also by the dry distillation of gum-ammoniac, from the aqueous extract of the bilberry plant, and generally from the extracts of all plants which contain tannic acid.[5] It is also obtained when filter paper, starch, or sugar is heated to 200—280°,[6] as well as by the dry distillation of wood; it therefore occurs in crude pyroligneous acid.[7]

Catechol is also found in the autumnal leaves of the Virginia creeper (*Ampelopsis hederacea*)[8] and in kino, the boiled juice of different species of Pterocarpus, Butea and Eucalyptus.[9]

It is likewise obtained when o-iodophenol[10] or o-phenolsulphonic acid[11] is fused with caustic potash. In order to prepare it according to the latter method, the proportion of 1 molecule of the acid to 24 molecules of caustic potash is taken and the melt heated for some time to 320—360°, when the yield amounts to 20 per cent. of the theoretical. By using less potash or by not heating so strongly, the yield is rendered smaller, as is also the case when the mass is heated to 400°.[12]

It is also obtained, together with quinol and quinone, by acting on a solution of phenol containing iron with hydrogen dioxide.[13]

---

[1] *Rep. Pharm.* **68**, 54.
[2] *Annalen,* **37**, 327 ; and also Wackenroder, *ibid.* 309.
[3] *J. Pr. Chem.* **52**, 450.   [4] *Ibid.* **55**, 65.
[5] Uloth, *Annalen,* **111**, 215.   [6] Hoppe-Seyler, *Ber.* **4**, 15.
[7] Pettenkofer, *Jahresb.* **1854**, 651 ; Buchner, *Annalen,* **96**, 188.
[8] Gorup-Besanez, *Ber.* **4**, 906.   [9] Flückiger, *ibid.* **5**, 1.
[10] Körner, *Zeitschr. Chem.* **1868**, 322.   [11] Kekulé, *ibid.* **1867**, 643.
[12] Degener, *J. Pr. Chem.* II. **20**, 304.
[13] Martinon, *Bull. Soc. Chim.* **43**, 155.

It is most readily prepared from its methyl ether, by warming it with fuming hydriodic acid, mixing the product with water and extracting with ether. After evaporating off the ether, the residual dark oil is purified by fractionation; the distillate solidifies to a cake, which may be further purified by recrystallization from benzene.[1]

Catechol is readily soluble in water, and crystallizes therefrom in thin prisms, and from benzene in broad plates. It is very soluble in alcohol and ether, melts at 104°, and boils at 240—245°. Its alkaline solution quickly turns brown in the air; it readily reduces solutions of the noble metals, and precipitates Fehling's solution on warming. On passing nitrogen trioxide through its ethereal solution it forms carboxytartronic or dihydroxytartaric acid (Part II. p. 266, this Part, p. 65). Ferric chloride produces in an aqueous solution of catechol an emerald green colour, which on addition of sodium carbonate, or better, bicarbonate, gives a beautiful violet red.[2]

The metallic compounds of catechol are very unstable, with the exception of the lead salt, $C_6H_4O_2Pb.$, which forms a white precipitate readily soluble in acetic acid.

1013 *Catechol monomethyl ether* or *Guaiacol*, $C_6H_4\begin{cases} OH \\ OCH_3 \end{cases}$, was first found in the distillation products of guaiacum,[3] and was afterwards noticed as an ingredient of beechwood-tar creosote. This is a mixture of phenols with the monomethyl ethers of catechol and its homologues, and will be further described under creosol, $C_6H_3(CH_3)OH(OCH_3)$ (Part IV. p. 32).

In order to prepare guaiacol, the crude compound, obtained from creosote by fractional distillation, is repeatedly shaken with moderately strong ammonia, washed and rectified. The oil is then dissolved in an equal volume of ether, and a small excess of concentrated alcoholic potash added to it. The potassium salt separates out, and is then recrystallized from alcohol and decomposed by dilute sulphuric acid.[4]

It is also obtained when equal molecules of catechol, caustic potash and potassium methyl sulphate are heated together,[5] or when vanillic acid, $C_6H_3(OCH_3)(OH)CO_2H$, is distilled with lime.[6]

[1] W. H. Perkin, jun. *Journ. Chem. Soc.* 1890, i. 587.

[2] Merz and Ris, *Ber.* **20**, 1190.

[3] UnVerdorben, *Pogg. Ann.* **8**, 402 ; *Ann. Chim. Phys.* III. **12**, 228 ; Sobrero, *Annalen*, **48**, 19 ; Völkel, *ibid.* **89**, 345.   [4] Hlasiwetz, *ibid.* **106**, 865.

[5] Gorup-Besanez, *ibid.*, **147**, 248.   [6] Tiemann, *Ber.* **8**, 1123.

Guaiacol is a strongly refractive liquid, having an aromatic smell, reminding one of Peru balsam. It boils at 200°, has at 13° a sp. gr. of 1·117, and is slightly soluble in water, but readily in alcohol; its solution gives with ferric chloride an emerald green colouration.

The metallic compounds of guaiacol are not very stable; the formation of potassium guaicate, $C_7H_7O_2K + 2H_2O$, is described above. By adding a little potash or potassium to guaiacol heated to 90°, the compound $C_7H_8O_2 + C_7H_7O_2K + H_2O$ is formed, which crystallizes from alcohol in shining prisms and dissolves in water with decomposition (Gorup-Besanez).

Catechol dimethyl ether, or *Veratrol*, $C_6H_4(OCH_3)_2$, was first obtained by distilling veratric acid, $C_6H_3(OCH_3)_2CO_2H$, with caustic baryta,[1] and is also formed when potassium guaicate is heated with methyl iodide.[2] It is a liquid with an aromatic odour, boils at 205—206°, and solidifies at 15° to a crystalline mass.

Catechol sulphuric acids are not known in the free state. On warming a solution of catechol in caustic potash with potassium disulphate, the salts $C_6H_4(SO_4K)_2$ and $C_6H_4(OH)SO_4K$ are formed. On treating the mixture with absolute alcohol, the first salt remains behind as a crystalline powder, and on evaporating the solution the second separates out in shining plates.[3]

These salts, or one of them, form a normal constituent of the urine of the horse and man.[4]

Catechol carbonate, $C_6H_4CO_3$, is obtained by the action of ethyl chlorocarbonate on a mixture of caustic potash and catechol or on the sodium salt:

$$C_6H_4{<}^{ONa}_{ONa} + 2ClCO.OC_2H_5 =$$

$$C_6H_4{<}^{O}_{O}{>}CO + CO(OC_2H_5)_2 + 2NaCl.$$

It crystallizes from alcohol or benzene in fine four-sided prisms, melting at 118°, and boils without decomposition between 225—230°.[5] By the action of acetyl chloride on catechol, the

[1] Merck, *Annalen*, **108**, 60; Kölle, *ibid.* **159**, 243.
[2] Marasse, *ibid.* **152**, 74.
[3] Baumann, *Ber.* **11**, 1913.
[4] Baumann, *Zeit. Physiol. Chem.* **1**, 244.
[5] Bender, *Ber.* **13**, 697; Wallach, *Annalen*, **226**, 34.

diacetate, $C_6H_4(OC_2H_3O)_2$, is formed, which crystallizes in needles.[1]

On heating guaiacol with acetic anhydride, aceto-guaiacol, $C_6H_4(OCH_3)OC_2H_3O$, a limpid liquid boiling at 235—240°,[2] is formed.

Tetrachlorocatechol, $C_6Cl_4(OH)_2$. To prepare this compound, chlorine is passed into an acetic acid solution of catechol or *o*-amidophenol, and the hexchlorodiketotetrahydrobenzene thus obtained treated with acetic acid and a solution of stannous chloride.[3] It crystallizes from alcohol in fine needles, which melt at 194—195°. On heating with acetic anhydride it yields a diacetyl compound $C_6Cl_4(O.C_2H_3O)_2$, which forms broad needles melting at 190°.

Tetrabromocatechol, $C_6Br_4(OH)_2$, is obtained by triturating together catechol and bromine,[4] as well as by heating protocatechuic acid, $C_6H_3(OH)_2CO_2H$, with bromine;[5] it crystallizes in colourless, transparent prisms which melt at 192—193°, and are insoluble in water. Its alcoholic solution is coloured dark blue by ferric chloride.

Nitrocatechol, $C_6H_3(NO_2)(OH)_2$, is obtained by the addition of sulphuric acid to an aqueous solution of catechol and potassium nitrate. It is readily soluble in water, and crystallizes from benzene in yellowish needles, melting at 157°. It dissolves in caustic potash, forming a splendid purple solution. This reaction is so delicate that nitrocatechol makes an excellent indicator for volumetric analysis.[6]

1014 *o-Hydroxyphenyl hydrosulphide*, $C_6H_4 \begin{Bmatrix} H \\ SH \end{Bmatrix}$, is obtained when sodium phenate is heated with sulphur to 180—200°.

$$2C_6H_5ONa + S = C_6H_5 \begin{Bmatrix} ONa \\ SNa \end{Bmatrix} + C_6H_5OH.$$

This body can be indirectly separated from the product by oxidation, the effect of which is to form dihydroxyphenyl disulphide.

$$2C_6H_4 \begin{matrix} OH \\ SH \end{matrix} + O = C_6H_4 \begin{matrix} OH \ HO \\ S——S \end{matrix} C_6H_4 + H_2O.$$

The latter is isolated by acidifying with sulphuric acid and

[1] Nachbaur, *Annalen*, **107**, 246. [2] Tiemann and Koppe, *ibid.* **14**, 2020.
[3] Zincke and Küster, *Ber.* **21**, 2729. [4] Hlasiwetz, *Annalen*, **142**, 250.
[5] Stenhouse, *Chem. News*, **29**, 95. [6] Benedikt, *Ber.* **11**, 362.

distilling with steam. The distillate is neutralised with soda and concentrated to obtain the salt, $(C_6H_4)_2S_2(OH)ONa$, which forms white crystalline crusts, dissolving in water and producing an intensely yellow solution. On addition of sulphuric acid and extraction with ether, the free dihydroxyphenyl disulphide is obtained; it is a thick oil, possessing a faint smell, and decomposing at 200°. It forms two series of salts; of these the normal ones are decomposed by carbon dioxide.

The dimethyl ether, $S_2(C_6H_4OCH_3)_2$, is obtained when the sodium salt is heated with caustic soda, methyl iodide, and methyl alcohol. It crystallizes from alcohol in odourless needles, melting at 119°. It is also formed by replacing the hydrogen of the hydroxyl in potassium *o*-phenolsulphonate, $C_6H_4(OH)SO_3K$, by methyl, converting the potassium anisoïlsulphonate so obtained, by means of phosphorus chloride, into the corresponding sulphonic chloride, and treating this, in alcoholic solution, with zinc dust. On oxidation with chromic acid, the methyl ether is reconverted into anisoïl-*o*-sulphonic acid, which, on fusion with caustic potash, yields catechol.

In order to prepare *o*-hydroxyphenyl hydrosulphide, sodium amalgam is added to a solution of the above-mentioned sodium salt.

$$
\begin{array}{l} C_6H_4 \diagdown \!\!\!\!\!\!\!\! \begin{array}{l} ONa \\ S \end{array} \\ \diagup S \\ C_6H_4 \diagdown \!\!\!\!\!\!\!\! \begin{array}{l} S \\ OH \end{array} \end{array} + 3NaOH + H_2 = 2C_6H_4 \diagdown \!\!\!\!\!\!\!\! \begin{array}{l} ONa \\ SNa \end{array} + 3H_2O.
$$

By the decomposition of the product with dilute sulphuric acid, *o*-hydroxyphenyl hydrosulphide is obtained as a strongly refractive liquid, having a penetrating smell and, like phenol, corroding the skin. It solidifies at a low temperature to a crystalline mass, similar to phenol, which melts at 5—6°. It is a tolerably strong acid, decomposing carbonates; its aqueous solution reddens litmus. On the addition of a little ferric chloride and carbonate of soda a very intense green colour is obtained, which, by adding caustic soda, is changed to a deep red. Oxidizing agents readily convert it again into dihydroxyphenyl disulphide, which is also formed by exposing an alkaline solution of the mercaptan to the air.[1]

[1] Haitinger, *Monatsh.* **4**, 165.

1015 It has already been mentioned in the introduction (p. 49) that p-hydroxybenzene readily loses two atoms of hydrogen on oxidation, and is converted into a compound termed quinone, which has probably the constitution

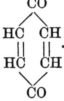

For a long time no corresponding compounds were known in the benzene series, containing the oxygen atoms in the ortho-position, although such a derivative of naphthalene was prepared by Stenhouse and Groves in 1877. Zincke has recently succeeded in obtaining halogen derivatives of benzene o-quinone by the oxidation of tetrachloro- and tetrabromocatechol.

$$
\begin{array}{c}
CO \\
\diagup\diagdown \\
Cl.C \quad CO \\
\| \quad\quad | \\
Cl.C \quad C.Cl \\
\diagdown\diagup \\
CCl
\end{array}
$$

Tetrachlorobenzene-o-quinone, , is prepared by treating tetrachloro-catechol with nitric acid,[1] and forms a deep red crystalline powder melting at 129—130°.

Tetrabromobenzene-o-quinone $C_6Br_4O_2$, is prepared in a similar manner from tetrabromocatechol. It forms thick garnet red prisms or tablets, which melt at 150–151°, and are soluble in alcohol, ether and benzene.[2]

m-DIHYDROXYBENZENE, RESORCIN, OR RESORCINOL, $C_6H_4(OH)_2$.

1016 This body was first obtained by Barth and Hlasiwetz by fusing galbanum or gum-ammoniac with caustic potash, and since it is very similar to orcin (dihydroxytoluene), its higher homologue, and is obtained from resins, it was named resorcin.[3] It is also obtained in a similar way from *Assafœtida,*[4] *Sagapenum*, and *Acaroïd-resin*.[5] It is also abundantly produced when impure brazilin, which separates as a crust from Brazil wood extract on standing, is subjected to dry distillation.[6]

[1] Zincke and Küster, *Ber.* **21**, 2730.   [2] Zincke, *ibid.* **20**, 1777.
[3] *Annalen*, **130**, 354.   [4] *Ibid.* **138**, 63.
[5] *Ibid.* **139**, 78.   [6] Kopp, *Ber.* **6**, 446.

Resorcinol is further obtained by fusing m-iodophenol,[1] phenol m-sulphonic acid [2] or m-benzenedisulphonic acid with caustic potash. It is however also formed from certain para- and ortho-compounds such as p-iodophenol, p-benzenedisulphonic acid,[3] and o-bromophenol,[4] the latter yielding also catechol. It has been usually supposed that intermolecular changes take place at the temperature employed, which is in agreement with the fact that p-iodophenol yields quinol at a low temperature, and resorcinol at a high temperature. It has however been pointed out by Nölting and Stricker [5] that this so-called intermolecular change is due to an alternate reduction and oxidation brought about by the caustic potash at a high temperature. Thus the formation of resorcinol from p-iodophenol may be supposed to take place as follows:

$$\text{I. } 2K\overset{.}{H}O = K_2O + H_2 + O$$

$$\text{II. } C_6H_4\begin{cases} OH\ (1) \\ I\quad\ (4) \end{cases} + O = C_6H_3\begin{cases} OH\ (1) \\ OH\ (3) \\ I\quad\ (4) \end{cases}$$

$$\text{III. } C_6H_3\begin{cases} OH\ (1) \\ OH\ (3) \\ I\quad\ (4) \end{cases} + H_2 = C_6H_4\begin{cases} OH\ (1) \\ OH\ (3) \end{cases} + HI.$$

On fusing phenol for a long time with an excess of caustic soda, hydrogen is evolved, and resorcinol, catechol, phloroglucinol, $C_6H_3(OH)_3$, and diresorcinol, $C_{12}H_6(OH)_4$,[6] are formed.

That resorcinol belongs to the meta-series is shown by the fact that it is easily obtained by the diazo-reaction from meta-amidophenol.[7]

Resorcinol is employed in the manufacture of different colouring matters. To prepare it on the large scale, 90 kilos of fuming sulphuric acid, of sp. gr. 2·244, are put into a cast-iron apparatus furnished with an agitator, and 24 kilos of pure benzene gradually added, the mixture gently heated for some hours, and the temperature then raised to 275°, in order to convert the benzene completely into the disulphonic acid. After cooling, the mass is poured into 2,000 kilos of water, heated to boiling, neutralized with milk of lime, and the gypsum removed by the filter press. The calculated quantity

[1] Körner, *Zeitsch. Chem.* **1868**, 322.
[2] Barth and Senhofer, *Ber.* **9**, 969.
[3] *Ibid.* **8**, 1482.
[4] Fittig and Mager, *ibid.* **7**, 1175 ; **8**, 365.
[5] *Ber.* **20**, 3022.
[6] Barth and Schreder, *ibid.* **12**, 417.
[7] Bantlin, *ibid.* **11**, 2101.

of soda is then added to the solution, the calcium carbonate separated by the filter press, the solution evaporated to dryness, and 60 kilos of the product thus obtained fused with 150 kilos of caustic soda in an iron vessel for eight to nine hours at 270°. The cooled melt is then dissolved in 500 kilos of boiling water and the solution boiled with hydrochloric acid as long as sulphur dioxide is given off. The cooled liquid is then systematically extracted with ether in a copper extraction-apparatus and the ether distilled off. The crude resorcinol thus obtained is heated up to 215°, in order to free it from water and adhering ether, the loss of which amounts on the whole to about 1 per cent. The residue, which contains 92 to 94 per cent. of resorcinol, forms the commercial product.[1] In order to prepare the pure compound, this crude product is distilled, when first water, then phenol, and, lastly, resorcinol comes over. This can be further purified by recrystallization from benzene, or by sublimation.

Resorcinol is very readily soluble in water, still more readily in alcohol and ether, but only with difficulty in cold benzene, and crystallizes in large rhombic prisms or tablets. It melts at 118°, boils at 276·5°, but readily sublimes at a lower temperature in lustrous silky needles. Its taste is intensely sweet, but causes subsequent irritation. In the warm state it reduces ammoniacal silver solution and Fehling's solution. It is distinguished from catechol by the fact that its solution produces a violet colour with ferric chloride, and is not precipitated by lead acetate. It forms a salt with two molecules of phenyl hydrazine[2] which crystallizes from benzene in fine white needles melting at 276°

In order to detect small quantities of resorcinol, it is heated for some minutes with an excess of phthalic anhydride almost to the boiling point of the latter, and the product dissolved in dilute caustic soda. If resorcinol be present, the dilute alkaline solution exhibits a fine green fluorescence, fluoresceïn being formed (Part IV. p. 458).

1017 *Resorcinol ether*, $C_{12}H_{10}O_3$ or $O(C_6H_4OH)_2$, is the name given to a body which is obtained by heating resorcinol with fuming hydrochloric acid to 180°, when it separates out like a resin.[3] It is also formed when resorcinol is heated with resor-

---

[1] Binschedler and Busch, *Jahresb.* **1878**, 1137, and 1184.
[2] Baeyer and Kochendörfer, *Ber.* **22**, 2194.
[3] Barth, *Annalen*, **164**, 122 ; *Ber.* **9**, 308.

cinoldisulphonic acid [1] as well as by the action of fuming sulphuric acid on resorcinol.[2] By dissolving the latter in a large excess of the acid an orange-coloured solution is obtained, which gradually darkens, and turns greenish-blue, green, and finally a beautiful blue. On then heating it to 100° it is coloured purple-red, and on the addition of water, yellow (Kopp). The resorcinol ether obtained by means of hydrochloric acid contains the compound $C_{24}H_{13}O_5$. To separate these bodies, the crude product is dissolved in strong alcohol, and an alcoholic solution of lead acetate added, the lead compound of the ether being precipitated, filtered off, dissolved in glacial acetic acid and decomposed by hydrochloric acid; or it may be covered with alcohol and treated with sulphuretted hydrogen gas.[3] Resorcinol ether is a resinous mass, or a fiery, brownish-red powder, showing when pressed a greenish metallic lustre. It dissolves in alkalis forming a deep red solution, which, when dilute, exhibits a beautiful green fluorescence. On fusing it with caustic alkali, a mixture of resorcinol with other bodies is obtained.

Resorcinol monomethyl ether, $C_6H_4(OH)OCH_3$, is formed, together with the dimethyl ether, when resorcinol is heated with caustic potash and potassium methylsulphate,[4] or when resorcinol and sodium are dissolved in methyl alcohol and then boiled with methyl iodide.[5] The monomethyl ether is an oily liquid, soluble with difficulty in cold water, but readily soluble in hot water; its solution is coloured light violet by ferric chloride. It is readily soluble in dilute caustic soda, and boils at 243—244°.

Resorcinol dimethyl ether, $C_6H_4(OCH_3)_2$, is a light mobile liquid, having an aromatic smell. It boils at 214°, is not soluble in caustic soda and is not coloured by ferric chloride.

Resorcinol diethyl carbonate, $C_6H_4(O.CO.C_2H_5)_2$, is obtained by the action of ethyl chloroformate on the sodium compound of resorcinol. It is a thick oily liquid boiling at 298—302°.[6]

Diacetoresorcinol, $C_6H_4(OC_2H_3O)_2$, is obtained by the action of acetyl chloride on resorcinol and is a strongly refractive liquid, boiling with slight decomposition at 278°.[7]

Thioresorcinol, $C_6H_4(SH)_2$, is obtained by the action of tin and hydrochloric acid on benzene *m*-disulphonic chloride,

[1] Hazura and Julius, *Monatsh.* **5**, 191.
[2] Kopp, *Ber.* **6**, 447; Annaheim, *ibid.* **10**, 976.
[3] Barth and Weidel, *ibid.* **10**, 1464.　　[4] Habermann, *ibid.* **10**, 867.
[5] Tiemann and Parrisino, *ibid.* **13**, 2362.　　[6] Wallach, *Annalen*, **226**, 84.
[7] Malin, *ibid.* **138**, 78; Nencki and Sieber, *J. Pr. Chem.* II. **23**, 149.

$C_6H_4(SO_2Cl)_2$,[1] and forms crystals, which have a penetrating odour. It melts at 27°, and boils at 243°. Its lead salt, $C_6H_4S_2Pb$ is a yellowish-red precipitate which on heating with cyanogen iodide and alcohol yields the thiocyanate, $C_6H_4(SCN)_2$, crystallizing in shining needles and melting at 54°.[2]

Thioresorcinol is sometimes found in commercial resorcinol.

CHLORINE SUBSTITUTION PRODUCTS OF RESORCINOL.

		Melting-point.	Boiling-point.
Monochlororesorcinol, $C_6H_3Cl(OH)_2$, { indistinct crystals		89°	256°
Dichlororesorcinol, $C_6H_2Cl_2(OH)_2$ rhombic prisms		77°	249°
Trichlororesorcinol, $C_6HCl_3(OH)_2$, fine needles	.	83°	—

1018 These bodies are obtained by the action of sulphuryl chloride on resorcinol. Trichlororesorcinol is also obtained when chlorine is passed through an aqueous solution of resorcinol. The aqueous solutions of the first two are coloured bluish-violet by ferric chloride.[3]

Pentachlororesorcinol, $C_6HCl_5O_2$, is obtained by the action of potassium chlorate and hydrochloric acid on resorcinol.[4] It crystallizes from carbon disulphide in flat prisms, melting at 92·5°. It may be heated to its boiling-point without decomposition, but a concentrated solution of acid potassium sulphite converts it into isotrichlororesorcinol, $C_6HCl_3(OH)_2$, which crystallizes in needles melting at 69°.[5]

Pentachlororesorcinol effloresces in the air, being converted into a modification which melts at 65°. This is obtained directly when pentachlororesorcinol is dissolved in hot water, and separates out on cooling in indistinct crystals.[6]

---

[1] Körner and Monselise, *Jahresb.* **1876**, 450 ; Pazschké, *J. Pr. Chem.* II. **2**, 418.

[2] Gabriel, *Ber.* **10**, 184.   [3] Reinhard, *J. Pr. Chem.* II. **17**, 321.

[4] Stenhouse, *Proc. Roy. Soc.* **20**, 72.   [5] Claessen, *Ber.* **11**, 1441.

Liebermann and Dittler, *Annalen*, **169**, 265.

BROMINE SUBSTITUTION PRODUCTS OF RESORCINOL.

		Melting-point.
Bromoresorcinol [1] $C_6H_3Br(OH)_2$, prisms . . .		91°
a-Dibromoresorcinol [2] $C_6H_2Br_2(OH)_2$, needles . . .		92—93°
β-Dibromoresorcinol [3] $C_6H_2Br_2(OH)_2$, long needles .		83—85°
γ-Dibromoresorcinol $C_6H_2.Br_2(OH)_2$		110—112°

1019 *Tribromoresorcinol*, $C_6HBr_3(OH)_2$, is obtained by the action of bromine water on resorcinol.[4] It crystallizes in small needles which melt at 104°, are soluble with difficulty in cold water, and readily in alcohol. A second tribromoresorcinol is also known, melting at 111°. It is obtained by the reduction of pentabromoresorcinol [5] or by the action of bromine on resorcinol dissolved in acetic acid.[6]

Tetrabromoresorcinol, $C_6Br_4(OH)_2$, is obtained by warming pentabromoresorcinol with sulphuric acid;[7] it crystallizes from dilute alcohol in needles melting at 167°.

Pentabromoresorcinol, $C_6HBr_5O_2$, separates out when a concentrated aqueous solution of resorcinol [8] is poured into bromine; it crystallizes from carbon disulphide in tetragonal tables melting at 113·5°.

Hydriodic acid and other reducing agents convert it into tribromoresorcinol.[9] On heating to 160° it gradually decomposes into bromine and tribromoresoquinone, $C_6HBr_3O_2$, crystallizing from alcohol in small orange-coloured needles, which are decomposed on heating.[10]

Hexbromoresorcinol, $C_6Br_6O_2$, is obtained by dissolving tetrabromoresorcinol in caustic potash, and then adding hydrochloric acid and bromine water. It forms monosymmetric

---

[1] Zehenter, *Monatsh.* **8**, 293.
[2] Baeyer, *Annalen*, **183**, 57; Hofmann *Ber.* **8**, 64.
[3] Zehenter, *Monatsh.* **2**, 478.
[4] Hlasiwetz and Barth, *Annalen*, **130**, 357.
[5] Stenhouse, *ibid.* **163**, 184.
[6] Colman and Perkin, Private Communication.
[7] Claassen, *Ber.* **11**, 1440.
[8] Stenhouse, *Proc. Roy. Soc.* **20**, 72.
[9] Benedikt, *Monatsh.* **1**, 351.
[10] Liebermann and Dittler, *Annalen*, **169**, 259.

crystals which melt at 136°, and by the action of tin and hydrochloric acid are again converted into tetrabromoresorcinol.[1]

To the foregoing compounds the constitutional formulæ $C_6HBr_3(OBr)_2$ and $C_6Br_4(OBr)_2$ have usually been assigned. The formation of tribromoresorcinol from this compound is explained by this formula but not the formation of tetrabromoresorcinol, and it is much more probable that they are derivatives of hydrobenzenes, having the formulæ :

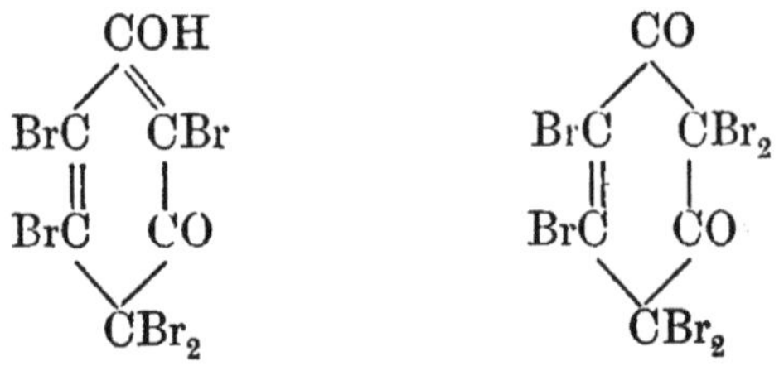

$$
\begin{array}{cc}
\text{COH} & \text{CO} \\
\text{BrC} \quad \text{CBr} & \text{BrC} \quad \text{CBr}_2 \\
\text{BrC} \quad \text{CO} & \text{BrC} \quad \text{CO} \\
\text{CBr}_2 & \text{CBr}_2
\end{array}
$$

Pentachlororesorcinol (p. 160) and hexchloro- and hexbromophenol (p. 130) have probably a similar constitution.

IODINE SUBSTITUTION PRODUCTS OF RESORCINOL.

1020 *Mono-iodoresorcinol*, $C_6H_3I(OH)_2$, was obtained by Stenhouse by gradually adding lead oxide to a solution of resorcinol and iodine in ether. It crystallizes in rhombohedral prisms melting at 67°.[2]

Tri-iodoresorcinol, $C_6HI_3(OH)_2$, is obtained when chloride of iodine is added to an aqueous solution of resorcinol,[3] or when this body is placed in a solution of potassium iodate and iodine in potassium iodide.[4] It crystallizes from carbon disulphide in needles melting at 154°.

NITRO-SUBSTITUTION PRODUCTS OF RESORCINOL.

1021 *Mononitroresorcinol*, $C_6H_3(NO_2)(OH)_2$, is obtained in two isomeric forms, together with other products, which will be described later on, when an ethereal solution of resorcinol is

[1] Benedikt, *Monatsh.* **1**, 366.
[2] Stenhouse, *Chem. News*, **26**, 279.
[3] Michael and Norton, *Ber.* **9** 1752.
[4] Claassen, *ibid.* **2**, 1442.

treated with nitric acid containing nitrous acid. They can readily be separated by distillation with steam. The volatile nitro-resorcinol, (OH : NO$_2$: OH = 1 : 2 : 3), crystallizes from dilute alcohol in orange-red prisms melting at 85°, and having a penetrating smell like that of *o*-nitrophenol.

The non-volatile nitroresorcinol, (1 : 4 : 3), forms lemon-coloured, hair-like needles melting at 115°. Its salts, which crystallize well, have a deep yellow to orange-red colour.[1]

Fuming sulphuric acid converts it into nitroresorcinol ether, $[C_6H_3(NO_2)OH]_2O$, which crystallizes in light rose-coloured needles, or in brown warty masses containing one molecule of water. It forms two barium salts, and is converted by concentrated nitric acid into trinitroresorcinol.[2]

Dinitroresorcinol, $C_6H_2(NO_2)_2(OH)_2$, (1 : 3 : 2 : 4), is formed by the action of nitric acid on dinitrosoresorcinol. It crystallizes from alcohol in small, light yellow plates which melt at 142°, and are converted by dilute nitric acid into trinitroresorcinol or styphnic acid.[3]

Isodinitroresorcinol, $C_6H_2(NO_2)_2(OH)_2$, (1 : 3 , 4 : 6), is obtained when diacetylresorcinol is converted by the action of concentrated nitric acid into the dinitro-compound, and this decomposed by hydrochloric acid.[4] It forms small, light-brown, shining plates, which melt at 212·5°, and are converted by a mixture of sulphuric acid and nitric acid into styphnic acid.[5]

Trinitroresorcinol or *Styphnic Acid*, $C_6H(NO_2)_3(OH)_2$.—On boiling logwood extract with nitric acid, Chevreul in 1808 obtained a crystalline body, which he considered to be a compound of an oil, or resinous matter with nitric acid; not only is its mode of preparation very nearly identical with that of Welter's Bitter (Picric Acid), but it also forms explosive salts like this body.[6] Erdmann then found, in 1846, that by the action of nitric acid on euxanthic acid, $C_{19}H_{16}O_{10}$, the magnesium salt of which occurs in commerce under the name Purrée or Indian yellow, an acid is obtained to which he gave the name of oxypicric acid, because it contained an atom of oxygen more than picric acid.[7] About this time Böttger and Will observed that when certain gum-resins or vegetable gums, as well as the extracts of different

[1] Flitz, *Ber.* **8**, 631.

[2] Weselsky and Benedikt, *Monatsh.* **1**, 887 ; Hazura and Julius, *ibid.* **5**, 188.

[3] Kostanecki and Feinstein, *Ber.* **21**, 3123 ; Benedikt and Hübl, *Monatsh.* **2**, 323.   [4] Typke, *Ber.* **16**, 551.

[5] Kostanecki and Feinstein, *ibid.* **21**, 3123.   [6] *Annalen*, **66**, 246 ; **73**, 43.

[7] *J. Pr. Chem.* **37**, 409.

dye-woods, bodies which we now know yield resorcinol on fusion
with potash, are boiled with nitric acid, an acid similar to picric
acid is obtained, which they named styphnic acid, because it
does not taste bitter, but astringent ($\sigma\tau\acute{v}\phi\nu o\varsigma$).[1] They per-
ceived that this body is identical with Chevreul's substance, and
supposed that this was also the case with oxypicric acid, a view
which Erdmann confirmed.[2] Schreder showed that it is trinitroresorcinol,[3] and Stenhouse afterwards obtained it by the
action of nitrosulphuric acid on resorcinol.[4] It is also formed
by the continued action of fuming nitric acid on m-nitrophenol.[5]

To prepare styphnic acid, finely-powdered resorcinol is dissolved gradually and with continual stirring, in 5 to 6 parts of
concentrated sulphuric acid warmed to about 40°; then cooled
down to 10—12°, and 2 to 2·5 times the theoretical quantity of
nitric acid added; at first concentrated acid mixed with 10 per
cent. by weight of water is employed, then concentrated acid
without admixture of water, and finally fuming acid; during
this operation the mixture must be continually agitated. It is
now allowed to stand over night, then brought into 1·5 to 2
volumes of cold water and passed through a vacuum filter to
separate the crystals from the mother-liquor. On evaporation a
further crop of crystals is obtained.[6]

Styphnic acid separates from its solution in dilute alcohol in
large, sulphur-coloured, hexagonal crystals, which melt at 175·5°,
and dissolve in 165 parts of water at 14°, more abundantly in
hot water, and readily in alcohol and ether. From its aqueous
solution, which has an acid reaction, it is precipitated by even a
small quantity of a strong acid (Stenhouse).

The Styphnates.—Styphnic acid forms two series of yellow
coloured salts, most of which are only slightly soluble, and, on
heating, explode more violently than the picrates (Böttger and
Will).

Dimethyl styphnate, $C_6H(NO_2)_3(OCH_3)_2$, was obtained by König
by the action of nitrosulphuric acid on resorcinol dimethyl
ether; it crystallizes in small plates, melting at 123—124.°[7]

Diethyl styphnate, $C_6H(NO_2)_3(OC_2H_5)_2$, is obtained by the
action of ethyl iodide on silver styphnate, and crystallizes in

[1] *Annalen,* **58**, 273.　　　[2] *J. Pr. Chem.* **38**, 355.
[3] *Annalen,* **158**, 244.　　　[4] *Chem. News,* **22**, 98.
[5] Lautlin, *Ber.* **11**, 2101; Henriques, *Annalen,* **215**, 321.
[6] Merz and Zetter, *Ber.* **12**, 2037.　　　[7] *Ber.* **11**, 1042.

long plates, which melt at 120·5°, and are quickly coloured orange-brown in the light (Stenhouse).

By the action of alcoholic ammonia it is easily converted into trinitro-*m*-diamidobenzene, $C_6H(NO_2)_3(NH_2)_2$, a yellow crystalline powder, which is only soluble with difficulty in the ordinary solvents, and on heating with dilute caustic soda is converted into sodium styphnate.

Styphnic acid is formed by the further nitration of both the adjacent and the symmetric dinitroresorcinol and must therefore have the formula:

$$\begin{array}{c} OH \\ NO_2 \diagup \diagdown NO_2 \\ H \diagdown \diagup OH \\ NO_2 \end{array}$$

Resorcinol indophane, $C_9H_4N_4O_6$.—The potassium salt, $C_9H_2K_2N_4O_6 + H_2O$, separates as a dark-brown, metallic-looking, crystalline mass, when warm solutions of potassium cyanide and potassium styphnate are mixed. It explodes on heating, and yields, on decomposition with dilute sulphuric acid, free resorcinol indophane, which forms needles having a metallic lustre and dissolving in water forming a bluish-violet solution.[1]

Tetranitroresorcinol, $C_6(NO_2)_4(OH)_2$, is obtained, as already stated, together with trinitrophenol and styphnic acid by the further nitration of γ-dinitrophenol. It is only slightly soluble in water, but readily in alcohol, and crystallizes in colourless or yellowish needles, which melt at 166° and sublime easily. It is however, still questionable whether this compound is a derivative of resorcinol.

p-DIHYDROXYBENZENE, HYDROQUINONE, OR QUINOL
$C_6H_4(OH)_2$.

1022 By the dry distillation of quinic acid, Caventou and Pelletier obtained a crystalline body, which they did not thoroughly examine, but named pyroquinic acid. Wöhler,[2] however, gave it the name of hydroquinone, because it is readily formed by the combination of hydrogen and quinone, $C_6H_4O_2$, which is an oxidation product of quinic acid and will be described later on.

---

[1] Schreder, *Annalen*, **163**, 298.　　　　[2] *Ibid.* **51**, 145; **65**, 349.

He found that this reduction can be performed with hydriodic acid and telluretted hydrogen, but that hydroquinone is best obtained when sulphur dioxide is passed through warm saturated solutions containing some undissolved quinone :

$$C_6H_4O_2 + 2H_2O + SO_2 = C_6H_6O_2 + SO_4H_2.$$

Besides these, almost all other reducing agents convert quinone into quinol.

Quinol is also formed by fusing p-iodophenol with caustic potash at a moderate temperature,[1] and by boiling p-diazobenzene sulphate with dilute sulphuric acid.[2]

The formation of quinol from succinic acid and acetoacetic acid is very interesting. By the action of sodium on the ethyl-salt of the former, ethyl succinyl succinate (Vol. III. Part II. p. 211) is obtained, which is also formed when ethyl bromacet-acetate is treated with sodium.[3]

$$
2\;
\begin{array}{c}
CH_3 \\
| \\
CO \\
| \\
CHBr \\
| \\
CO_2.C_2H_5
\end{array}
+ 2Na =
\begin{array}{c}
CO_2.C_2H_5 \\
| \\
CH_2-CH \\
| \quad\; | \\
CO \quad CO \\
| \quad\; | \\
CH-CH_2 \\
| \\
CO_2C_2H_5.
\end{array}
+ H_2 + 2NaBr.
$$

By the action of bromine this is converted into the diethyl-salt of quinoldicarboxylic acid, $C_6H_4O_2(CO_2H)_2$. The free acid crystallizes from hot water in hair-like needles. Its solution is coloured pure blue by ferric chloride. On dry distillation the acid is decomposed into quinol and carbon dioxide.[4]

Quinol is also found in the distillation products of the salts of succinic acid.[5]

For the preparation of quinol Nietzki[6] recommends the following method :—1 part of aniline is dissolved in 25 parts of water and 8 parts of sulphuric acid, and to the carefully cooled liquid a concentrated solution of sodium bichromate added in small

[1] Körner, *Zeitsch. Chem.* **1866**, 622 and 731.
[2] Weselsky and Schuler, *Ber.* **9**, 1159.
[3] Duisberg, *ibid.* **16**, 133.
[4] Herrmann, *ibid.* **16**, 1411 ; *Annalen*, **211**, 308.
[5] Richter, *J. Pr. Chem.* II. **20**, 207.
[6] *Ber.* **19**, 1467.

portions at a time, cooling well between each addition. If potassium bichromate be used it must be added in the form of powder. The most suitable temperature for the reaction is 5—10°. The mixture first assumes a dark-green colouration, which towards the end of the operation changes to a deep blue-black tinge; after further addition the precipitate for the most part disappears, and a cloudy brown liquid is obtained, the chief constituents of which are quinone and quinhydrone. To convert them into quinol, sulphur dioxide is passed into the solution and the whole extracted with ether. The raw quinol thus obtained amounts to about 85 per cent. of the theoretical yield.

In place of the above procedure Schniter[1] recommends that one-third of the bichromate necessary should be added at once, and the remainder after 12—24 hours.

The raw quinol is best purified by distillation, and recrystallizing the distillate from the smallest possible quantity of hot water. It forms rhombohedral prisms, which sublime on heating in monosymmetric plates.

These, on recrystallization from hot water, reproduce crystals of the former kind.[2] Quinol is therefore dimorphous; it has a slightly sweet taste, melts at 169°, and distils without decomposition. It is readily soluble in hot water, alcohol, and ether, and slightly in cold benzene. On fusing with caustic potash it remains unchanged; if its vapour be passed through a tube at a low red heat it is decomposed into hydrogen and quinone.[3] The latter is also very easily formed by the action of oxidizing agents for which reason quinol reduces silver nitrate solution on warming, and Fehling's solution even in the cold. It is on this account largely used as a developer in photography. Ferric chloride also easily oxidizes it, by means of which reaction it can be readily distinguished from its isomerides. It also differs from catechol in not being precipitated by lead acetate. When dissolved in a hot solution of this salt, the compound $2[C_6H_6O_2 + (C_6H_5O_2)_2Pb] + 3H_2O$ separates out on cooling in oblique rhombic prisms (Wöhler). When sulphuretted hydrogen is passed through a cold saturated solution of quinol, the solution gently warmed until crystals appear, and the gas then allowed to stream through until these are redissolved, the compound $(C_6H_6O_2)_3SH_2$ separates out on cooling in colourless, transparent rhombohedra, which are odourless, and stable in the dry state. Cold water decomposes

---

[1] *Ber.* **20**, 2283.   [2] Lehmann, *Jahresb.* **1877**, 566.
[3] Hlasiwetz, *Annalen*, **175**, 68 ; **177**, 336 ; Hesse, *ibid.* **114**, 297.

them slowly, hot water quickly, into their constituents. · By employing a saturated solution warmed to 40°, long, colourless prisms of the composition $(C_6H_6O_2)_2SH_2$ are obtained, which behave like the preceding compound.[1] On passing sulphur dioxide through a solution of quinol, yellow rhombohedra of the composition $(C_6H_6O_2)_3SO_2$, are formed, which decompose on heating.[2]

1023 *Quinol methyl ether*, $C_6H_4(OCH_3)OH$, was first obtained together with quinol, as a decomposition product of *arbutin*. It is formed together with the dimethyl ether by heating quinol with caustic potash and potassium methyl sulphate to 170°.[3] It is more easily obtained by digesting 1 pt. of potash, 2 pts. quinol, 3 pts. methyl iodide, and some methyl alcohol in a flask connected with an inverted condenser, until the alkaline reaction disappears. The methyl alcohol is evaporated off and the residue distilled in a current of steam, when the dimethyl ether comes over. The residue is then extracted with ether and the latter distilled off, the monomethyl ether, together with some quinol, remaining behind. These are then separated by cold benzene, in which the methyl ether is readily soluble ; after the removal of the benzene this is distilled.[4] It crystallizes in rhombic plates or prismatic tables melting at 53° and boiling at 243°. It is not oxidized by ferric chloride, but reduces silver nitrate solution on warming.

Quinol dimethyl ether, $C_6H_4(OCH_3)_2$. is best obtained by warming 10 pts. of quinol, 12 pts. of caustic potash, and 30 pts. of methyl iodide diluted with double the volume of wood spirit, under an extra pressure of 200 mm. in an apparatus connected with an inverted condenser. On recrystallization from methyl alcohol or ordinary alcohol, it is obtained in splendid plates melting at 56°.[5] It behaves towards ferric chloride and silver nitrate like the preceding compound. When 15 to 20 volumes of ether are added to a concentrated solution of the methyl ether and caustic potash, the phenate, $C_6H_4(OCH_3)OK$, separates out as a crystalline powder, or, when the materials are not anhydrous, in crystalline tablets.[6]

Quinol ethyl ether, $C_6H_4(OC_2H_5)OH$, is readily soluble in hot

---

[1] Wöhler, *Annalen*, **69**, 294.
[2] Clemm, *ibid.* **110**, 357 ; see also Hesse, *ibid.* **114**, 300.
[3] Hlasiwetz and Habermann, *ibid.* **177**, 338.
[4] Hesse, *ibid.* **200**, 254 ; Tiemann, *Ber.* **14**, 1989.
[5] Mühlhäuser, *Annalen*, **207**, 252.
[6] Michael, *Am. Chem. Journ.* **5**, 176.

water, and crystallizes in very thin, satin-like needles; it melts at 66° and boils at 246—247°.[1]

Quinol diethyl ether, $C_6H_4(OC_2H_5)_2$, forms large thin plates, which melt at 72° and are volatile with steam.[2]

Fiala[3] has prepared the following mixed ethers of quinol; these bodies possess an aromatic odour resembling fennel, and have a burning taste:

		Melting-point.
$C_6H_4\begin{cases} OCH_3 \\ OC_2H_5 \end{cases}$	crystals.	39°
$C_6H_4\begin{cases} OCH_3 \\ OC_3H_7 \end{cases}$	crystals.	26°
$C_6H_4\begin{cases} OCH_3 \\ OC_2H_3(CH_3)_2 \end{cases}$	liquid.	—

1024 *Arbutin*, $C_{12}H_{16}O_7$.—This glucoside was found by Kawalier in the leaves of the bearberry (*Arbutus* s. *Arctostaphylos uva-ursi*), and he showed that it is decomposed by emulsin into glucose and arctuvin.[4] Strecker, who recognised the latter compound as quinol, proved the composition of arbutin, and also effected its decomposition by boiling with dilute sulphuric acid.[5] He also found it in the leaves of the Wintergreen (*Pyrola umbellata*).[6] In order to prepare it, the leaves are boiled with water, lead acetate added to the solution to precipitate tannic acid, &c., and the filtrate treated with sulphuretted hydrogen and evaporated. Arbutin crystallizes from boiling water in long silky needles containing half a molecule of water of crystallization, which is driven off at 100°. It has a bitter taste, and gives a light blue colour with ferric chloride. By heating it with acetic anhydride, the pentacetate, $C_{12}H_{11}(C_2H_3O)_5O_7$, which crystallizes from hot alcohol in needles or plates, is obtained.

These different reactions show that arbutin is an ether of glucose, and at the same time a phenol: its constitution is therefore expressed by the following formula (Schiff):[7]

$$CHO$$
$$|$$
$$(CH.OH)_4$$
$$|$$
$$CH_2.OC_6H_4.OH.$$

[1] Hantzsch, *J. Pr. Chem.* II. **22**, 246; Wichelhaus, *Ber.* **12**, 1501·
[2] Rakowski, *Neues Handwörterb.* **2**, 560.   [3] *Monatsh.* **5**, 232.
[4] *Annalen*, **84**, 356.   [5] *Ibid.* **107**, 228.
[6] Zwenger and Himmelmann, *ibid.* **129**, 205.   [7] Schiff, *ibid.* **154**, 237.

Hlasiwetz and Habermann proposed for arbutin the formula $C_{25}H_{34}O_{14}$, because they found that on decomposition it yielded equal molecules of quinol and quinol methyl ether.[1] Arbutin is, however, more probably, a mixture of the glucosides of both compounds,[2] and, indeed, the analyses of different preparations show varying relations. By the fractional crystallization of such a mixture, Schiff succeeded in obtaining a pure normal arbutin crystallizing in needles, 2—3 centimetres in length, which, after being dried at 110—115°, melt at 165—166°.[3]

On the other hand, he could not then obtain a *methylarbutin*, $C_{12}H_{15}(CH_3)O_7$, free from arbutin. Michael, however, obtained it artificially, by allowing a solution of 11 parts of the above-mentioned potassium compound of quinol methyl ether, and 25 parts of aceto-chlorohydrose in absolute alcohol to stand for several days (see Phenolglucoside, p. 120). It crystallizes from water in long, silky needles, containing half a molecule of water of crystallization, and melting in the anhydrous state at 168—169°. Methylarbutin gives no colouration with ferric chloride; it is easily decomposed on heating with acids or in presence of emulsin.[4]

Schiff then obtained this compound by heating a solution of ordinary arbutin, methyl iodide, and caustic potash, in methyl alcohol. It contained one molecule of water of crystallization, and melted in the anhydrous state at 175—176°. He also found that mixtures of both glucosides melt lower than either of them separately, and believes that the true melting-point of pure arbutin is somewhat higher than 190·5°, the temperature which he had previously observed.[5]

Diacetoquinol, $C_6H_4(OC_2H_3O)_2$, is obtained by heating quinol with acetic anhydride or acetyl chloride, and crystallizes in tables or plates, melting at 123—124°, and readily subliming in needles It is easily soluble in benzene and ether, and slightly in alcohol and hot water.[6]

Quinol ethylcarbonate, $C_6H_4(O.CO_2C_2H_5)_2$, is obtained by the action of ethyl chlorocarbonate on a mixture of caustic potash and quinol ; it crystallizes from absolute alcohol in large needles;

[1] *Annalen*, **177**, 342.   [2] Fittig, *Org. Chem.* 2te. Aufl. 638.
[3] *Annalen*, **206**, 159.   [4] *Am. Chem. Journ.* **5**, 176 ; *Ber.* **14**, 2079.
[5] *Ber.* **15**, 1841 ; see also *Annalen*, **221**, 365.
[6] Rakowski, *Neues Handwörterb.* **2**, 560 ; Hesse, *Annalen*, **200**, 244 ; Nietzki, *Ber.* **11**, 470.

which melt at 101°, boil at 310°, and are not attacked by alkalis or acids even on warming.[1]

Acetonequinol, $C_6H_6O_2.C_3H_6O$, is obtained by the direct combination of its constituents and forms beautiful, transparent, monosymmetric crystals, which give off acetone in the air and become opaque without falling to powder. This body has an analogous composition to quinhydrone (see p. 185).[2]

Thioquinol, $C_6H_4(SH)_2$, may be obtained by the action of tin and hydrochloric acid on *p*-benzenedisulphonic chloride; it crystallizes in small six-sided plates, melting at 98°. Its lead salt, $C_6H_4S_2Pb$, is a yellowish-red precipitate.[3]

CHLORINE SUBSTITUTION PRODUCTS OF QUINOL.

1025 These bodies are not formed by the direct action of chlorine on quinol, because the latter is thereby converted into quinone. They can, however, be obtained by reducing the substitution products of quinone, as well as by treating them with hydrochloric acid.

Chloroquinol, $C_6H_3Cl(OH)_2$, is obtained by treating quinone with concentrated hydrochloric acid (Wöhler),[4] and by the action of sulphurous acid on chloroquinone (Städeler).[5] It is readily soluble in water and crystallizes in prisms melting at 98°.[6]

a-Dichloroquinol, $C_6H_2Cl_2(OH)_2$ (Cl : Cl = 2 : 5), was prepared by Städeler from the corresponding dichloroquinone; it is also obtained by the combination of monochloroquinone with hydrochloric acid (Levy and Schultz). It is much more readily soluble in hot water than in cold, and crystallizes in very long needles, or short, thick prisms, melting at 166°.

β-Dichloroquinol, (Cl : Cl = 2 : 6), is obtained by reducing *β*-dichloroquinone. It crystallizes from dilute alcohol in small yellow plates, melting at 157—158°.[7]

[1] Bender, *Ber.* **13**, 697 ; Wallach, *Annalen*, **226**, 85.
[2] Habermann, *Monatsh.* **5**, 329.
[3] Körner and Monselise, *Jahresb.* **1876**, 450.
[4] *Annalen*, **51**, 155.
[5] *Ibid.* **69**, 307.
[6] Levy and Schultz, *Ber.* **13**, 1427.
[7] Faust, *Annalen*, Suppl. **6**, 154.

Trichloroquinol, $C_6HCl_3(OH)_2$, is formed by the action of sulphurous acid on trichloroquinone,[1] and, together with tri-chloroquinone and other products, when benzene is treated with potassium chlorate and dilute sulphuric acid.[2] It is slightly soluble in water, readily in alcohol, and crystallizes in prisms, melting at 134°.

Tetrachloroquinol, $C_6Cl_4(OH)_2$, is obtained by the reduction of tetrachloroquinone (Städeler, Gräbe). It forms plates which are insoluble in water, but which readily dissolve in alcohol and alkalis, and reduce silver solution. On heating in a current of air it sublimes in long flat needles, with partial decomposition.

Its dimethyl ether, $C_6Cl_4(OCH_3)_2$, is formed by treating a solution of quinol dimethyl ether in acetic acid with chlorine. It crystallizes in needles, which melt at 153—154° and sublime without decomposition.

BROMINE AND IODINE SUBSTITUTION PRODUCTS OF QUINOL.

1026 These are formed in a similar way to the corresponding chlorine substitution products, but can also be obtained by the direct bromination of quinol.

Bromoquinol, $C_6H_3Br(OH)_2$, is obtained by the combination of quinone and hydrobromic acid,[4] as well as by acting upon an ethereal solution of quinol with a solution of bromine in chloroform.[5] It is very readily soluble in water, and crystallizes from light petroleum in lustrous silky plates, melting at 110—111°.

Dibromoquinol, $C_6H_2Br_2(OH)_2$, is obtained by the action of bromine on a warm solution of quinol in acetic acid,[6] and is also formed when quinone (Wichelhaus) or bromoquinone (Sarauw) is heated with hydrobromic acid. It is scarcely soluble in cold water, and crystallizes from a boiling solution in long needles melting at 186°.

Tribromoquinol, $C_6HBr_3(OH)_2$, is formed by the action of

[1] Städeler, Gräbe, *Annalen*, **146**, 25 ; Stenhouse, *Journ. Chem. Soc.* **21**, 146.
[2] Krafft, *Ber.* **10**, 797.
[3] Habermann, *ibid.* **11**, 1035.   [4] Wichelhaus, *ibid.* **12**, 1504.
[5] Sarauw, *Annalen*, **209**, 105.   [6] Benedikt, *Monatsh.* **1**, 345.

hydrobromic acid on dibromoquinol, as well as by that of bromine on quinol or quinone (Sarauw).

$$C_6H_4O_2 + 2Br_2 = C_6H_3Br_3O_2 + HBr.$$

It is very slightly soluble in cold, readily in hot water, and crystallizes in silky needles melting at 136°.

Tetrabromoquinol, $C_6Br_4(OH)_2$, is obtained by the action of sulphurous acid on tetrabromoquinone,[1] or by treating it with hydriodic acid and phosphorus.[2] It is further obtained by the combination of tribromoquinol with hydrobromic acid, and by the action of bromine on a solution of quinol in acetic acid (Sarauw).

Melting-point.
Diiodoquinol,[3] $C_6H_2I_2(OH)_2$, silky needles . . . .   142·5°

NITRO-SUBSTITUTION PRODUCTS OF QUINOL.

1027 *Nitroquinol*, $C_6H_3(NO_2)(OH)_2$, is not known in the free state ; but several of its ethers have been obtained by the action of nitric acid on quinol ether.

Melting-point.

Monomethyl ether, $C_6H_3(NO_2)(OCH_3)OH$, { orange yellow needles } 83°
Dimethyl ether, $C_6H_3(NO_2)(OCH_3)_2$, golden yellow needles 71·5°
Monoethyl ether, $C_6H_3(NO_2)(OC_2H_5)OH$, deep yellow needles 83°
Diethyl ether, $C_6H_3(NO_2)(OC_2H_5)_2$, golden yellow needles 49°

Dinitroquinol, $2C_6H_2(NO_2)_2(OH)_2 + 3H_2O$, was obtained by Strecker by boiling dinitro-arbutin with dilute sulphuric acid,[4] and by Nietzki by treating the diacetate, described below, with cold caustic soda.[5] It crystallizes from hot water in fine flat needles having a golden lustre, which lose water of crystal-lization at 100° with disintegration, and then turn brown and melt at 135—136°. It colours the skin a carmine red. Its yellow solution is coloured first blood-red, and then bluish-violet by the gradual addition of an alkali, whilst its ammoniacal solution is coloured purple-red on boiling, leaving metallic-green crystals on evaporation. The arrangement of the groups is as follows :[6]

$$OH.NO_2.OH.NO_2 = 1.2.4.6.$$

[1] Stenhouse, *Annalen*, **91**, 310.   [2] Stenhouse, *Journ. Chem. Soc.* **23**, 11.
[3] Metzeler, *Ber.* **21**, 2555.   [4] *Annalen*, **118**, 293.   [5] *Ber.* **11**, 470.
[6] Nietzki and Preusser, *ibid.* **20**, 799.

Its ethers are obtained similarly to those of nitroquinone.[1]

Melting-point.

Monomethyl ether, $C_6H_2(NO_2)_2(OCH_3)OH,$ $\left\{ \begin{array}{c} \text{greenish} \\ \text{needles} \end{array} \right\}$ 102°

Dimethyl ether, $C_6H_2(NO_2)_2(OCH_3)_2$ $\left\{ \begin{array}{c} \text{yellow} \\ \text{crystals} \end{array} \right\}$ 169—170°

Monoethyl ether, $C_6H_2(NO_2)_2(OC_2H_5)OH,$ $\left\{ \begin{array}{c} \text{brownish} \\ \text{yellow} \\ \text{needles} \end{array} \right\}$ 71°

a-Diethyl ether $\left. \begin{array}{l} \\ \beta\text{-Diethyl ether} \end{array} \right\}$ $C_6H_2(NO_2)_2(OC_2H_5)_2$ $\left\{ \begin{array}{l} \text{lemon yellow plates 130°} \\ \quad\text{,,} \quad\quad \text{,,} \quad\quad \text{,,} \quad 176° \end{array} \right.$

Diacetodinitroquinol, $C_6H_2(NO_2)_2(OC_2H_3O)_2$, is obtained by dissolving quinol diacetate in fuming nitric acid,[2] and crystallizes from alcohol in sulphur-coloured needles melting at 96°.

Trinitroquinol, $C_6H(NO_2)_3(OH)_2$.—The dimethyl ether of this substance is obtained by adding a solution of quinol dimethyl ether in acetic acid to a well-cooled mixture of nitric and sulphuric acids; it crystallizes from alcohol in long, yellow needles which melt at 100—101° (Habermann).

The diethyl ether is obtained by the nitration of both the dinitroquinol diethyl ethers, and forms long, straw-coloured needles melting at 130° (Nietzki).

QUINONE, OR BENZOQUINONE, $C_6H_4O_2$.

1028 Woskresensky obtained this body by oxidizing quinic acid with manganese dioxide and sulphuric acid, and named it quinoyl,[3] a name which was changed by Berzelius to that now used. It is also obtained by the oxidation of quercitol, $C_6H_7(OH)_5$, (*vide* p. 207),[4] caffetannic acid, extract of coffee-beans, the leaves of the coffee shrub, the holly, and other plants,[5] as well as several para-disubstitution products of benzene, such as p-diamidobenzene,[6] p-amidophenol,[7] p-amidobenzene-

---

[1] Weselsky and Benedikt, *Monatsh.* **2**, 369; Mühlheimer, *Annalen,* **207**, 253; Habermann, *Ber.* **11**, 1037; Nietzki, *ibid.* **11**, 1448; **12**, 39; *Annalen,* **215**, 125.   [2] Nietzki, *Ber.* **11**, 470.   [3] *Annalen,* **27**, 268.

[4] Prunier, *Compt. Rend.* **82**, 1113.

[5] Stenhouse, *Mem. Chem. Soc.* **2**, 226; *Annalen,* **89**, 244.

[6] Hofmann, *Jahresb.* **1863**, 415.

[7] Körner, Kekulé's *Org. Chem.* **3**, 103; Andresen, *J. Pr. Chem.* II. **23**, 173.

sulphonic acid (sulphanilic acid),[1] phenol-p-sulphonic acid,[2] &c. It is also formed by the oxidation of aniline (Hofmann), in which way it is best obtained. The method has already been described under quinol (p. 166). The quinone may be extracted direct from the solution of quinone and quinhydrone with ether. It is however preferable to convert the whole into quinol as there described, dissolve the latter in as small quantity of water as possible, and add sulphuric acid, and bichromate, cooling well during the operation.[3] The separated quinone may be filtered off and the mother liquor extracted with ether, from which quinone separates on evaporation in golden-yellow plates.[4] It can also be obtained by the direct oxidation of benzene. When 4 parts of this are gently heated with 1 part of chromium oxychloride, CrO_2Cl_2, hydrochloric acid gas is evolved, and a brown precipitate separates out, which has probably the formula $C_6H_4(OCrOCl)_2$, since it is decomposed by water into chromium trioxide, hydrochloric acid, and quinone, which remains dissolved in the excess of benzene.[5]

Wöhler, who first closely examined the quinone prepared from quinic acid, describes it as follows:[6] "There are probably few substances which have so great a power of crystallization as this. When even small quantities of it are sublimed, crystals are obtained an inch in length. After being melted it solidifies to a crystalline mass. It dissolves in quantity in boiling water with a reddish-yellow colour, and, when this solution is cooled crystallizes out in long but less transparent prisms, which are somewhat darker and of a less beautiful yellow than those obtained by sublimation.

"Its solution stains the skin permanently brown. It is so volatile that, even at the ordinary temperature, it sublimes from one side of the vessel to the other. Its strong odour, which irritates both the eyes and nose, causes an after-effect similar to that produced by iodine or chlorine."

Quinone is readily soluble in hot alcohol and light petroleum, crystallizing from the latter in beautiful yellow prisms,[7] which melt at 115·7°.[8] Hofmann found its vapour density to be 3·72-3·79, by means of which determination its then somewhat doubtful molecular weight was confirmed.[9] Its alkaline solution

[1] Ador and Meyer, *Annalen*, **159**, 7. [2] Schreder, *Ber.* **8**, 760.
[3] Nietzki, *ibid.* **20**, 1468. [4] Nietzki, *ibid.* **10**, 1934.
[5] Etard, *Ann. Chim. Phys.* V. **22**, 270. [6] *Annalen*, **51**, 148 ; **65**, 349.
[7] Hesse, *ibid.* **200**, 240. [8] Hesse, *ibid.* **114**, 300. [9] *Ber.* **3**, 583.

quickly turns dark brown, and its aqueous solution a dark
yellowish-red in the air, and the latter deposits a blackish-
brown substance (Wöhler). In presence of sodium acetate
this decomposition takes place more rapidly, particularly on
warming, the acetate, however, remaining unchanged. At the
same time a certain quantity of quinol is obtained, which can
be isolated by extracting with ether.[1] Its alcoholic solution
after exposure for five months to direct sunlight was almost
completely converted into quinol,[2] and as already stated, it is
easily converted into the latter substance by reducing agents.
It acts therefore as a strong oxidizing agent, and decomposes
hydriodic acid with separation of iodine. It is not poisonous, in
spite of its strong smell and its action on the living skin ; 0·5
and even 1 grm. given to a dog produced no effect. It could
not be found in the urine, and what had become of it could
not be ascertained.[3]

To detect quinone dissolved in water, a few drops of a saturated
solution of hydrocaerolignone, $C_{12}H_4(OCH_3)_2(OH)_2$. are added.
The solution immediately turns a yellowish-red; and then deposits
steel-blue, iridescent needles of caerolignon, $C_{12}H_4(OCH_3)_4O_2$, the
solution becoming decolourized. In this way 1 part of quinone
in 200,000 parts of water can be detected. In more dilute solutions the separation of crystals does not take place, but the
colouration is still produced by 1 part of quinone in 1,000,000
parts of water.[4]

CHLORINE SUBSTITUTION PRODUCTS OF QUINONE.

1029 *Monochloroquinone*, $C_6H_3ClO_2$, is obtained by distilling
a salt of quinic acid with common salt, manganese dioxide and
dilute sulphuric acid,[5] as well as by oxidizing chloroquinol[6] or
chloramidophenol,[7] with an ice-cold solution of potassium bichromate in dilute sulphuric acid. It is readily soluble in water
alcohol, and ether, and forms yellowish-red rhombic crystals,
smelling like quinone, melting at 57°, and volatilizing at the
ordinary temperature.

[1] Hesse, *Annalen,* **220,** 365.   [2] Ciamician, *Gazzetta* **16,** 111.
[3] Wöhler and Frerichs, *Annalen* **65,** 343.   [4] Liebermann, *Ber.* **10,** 1615.
[5] Städeler, *Annalen,* **69,** 302   [6] Levy and Schultz, *ibid.* **210.**
144.   [7] Kollrepp, *ibid.* **234,** 14.

p-Dichloroquinone, $C_6H_2Cl_2O_2$(Cl : Cl = 2 : 5), was obtained by Städeler, together with the preceding compound ; Carius prepared it by the action of chlorous acid on benzene.[1] It is also obtained when *a*-dichloroquinol is oxidized with dilute nitric acid, or *p*-dichloraniline with a solution of chromic acid (Levy and Schultz). It is best prepared by treating quinone twice in succession with hydrochloric acid, adding water and dilute sulphuric acid, and then, with the usual precautions potassium bichromate.[2] It crystallizes in dark yellow, mono-symmetric tables melting at 159°, which are volatile with steam, insoluble in water, scarcely soluble in cold alcohol, but readily in boiling alcohol. On treatment with bromine a molecular change takes place, and a *m*-dichloro-*m*-dibromoquinone is obtained.[3]

m-Dichloroquinone (Cl : Cl = 2 : 6) is obtained by the action of cold fuming nitric acid on trichlorophenol,[4] or by passing nitrogen trioxide through its aqueous solution.[5] It is slightly soluble in water and cold alcohol, and crystallizes from boiling alcohol or light petroleum in large rhombic straw-coloured prisms, melting at 120°, and subliming readily. It is also obtained by the oxidation of *m*-dichloro-*p*-diamidobenzene,[6] and of dichloro-*p*-amidophenol.[7]

Trichloroquinone, $C_6HCl_3O_2$.—Woskresensky, by the action of chlorine on quinone, obtained a compound of this composition, which he named chloroquinoyl, but according to his statement, which is supported by the experience since gained concerning chlorinated quinones, this was a mixture. Trichloroquinone was first obtained pure, together with tetrachloroquinone and the above-mentioned compound, by Städeler, by the chlorination of quinic acid. By the action of hydrochloric acid and potassium chlorate on many aromatic substances, Gräbe[8] has shown that a mixture of trichloroquinone and tetrachloroquinone is obtained, whereas it was formerly believed that only the latter compound is produced.

In order to prepare trichloroquinone, Gräbe's process modified by Knapp and Schultz is employed.[9] 1 part of phenol is dissolved in an equal weight of sulphuric acid at 100°. The

[1] *Annalen*, **143**, 316. [2] Hantzsch and Schniter, *Ber.* **20**, 2279.
[3] *Ibid.* p. 2280. [4] Faust, *Annalen*, **149**, 153.
[5] Weselsky, *Ber.* **3**, 646. [6] Levy, *ibid.* **16**, 1444.
[7] Kollrepp, *Annalen*, **234**, 14. [8] *Ibid.* **146**, 1.
[9] *Ibid.* **210**, 174.

phenolsulphonic acid thus obtained is brought into a hot aqueous solution of 4 parts of potassium chlorate, and an excess of crude hydrochloric acid added. An energetic reaction soon commences, which is finished after standing twenty-four hours by passing in steam. The mixture of trichloroquinone and tetrachloroquinone is washed with hot water and cold alcohol, then suspended in water and the liquid saturated with sulphur dioxide, and allowed to stand until the crystals have become colourless. Boiling water then only extracts trichloroquinol, while tetrachloroquinol remains behind. By the addition of fuming nitric acid to the hot solution of the former, trichloroquinone separates out. According to Stenhouse, it is better to dissolve in hot water containing sulphuric acid and add potassium bichromate to it.[1] The trichloroquinone is then purified by recrystallization from alcohol.

It is also obtained by the action of chromium oxychloride on benzene.[2]

$$4CrO_2Cl_2 + C_6H_6 = C_6HCl_3O_2 + 2Cr_2O_3 + 5HCl$$

It is further obtained, together with tetrachloroquinone, when bleaching powder is added to a boiling solution of p-amido-phenol[3] in hydrochloric acid, or by decomposing trichloramido-phenol with bromine water,[4] and crystallizes in large yellow plates melting at 165—166°, and is insoluble in cold water, slightly soluble in cold, and readily in hot alcohol.

1030 *Tetrachloroquinone* or *Chloranil*, $C_6Cl_4O_2$.—Erdmann first obtained this compound by passing chlorine through an alcoholic solution of chlorisatin, and termed it *chloranil*,[5] a name which is still employed. Fritzsche obtained it by the action of hydro-chloric acid and potassium chlorate on aniline,[6] and Hofmann, from phenol, quinone, chloraniline, salicylic acid, salicyl aldehyde, isatin, &c.[7] It has since been frequently observed as a final product of the action of the above reagents on aromatic bodies. That the substance thus obtained always contains trichloro-quinone has already been stated. It is also obtained when symmetrical tetrachlorobenzene is oxidized with concentrated nitric acid, but is not formed from the isomerides of this com-pound.[8] Pentachlorophenol is similarly converted into chloranil.[9]

[1] *Journ. Chem. Soc.* **21**, 149.   [2] *Carstanjen, Ber.* **2**, 633.
[3] Schmitt and Andresen, *J. Pr. Chem.* II. **23**, 436.
[4] *Ibid.* II. **24**, 434.   [5] *Annalen,* **48**, 309.
[6] *Neues Handwörterbuch.* **2**, 561.   [7] *Annalen,* **52**, 57.
[8] Beilstein and Kurbatow, *ibid.* **192**, 206.   [9] Merz and Weith, *Ber.* **5**, 460.

Its preparation from phenol has been already described ; it is isolated as tetrachloroquinol, and this is oxidized by nitric acid, and repeatedly extracted with boiling alcohol, in order to remove any admixed trichloroquinone. If it be required to convert the trichloroquinone, formed at the same time, into chloranil, the solution must be boiled for a long time with concentrated hydrochloric acid, and the tetrachloroquinol so obtained, oxidized with fuming nitric acid (Knapp and Schultz).

The mixture obtained by the action of hydrochloric acid and potassium chlorate on phenol can, according to Stenhouse, also be converted into chloranil by dissolving it in an equal weight of water, adding half its weight of iodine, and passing in chlorine, until it is only slowly absorbed ; the chloranil formed is then distilled off.

Chloranil is employed in the colour industry as an oxidising agent, and is obtained on the large scale as stated above, or by the oxidation of trichlorophenol with potassium bichromate and sulphuric acid.

Chloranil is insoluble in water, very slightly soluble in cold, only sparingly in boiling alcohol, and somewhat more easily in ether. It crystallizes in small gold-coloured plates, or from boiling benzene in light-yellow, transparent prisms ;[1] it begins to vapourize at 150° and sublimes rapidly at 210°. It melts at a high temperature and boils with partial decomposition. It is not attacked by concentrated sulphuric acid, nitric acid, or aqua regia, and this explains why it is so frequently obtained as a final product of the action of oxidizing and chlorinating agents on a large number of bodies. By heating it with phosphorus pentachloride to 180°, hexchlorobenzene is obtained (Gräbe) :

$$C_6Cl_4O_2 + 2PCl_5 = C_6Cl_6 + 2POCl_3 + Cl_2.$$

By heating with acetyl chloride to 160—180° it is converted into di-acetotetrachloroquinol :

$$C_6Cl_4O_2 + 2C_2H_3OCl = C_6Cl_4(OC_2H_3O)_2 + Cl_2.$$

It has already been stated, that reducing agents convert it into tetrachloroquinol ; this also takes place on boiling with concentrated hydrochloric acid, or still more readily with hydrobromic acid :[2]

$$C_6Cl_4O_2 + 2HBr = C_6Cl_4(OH)_2 + Br_2.$$

[1] Levy and Schultz, Annalen, 210, 154.
[2] Levy and Schultz, Ber. 12, 1430 ; Sarauw, Annalen, 209, 125.

By adding chloranil to a dilute solution of acid potassium sulphite, the potassium salt of dichloroquinoldisulphonic acid is formed:[1]

$$C_6Cl_4O_2 + 3SO_3KH + H_2O = C_6Cl_2(OH)_2(SO_3K)_2 + 2HCl + SO_4KH$$

This crystallizes in colourless tables, but the free acid is only known in aqueous solution. Both the latter and the solutions of its salts, are coloured indigo-blue by ferric chloride.

On treating chloranil with a concentrated solution of acid, or better, normal potassium sulphite, the potassium salt of *thiochronic acid* is the chief product, the preceding compound being also formed:[2]

$$C_6Cl_4O_2 + 5SO_3K_2 + H_2O = C_6(OH)(SO_4K)(SO_3K)_4 + 4KCl + KOH$$

It crystallizes from hot water in yellow rhombic prisms containing 4 molecules of water, and its solution is coloured a deep brownish red by ferric chloride. The free acid is only known in aqueous solution, which decomposes when gently heated.

On heating the potassium salt with water to 130—140°, potassium quinoldisulphonate, $C_6H_2(OH)_2(SO_3K)_2 + 4H_2O$, is obtained, together with acid potassium sulphate. The free acid crystallizes in thick tablets, deliquesces in the air, and is coloured blue by ferric chloride (Gräbe).

Hesse obtained an acid isomeric with the above by the action of fuming sulphuric acid on quinic acid. It forms a syrup which readily dissolves in water; its salts are coloured deep blue by ferric chloride.[3]

BROMINE AND IODINE SUBSTITUTION PRODUCTS OF QUINONE.

	Melting-point.
Monobromoquinone,[4] $C_6H_3BrO_2$, yellow tables or needles	55—56°
a-Dibromoquinone,[5] $C_6H_2Br_2O_2$, { small lustrous golden plates }	188°
β-Dibromoquinone,[6] $C_6H_2Br_2O_2$, lustrous yellow plates	122

[1] Hesse, *Annalen*, **114**, 324 ; Greiff, *Jahresb.* **1863**, 392.
[2] Hesse, *loc. cit.* ; Gräbe, *Annalen*, **146**, 40.
[3] Hesse, *ibid.* **110**, 195.   [4] Sarauw, *ibid.* **209**, 102, 106.
[5] Sarauw ; Benedikt, *Monatsh.* **1**, 134.
[6] Levy and Schultz, *Annalen*, **210**, 158.

Melting-point.

γ-Dibromoquinone,[1] $C_6H_2Br_2O_2$, { yellow fibrous crystals or needles }   76°

Tribromoquinone,[2] $C_6HBr_3O_2$, lustrous golden plates   147°

Tetrabromoquinone or Bromanil,[3] $C_6Br_4O_2$, { golden rhombic plates or thick tablets }   —

Melting-point.

a-Di-iodoquinone,[4] $C_6H_2I_2O_2$, small golden plates   177—179°

β-Di-iodoquinone,[5] $C_6H_2I_2O_2$, yellow needles . . 157—159°

NITRO-SUBSTITUTION PRODUCTS OF QUINONE.

1031 *Nitroquinone*, $C_6H_3(NO_2)O_2$.—Chromium oxychloride acts on nitrobenzene just as on benzene (p. 175), with formation of the compound $C_6H_3(NO_2)(OCrOCl)_2$; this is decomposed by water as follows :

$$C_6H_3(NO_2)(OCrOCl)_2 + H_2O = C_6H_3(NO_2)O_2 + Cr_2O_3 + 2HCl.$$

The product is shaken with dilute caustic soda, filtered, and the nitroquinone precipitated from the filtrate by hydrochloric acid. It forms small yellowish brown plates which melt at 232°, and are tolerably soluble in hot water and readily in alcohol.[6]

QUINHYDRONES.

Quinhydrone, $C_{12}H_{10}O_4 = C_6H_4O_2 + C_6H_4(OH)_2$.

1032 This compound, which Wöhler described as " green hydroquinone," is formed by the partial reduction of quinone or the cautious oxidation of quinol. He makes the following remarks: "In all cases in which the green compound is

---

[1] Böhmer, *J. Pr. Chem.* II. **24**, 264.   [2] Sarauw, *loc. cit.*
[3] Stenhouse, *Annalen*, **91**, 307 ; *Journ. Chem. Soc.* **23**, 10 ; Sarauw.
[4] Seifert, *J. Pr. Chem.* II. **28**, 437.   [5] Metzeler, *Ber.* **21**, 2555.
[6] Etard, *Ann. Chim. Phys.* V. **22**, 272.

obtained, it separates out in the crystalline state, the liquid being momentarily coloured a deep red, and then suddenly becoming full of the most beautiful green metallic prisms, which, even with small quantities, are frequently an inch in length."

" It is most readily obtained by adding ferric chloride to a solution of the colourless hydroquinone."

" It is prepared from quinone by mixing a saturated solution with sulphurous acid, which in order to obtain large crystals, must be added all at once, but only in such quantity that some quinone still remains unchanged; otherwise the action will proceed further with formation of the colourless hydroquinone."

" The most remarkable method of formation of green hydro-quinone is by the mutual action of the colourless compound and quinone; when their solutions are mixed, they instantaneously combine forming the green crystals, no other compound being produced." " Green hydroquinone is one of the most beautiful substances which organic chemistry has produced. It is very similar to murexide, but excels it in lustre and beauty of colour. In this respect it bears the greatest resemblance to the metallic green of the rose-chafer, or of the feathers of the humming-bird. The crystals are generally very fine, and often very long. Under a high magnifying power the finer ones are seen to be transparent showing a reddish-brown colour. It has a sharp taste and a weak quinone-like odour; it readily fuses to a brown liquid, and at the same time a portion of it sublimes in small green plates, whilst another portion decomposes with formation of quinone, which sublimes in its characteristic yellow crystals. It is slightly soluble in cold water, and dissolves in greater quantity in hot water forming a brownish red solution, from which it again crystallizes out on cooling. The green hydroquinone is readily soluble in alcohol and ether with a yellow colour; on evaporation it retains its green metallic lustre, and, in the crystalline condition, presents, especially on white porcelain, a very striking and beautiful appearance."

It crystallizes from hot glacial acetic acid in greenish black tablets or prisms (Hesse).

Quinhydrone is a compound consisting of equal molecules of quinone and quinol. On boiling with water it is decomposed into its constituents, quinone volatilizing and quinol remaining behind. Oxidizing agents convert it into the first, and reducing agents into the second. " It dissolves in ammonia with a deep green colour, which in presence of air immediately changes into a dark, brownish red. On evaporation it remains behind as a brown, amorphous mass" (Wöhler).

Quinhydrone dimethyl ether, $C_{18}H_{14}O_6(CH_3)_2 = C_6H_4O_2 + 2C_6H_4$ $(OCH_3)OH$, is formed by the combination of quinone with quinol methyl ether,[1] when the solutions of these in hot light petroleum are mixed.[2] On cooling, the compound separates out in beautiful greenish black prisms with a metallic lustre, which become reddish brown in the light. The ether dissolves in warm water with decomposition, and on reduction with sulphurous acid gives two molecules of quinol methyl ether for each molecule of quinol.

1033 *Chlorine substitution products of quinhydrone* are not formed by the combination of a chlorinated quinol with quinone, nor of a chlorinated quinone with quinol, quinhydrone being always formed when these are brought together.[3]

$$C_6H_4O_2 + 2C_6H_5ClO_2 = C_{12}H_{10}O_4 + C_6H_4Cl_2O_2.$$
$$C_6H_3ClO_2 + 2C_6H_6O_2 = C_{12}H_{10}O_4 + C_6H_5ClO_2.$$
$$2C_6H_4O_2 + C_6H_4Cl_2O_2 = C_{12}H_{10}O_4 + C_6H_2Cl_2O_2.$$

Chlorinated quinones, on the contrary, combine with chlorinated quinols.

Dichloroquinhydrone or *brown chloroquinol,* $C_{12}H_8Cl_2O_4$, was obtained by Wöhler by the oxidation of chloroquinol with ferric chloride. It is also obtained by the combination of chloroquinone with chloroquinol, and is the first product of the action of hydrochloric acid on quinone (Städeler). It separates out at first as an oil and solidifies after some time to a greenish brown crystalline mass which, when left in a glass tube, sublimes in fine, long, brown needles, which stain the skin a dark purplered colour.

Tetrachloroquinhydrone or *violet bichloroquinol,* $C_{12}H_6Cl_4O_4 + 2H_2O$, is obtained by warming dichloroquinol with a solution of dichloroquinone, as well as by the oxidation of the former with ferric chloride (Städeler). It crystallizes in small violet prisms, or long, flat, dark green needles, which lose their water of crystallization over sulphuric acid, or at 70°. At 120° they melt and decompose into their constituents. It is insoluble in water, and dissolves in alcohol forming a yellow solution, whilst its solution in ammonia is green.

[1] Wichelhaus, *Ber.* **12,** 1501 ; Hesse, *Annalen,* **200,** 254.
[2] Nietzki, *Ber.* **12,** 1982.   [3] Wichelhaus, *ibid.* **12,** 1503.

Hexchloroquinhydrone or *yellow trichloroquinol*, $C_{12}H_4Cl_6O_4$, is formed, according to Woskresensky, by passing chlorine over quinone, or by the oxidation of trichloroquinol (Städeler).[1] It crystallizes in small, yellow, lustrous plates, which are slightly soluble in boiling water and more readily in boiling alcohol. It melts and sublimes a few degrees above 100°, and possesses a penetrating aromatic odour.

By the action of an insufficient quantity of cold nitric acid on trichloroquinol, Gräbe obtained long, black needles, which he considered to be hexchloroquinhydrone, but which are more probably a hydrate of this.

Resorcinolquinone, $C_{12}H_{10}O_4 = C_6H_4O_2 + C_6H_4(OH)_2$, is isomeric with quinhydrone, and is obtained by dissolving equal molecules of resorcinol and quinone in warm benzene. It forms almost black needles, which in transmitted light appear of a garnet-red colour, and have a green surface lustre. They melt at 90° and are readily soluble in water and alcohol.[2]

Phenoquinone, $C_{18}H_{16}O_4 = C_6H_4O_2 + 2C_6H_5.OH$, was first obtained by Wichelhaus by the oxidation of phenol in aqueous solution ; it is also obtained by the direct combination of quinone with phenol,[3] when the solutions of these in hot light petroleum are mixed.[4] It crystallizes in splendid red needles, the broad faces of which show a green lustre. They have a slightly sharp smell, melt at 71°, and are very volatile. Phenoquinone is soluble in cold water, but more readily in alcohol and ether, as well as in light petroleum, by means of which it may be separated from quinone and quinhydrone. It is split up by alkalis or acids into quinone and phenol. On the addition of potash the red needles become blue, and with baryta or ammonia, green.

1034 *Constitution of the Quinhydrones.* Various views have been advanced concerning the constitution of these compounds, which are formed by the direct combination of phenols with quinone. Quinhydrone was first looked upon as a compound of equal molecules of quinone and quinol, and Gräbe gave the following constitutional formula : [5]

$$C_6H_4\!\!\begin{array}{c} \diagup OH \quad HO \diagdown \\ \diagdown O \!\!-\!\!-\!\!-\!\! O \diagup \end{array}\!\!C_6H_4$$

[1] *J. Pr. Chem.* **18**, 419.   [2] Nietzki, *Ber.* **12**, 1982.
[3] Wichelhaus, *ibid.* **5**, 248, 346.
[4] Nietzki, *ibid.* **12**, 1981 ; Hesse, *Annalen*, **200**, 251.
[5] Nietzki, *ibid.* **146**, 61.

Wichelhaus opposed this view; he explains the formation of phenoquinone according to the following equation:

$$C_6H_4\!\!<\!\!\genfrac{}{}{0pt}{}{O}{O}\!\!> + 2HO.C_6H_5 = C_6H_4\!\!<\!\!\genfrac{}{}{0pt}{}{O.OC_6H_5}{O.OC_6H_5} + H_2.$$

Quinhydrone and its methyl ether he supposed to be formed in a similar manner:

$$C_6H_4\!\!<\!\!\genfrac{}{}{0pt}{}{O}{O}\!\!> + 2HO.C_6H_4.OH = C_6H_4\!\!<\!\!\genfrac{}{}{0pt}{}{O.O.C_6H_4.OH}{O.O.C_6H_4.OH} + H_2.$$

The hydrogen formed would act as a reducing agent, changing, for example, a portion of the quinone into quinol.[1] Liebermann however, showed by quantitative researches that quinhydrone is a compound of equal molecules of its constituents;[2] Nietzki then found that quinone can be volumetrically determined by sulphurous acid and iodine solution. He thus obtained the formula $C_{12}H_{10}O_4 = C_6H_4O_2 + C_6H_6O_2$ for quinhydrone,[3] and $C_{18}H_{16}O_4 = C_6H_4O_2 + 2C_6H_6O$ for phenoquinone.[4] He also found that if phenol and quinone are dissolved in the correct proportions in petroleum light, they combine completely to form phenoquinone, an observation which was also made by Hesse. The latter found a further proof of the accuracy of the old formula for quinhydrone in the fact that when it is heated with acetic anhydride, quinone is set free, and 85 to 87 per cent. of diaceto-quinol is formed, instead of the theoretical 88·9, while according to the formula of Wichelhaus only 59·5 per cent. should be formed. Finally, he proved that quinhydrone dimethyl ether is a compound of one molecule of quinone with two molecules of quinol methyl ether.[5]

It follows, therefore, that one molecule of quinone combines with one molecule of a phenol, which contains two hydroxyls, and with two molecules of one which contains only one hydroxyl, or with their ethers.

Quinone tetrahydride, $C_6H_8O_2$, is obtained by heating succino-succinic acid (Part II. p. 211), and crystallizes on the gradual evaporation of its aqueous solution in short, flat, lustrous prisms,

[1] Hesse, *Ber.* **10**, 1781, 2005; **12**, 1500. [2] *Ibid.* **10**, 1614; 2000.
[3] *Ibid.* **10**, 2003. [4] *Ibid.* **12**, 1279.
[5] *Annalen*, **200**, 248.

having a peculiar, faint odour, and a cooling taste, and melting at 75°. Bromine converts it into bromanil:

$$
\begin{array}{ccc}
& \text{CO} & \\
& \diagup\diagdown & \\
\text{H}_2\text{C} & & \text{CH}_2 \\
| & & | \\
\text{H}_2\text{C} & & \text{CH}_2 \\
& \diagdown\diagup & \\
& \text{CO} &
\end{array}
+ 6\text{Br}_2 =
\begin{array}{ccc}
& \text{CO} & \\
& \diagup\diagdown & \\
\text{BrC} & & \text{CBr} \\
|| & & || \\
\text{BrC} & & \text{CBr} \\
& \diagdown\diagup & \\
& \text{CO} &
\end{array}
+ 8\text{HBr}
$$

1035 *The Constitution of Quinone.*—It has already been stated that the constitution of quinone is expressed by one of the following formulæ (p. 49):

$$
\begin{array}{ccc}
& \text{C} & \\
\diagup\!\!\diagup & | & \diagdown \\
\text{HC} & \text{O} & \text{CH} \\
| & | & || \\
\text{HC} & \text{O} & \text{CH} \\
\diagdown\!\!\diagdown & | & \diagup \\
& \text{C} &
\end{array}
\qquad
\begin{array}{ccc}
& \text{CO} & \\
& \diagup\diagdown & \\
\text{HC} & & \text{CH} \\
|| & & || \\
\text{HC} & & \text{CH} \\
& \diagdown\diagup & \\
& \text{CO} &
\end{array}
$$

According to the former quinone is a peroxide, and this is the view which until within the last few years was almost universally adopted. It readily explains the conversion of quinone into quinol and *vice versâ*, and is also supported by the close relation between quinonechlorimide and the indophenols (described in a later volume). On the other hand, it does not explain so satisfactorily the formation of quinone dioxime from quinone and hydroxylamine, a reaction which is characteristic of all ketones (Part II. p. 14) and is therefore in favour of the second formula. The ketone formula likewise most readily explains the formation of β-trichloracetylacrylic acid from quinone and chlorous acid, and also the formation of bromanil from succinosuccinic acid. A recent synthesis of xyloquinone from diacetyl by treatment with caustic soda (Part II. p. 185) can be readily understood if the ketone formula be assumed, as shown by the following equations:

$$
(1)\quad
\begin{array}{c}
\text{CH}_3.\text{CO.CO.CH}_3 \\
\\
\text{CH}_3.\text{CO.CO.CH}_3
\end{array}
=
\begin{array}{c}
\text{CH}_3.\text{C·CO.CH}_3 \\
|| \\
\text{CH.CO.CO.CH}_3
\end{array}
+ \text{H}_2\text{O}
$$

$$
(2)\quad
\begin{array}{c}
\text{CH}_3.\text{C.CO.CH}_3 \\
|| \\
\text{CH.CO.CO.CH}_3
\end{array}
=
\begin{array}{c}
\text{CH}_3.\text{C . CO . CH} \\
|| \qquad || \\
\text{CH.CO.C.CH}_3
\end{array}
+ \text{H}_2\text{O}
$$

whereas, with the peroxide formula no such simple explanation is possible. Further, Nef has lately shown [1] that quinone readily combines with bromine in chloroform solution, forming two additive compounds, $C_6H_4O_2Br_2$, and $C_6H_4O_2Br_4$. The first compound, *quinonedibromide*, crystallizes in flat yellow needles, which melt at 86°. It was first obtained by Sarauw [2] who, however, believed it to be an isomeric dibromoquinol. It readily loses hydrobromic acid on boiling with water, forming bromoquinone, $C_6H_3BrO_2$,

Quinone tetrabromide forms colourless lustrous scales, which become yellow at 110°, melt with evolution of hydrogen bromide at 170—175°, and on boiling with water yield a mixture of *m*- and *p*-dibromoquinone.

Again, when quinone is treated with phosphorus pentachloride or similar reagents, the two atoms of oxygen are replaced by two monovalent atoms, and simple derivatives of benzene are formed. This result would naturally follow from a compound having the first formula, whereas from the second we should expect to obtain derivatives of di-hydrobenzene.

The constitution of quinone must therefore be at present regarded as an open question. Perhaps the most probable conclusion is that free quinone has the second or ketone formula, and that its derivatives exist sometimes in the one form and sometimes in the other. The diketone formula will be employed as a general rule in the sequel.

A number of similar cases of compounds existing in two forms have also been observed. Thus phloroglucinol, which is described below (p. 193), behaves in some of its reactions as *s*-trihydroxybenzene (Fig. A), and in others as triketohexamethylene (Fig. B).

$$
\begin{array}{cc}
\text{C.OH} & \text{CO} \\
\text{HC} \diagup \diagdown \text{CH} & \text{H}_2\text{C} \diagup \diagdown \text{CH}_2 \\
\text{HO.C} \diagdown \diagup \text{C.OH} & \text{OC} \diagdown \diagup \text{CO} \\
\text{CH} & \text{CH}_2 \\
\text{FIG. A.} & \text{FIG. B.}
\end{array}
$$

Among the simplest examples are nitrous acid and cyanic acid, which are extremely unstable in the free state, but yield stable modifications of two kinds:

[1] *J. Pr. Chem.* II. **42**, 182.   [2] *Annalen*, **209**, 107.

Methyl Nitrite.

$$CH_3.O—N{=}O$$

Nitromethane.

$$CH_3—N{<}\begin{smallmatrix}O\\\ \\O\end{smallmatrix}{>}$$

Methyl Cyanate.

$$CH_3—O—C \equiv N$$

Methyl Isocyanate.

$$CH_3—N{=}C{=}O$$

The constitution of phloroglucinol and similar compounds can therefore be represented with equal accuracy by different structural, or as Laar[1] names them, tautomeric formulæ. The atoms are regarded as in continual motion within the molecule, and one form is converted into the other when the light and rapidly moving atoms of hydrogen are more strongly attracted by one or other of the remaining atoms. If, however, the hydrogen be replaced by a heavier atom or radical, the latter no longer escapes from the sphere of attraction, and the mobile form is converted into a stable one.

The oximes of quinone mentioned above will be described after amidophenol, and together with the nitroso-derivatives to which they are closely allied.

TRIHYDROXYBENZENES.

1036 We are acquainted with the three theoretically possible compounds :

Pyrogallol.

Phloroglucinol.

Hydroxyquinol.

Pyrogallol is obtained by the separation of carbon dioxide from gallic acid, $C_6H_2(OH)_3CO_2H$, which, as will be shown later contains the three hydroxyls in the adjacent position. Hydroxyquinol is obtained from quinol, which, as a para-compound, can only yield one trihydroxybenzene. In phloroglucinol, therefore, the hydroxyls must occupy the symmetrical position, which is confirmed by its synthesis and general behaviour.

[1] *Ber.* **18**, 468 ; **19**, 730.

PYROGALLOL, $C_6H_3(OH)_3$.

1037 Scheele, who was the first to obtain pure gallic acid, $C_6H_2(OH)_3CO_2H$, in 1786, showed that on dry distillation it yields a sublimate which, like gallic acid, forms a precipitate with ferrous sulphate. He mentions this as very remarkable,[1] whence it would appear that he considered the bodies to be different. Later chemists, nevertheless, considered this sublimate to be pure gallic acid, until Braconnot, in 1831, showed that sublimed is in reality different from ordinary gallic acid,[2] in which conclusion he was confirmed by Pelouze in 1838.[3] It was then called pyrogallic acid, which name is still frequently employed.

In order to prepare pyrogallol in small quantities, gallic acid, dried at 100°, is placed in a tubular retort and heated in an oil bath at 210—220° but no higher, carbon dioxide being passed through the apparatus; a yield of 30 per cent. is thus obtained. To obtain larger quantities, gallic acid is heated with 2 to 3 parts of water for half an hour in an autoclave to about 210—220°. A paper ring is placed between the vessel and its cover to allow the carbon dioxide to escape gradually. The solution is boiled with animal charcoal, filtered and concentrated, and the pyrogallol which crystallizes out on cooling distilled *in vacuo* to render it perfectly pure.[4] The yield is almost theoretical.

Pyrogallol crystallizes in thin plates, or white lustrous needles, melting at 115° and boiling with slight decomposition at 210°. It is readily soluble in water, alcohol and ether, tastes bitter and is poisonous; 2 to 4 grams given to a dog, kill it, producing the same symptoms as phosphorus poisoning.[5] Its alkaline solution rapidly turns brown and black in the air, and then contains carbonic acid, acetic acid, and black humus-like bodies. Oxygen is most rapidly absorbed by it when a solution of 0·25 grms. of pyrogallol in 10 cc. of caustic potash of specific gravity 1·05[6] is used. Its alkaline solution is therefore used in gas analysis (Liebig), but the fact that carbonic oxide is always given off has to be borne in mind,[7] the quantity of this being

---

[1] *Opusc.* **2**, 226.   [2] *Ann. Chim. Phys.* **46**, 206.
[3] *Ibid.* **54**, 378.
[4] De Luynes and Esperandieu, *Annalen*, **138**, 60.
[5] Personne, *Zeitsch. Chem.* **1869**, 728.
[6] Weyl and Zeitler, *Annalen*, **205**, 264.
[7] Calvert and Clöez, *ibid.* **130**, 248.

larger when pure oxygen is absorbed, than when it is absorbed from the air.[1]

Pyrogallol reduces solutions of gold, mercury, and silver. A perfectly pure solution of a ferrous salt produces a white tubidity in a solution of pyrogallol, but if the smallest trace of a ferric salt be present the liquid is coloured blue. Ferric chloride produces a red colour (see pyrogalloquinone). A very delicate reaction of pyrogallol is, that its aqueous solution is coloured brown by nitrous acid; it can hence be employed as a test for the latter compound.[2] From its property of reducing silver salts, pyrogallol is employed as a developer in photography. Since its price is seven times that of gallic acid, a solution available for dry plate work can be cheaply prepared by heating 10 grms. of gallic acid and 30 grms. of glycerol to 190—200° as long as carbon dioxide is evolved. The theoretical yield is thus obtained. After cooling, the residue is dissolved in 1 litre of water.[3] Pyrogallol is also employed as a hair dye and for different analytical purposes. It is further made use of in the manufacture of some colouring matters, such as galleïn and coëruleïn.

1038 *Pyrogallol dimethyl ether*, $C_6H_3(OCH_3)_2OH$, is found in beechwood creasote, and is formed when one molecule of pyrogallol, two molecules of caustic potash and two molecules of methyl iodide, are dissolved in absolute alcohol and heated to 150—160°. It crystallizes from boiling water in white prisms, melting at 51—52° and boiling at 253°. It forms fine, crystalline salts with the alkalis, which do not become black in the air.[4]

Pyrogallol trimethyl ether, $C_6H_3(OCH_3)_3$, is prepared by heating pyrogallol with more than three molecules of methyl iodide and caustic potash till the mixture has a neutral reaction, adding an excess of alkali, and distilling with steam. It comes over as an oil, which speedily solidifies. It melts at 47°, boils at 235° (corr.), and is insoluble in alkalis.[5]

The ethyl ethers of pyrogallol are obtained when pyrogallol is heated with caustic potash, ethyl iodide and absolute alcohol to 100°.[6] After evaporating off the alcohol the residue is distilled with steam. The triethyl ether comes over together with some

[1] Boussingault, *Annalen*, **130**, 249.

[2] Schönbein, *Zeitschr. Analyt. Chem.* **1**, 319.

[3] Thorpe, *Chem. News*, **43**, 109.

[4] Hofmann, *Ber.* **11**, 333.

[5] Will, *ibid.* **21**, 607.

[6] Benedikt, *ibid.* **9**, 125; Hofmann, *ibid.* **11**, 798.

diethyl ether, and is separated from it by means of caustic potash, in which the first compound is insoluble. In order to separate the residual mono-ethyl ether from any adhering diethyl ether, the latter is extracted with cold benzene.[1]

Pyrogallol mono-ethyl ether, $C_6H_3(OC_2H_5)(OH)_2$, is tolerably soluble in cold, readily in hot water and in alcohol, but very slightly in cold benzene. It crystallizes in needles melting at 95°. Its aqueous solution is coloured bluish violet by ferrous sulphate, and its alkaline solution turns brown in the air.

Pyrogallol diethyl ether, $C_6H_3(OC_2H_5)_2(OH)$, is readily soluble in cold benzene, slightly in cold dilute alcohol, and readily in hot. It forms crystals melting at 79° and boiling at 262°, and its alkaline solution is not coloured brown in the air.

Pyrogallol triethyl ether, $C_6H_3(OC_2H_5)_3$, crystallizes from alcohol in fine needles, melting at 39° and boiling at about 250°.

Triacetopyrogallol, $C_6H_3(OC_2H_3O)_3$, is obtained by the action of acetyl chloride on pyrogallol, and forms crystals which can be sublimed and are almost insoluble in water.[2]

Purpurogallin, $C_{20}H_{16}O_9$. Girard obtained this compound by oxidizing pyrogallol with silver nitrate or potassium permanganate and sulphuric acid.[3] According to Wichelhaus, who prepared it by the action of chromic acid solution on pyrogallol, it has the formula $C_{18}H_{14}O_9$.[4] He also obtained a compound by the action of quinone on pyrogallol, and termed it pyrogalloquinone, which has been proved to be identical with purpurogallin.[5] This body was then observed by Struve in his investigations on the action of nascent oxygen on pyrogallol. He found that it is formed when a dilute solution of pyrogallol is brought into contact with saliva, malt extract, &c., and allowed to stand in presence of air. It is more readily obtained by using a dilute solution of gum arabic,[6] and has also been prepared by exposing a solution of pyrogallol and sodium phosphate to the action of the air.[7] It was then investigated by Clermont and Chatand who confirmed the formula given by Girard.[8]

It is best prepared, according to Nietzki and Steinmann,[9] by treating a solution of pyrogallol containing acetic acid with

---

[1] Benedikt and Weselsky, *Monatsh.* **2**, 212.

[2] Nachbaur, *Annalen,* **107**, 244.   [3] *Ber.* **2**, 562.

[4] *Ibid.* **5**, 848.

[5] Loew, *J. Pr. Chem.* II. **15**, 322 ; Clermont, *Compt. Rend.* **102**, 1072 ; Nietzki and Steinmann, *Ber.* **20**, 1278.

[6] *Annalen,* **163**, 162.   [7] Loew, *loc. cit.*

[8] *Compt. Rend.* **94**, 1189, 1254, 1362.   [9] *Loc. cit.*

sodium nitrate as long as nitrogen is evolved in quantity, and recrystallizing the precipitated substance from acetic acid. It is further formed by oxidizing pyrogallol with potassium ferricyanide, and by the action of potassium nitrite on gallic acid.[1]

Purpurogallin crystallizes from alcohol in brown, velvet-like needles, melting at 256°. It forms a sodium salt, $C_{20}H_{12}Na_4O_9$, which crystallizes with difficulty in deliquescent needles, and gives with barium chloride an almost insoluble precipitate of $C_{20}H_{12}Ba_2O_9$. By heating it with acetic anhydride, the acetate, $C_{20}H_{12}O_9(C_2H_3O)_4$, is obtained, crystallizing from alcohol in lustrous, brown needles, melting at 186°. By heating with concentrated hydriodic acid, purpurogallin is converted into the hydrocarbon $C_{10}H_{14}$, which boils at 195° and undergoes polymerisation.[2] When distilled with zinc dust it gives considerable quantities of naphthalene, and is probably, therefore, a derivative of this hydrocarbon.[3]

Trichloropyrogallol, $C_6Cl_3(OH)_3$, is obtained by passing dry chlorine through a well-cooled mixture of pyrogallol and acetic acid, and crystallizes in fine needles containing 3 molecules of water. On adding baryta water to its ethereal solution a blue colouration is produced.[4]

By the action of chlorine on pyrogallol in acetic acid solution, Stenhouse and Groves obtained two compounds of unknown constitution, $C_{18}H_7Cl_{11}O_{10}$ and $C_{18}H_6Cl_{12}O_{12} + 2H_2O$, to which they gave the names *mairogallol* and *leukogallol* respectively. Mairogallol forms rhombic prisms melting with decomposition at 190°, and insoluble in water, whilst leukogallol separates in crusts of fine needles which melt with decomposition at 104°, and dissolve readily in water and alcohol.[5] Both of these compounds yield trichloropyrogallol on reduction with zinc dust and sulphuric acid.[6]

Tribromopyrogallol, $C_3Br_3(OH)_3$, is formed by triturating a mixture of bromine and pyrogallol, and crystallizes from hot water in lustrous, flat, rhombic needles.[7]

Nitropyrogallol, $C_6H_2(NO_2)(OH)_3 + H_2O$, is obtained by passing nitrogen trioxide through a solution of pyrogallol in ten times its weight of ether until carbon dioxide commences to come off. It crystallizes from boiling water in long, thin,

[1] Hooker, *Ber.* **20**, 3259.   [2] Clermont and Chatand, *loc. cit.*
[3] Nietzki and Steinmann, *loc. cit.*   [4] Webster, *Journ. Chem. Soc.* 1884, **1**, 205.
[5] *Annalen*, **179**, 237.   [6] Hantzsch and Schniter, *Ber.* **20**, 2036.
[7] Hlasiwetz, *Annalen*, **142**, 250.

brownish yellow needles or thick rhombic prisms, which become anhydrous at 100°, and melt with decomposition at 205°. Its solution is coloured deep red by lime water, and green by ferric chloride.

PHLOROGLUCINOL, $C_6H_3(OH)_3 + 2H_2O$.

1039 Hlasiwetz first prepared this body by heating phloretin, $C_{15}H_{14}O_5$, with caustic potash. This substance is thus decomposed into phloretinic acid, $C_6H_4(OH)C_2H_4.CO_2H$, and phloroglucinol, which obtained this name on account of its exceedingly sweet taste.[1] It is also formed from various glucosides, plant-extracts, resins, &c., such as quercetin, maclurin, catechin, kino, gamboge, dragon's blood and others, by fusing them with caustic potash.[2] It is likewise obtained by the action of sodium amalgam or caustic potash on naringin[3] and on morin,[4] $C_{12}H_8O_5$, which occurs together with maclurin, $C_{13}H_{10}O_6$, in fustic. The latter compound, which is obtained in the impure state as refuse in the preparation of fustic extract, is well adapted for the preparation of phloroglucinol.[5] It is fused with 3 parts of caustic potash and some water until the mass becomes pulpy. After cooling it is dissolved in water, acidified with sulphuric acid, and extracted with ether; on distilling off the latter, the residue is dissolved in water and lead acetate added to precipitate the protocatechuic acid. The solution is then treated with hydrogen sulphide, and the phloroglucinol obtained from the filtrate either by evaporation or by extraction with ether (Hlasiwetz and Pfaundler).

It can also be advantageously prepared from resorcinol by fusing this with a tolerably large excess of caustic soda, until the violent evolution of gas, which commences after some time, moderates, and the mass has become light chocolate-coloured. The solution is acidified with sulphuric acid, extracted with ether, the phloroglucinol obtained by evaporation, and recrystallized.[6] According to Tiemann and Will, the adhering resorcinol may be removed by heating the phloroglucinol to 100° and moistening at intervals, the resorcinol being thus sublimed. They also

---

[1] *Annalen*, **96**, 118.
[2] Hlasiwetz, *ibid.* **112**, 98 ; **134**, 118 ; **138**, 190 ; Hlasiwetz and Pfaundler, **127**, 357 ; Hlasiwetz and Barth, **134**, 283 ; **138**, 68.
[3] Will, *Ber.* **18**, 1322.   [4] Hlasiwetz, *Annalen*, **143**, 297.
[5] Benedikt, *ibid.* **185**, 114.   [6] Barth and Schreder, *Ber.* **12**, 503.

noticed the curious facts that ether dissolves phloroglucinol more readily from a neutral than from an acid solution, and that if the solution be saturated with common salt, a large portion of the phloroglucinol separates out.[1] Phloroglucinol is also obtained by fusing benzenetrisulphonic acid with an excess of caustic soda.[2]

Baeyer has also succeeded in obtaining phloroglucinol synthetically by heating ethyl malonate and ethyl sodiomalonate together at 145°. The sodium compound of ethylphloroglucinol-tricarboxylate is formed, and yields the free ethyl salt on treatment with dilute acids. (See Part **V.** p. 143). This has the constitution

$$CO_2C_2H_5.HC \begin{array}{c} CO \\ \diagup \diagdown \\ \diagdown \diagup \\ CO \end{array} CH.CO_2C_2H_5$$
$$\underset{CH.CO_2C_2H_5}{CO}$$

and yields phloroglucinol on fusion with caustic potash.[3] From this it would appear that the latter has the constitution

$$H_2C \begin{array}{c} CO \\ \diagup \diagdown \\ \diagdown \diagup \\ OC \end{array} CH_2$$
$$\underset{CH_2}{CO}$$

and is therefore *triketohexamethylene*. In order to ascertain whether this really is the case, Baeyer treated phloroglucinol with hydroxylamine, and obtained the *trioxime* $C_6H_6(NOH)_3$ as a sandy crystalline powder which blackens at 140°, and explodes violently at 155°. Phloroglucinol therefore behaves towards hydroxylamine as a triketone, whereas in a large number of reactions it behaves as a phenol. Reference has already been made to this point under quinone (p. 187), and as there explained, it is supposed that phloroglucinol and other similar compounds can exist in two *tautomeric* forms :

$$\begin{array}{c} C.OH \\ \diagup \diagdown \\ HC \qquad CH \\ \| \qquad | \\ HO.C \qquad C.OH \\ \diagdown \diagup \\ CH \end{array} \quad \text{and} \quad \begin{array}{c} CO \\ \diagup \diagdown \\ H_2C \qquad CH_2 \\ | \qquad | \\ OC \qquad CO \\ \diagdown \diagup \\ CH_2 \end{array}$$

---

[1] *Ber.* **14**, 954.　　　　　　　[2] *Ibid.* **12**, 422
[3] Baeyer, *ibid.* **18**, 3454 ; **19**, 159.

Phloroglucinol is readily soluble in water, alcohol, and ether, and crystallizes in small plates or rhombic tablets, which lose their water of crystallization at 100°, and melt at 217—218°. At a higher temperature it partly sublimes without decomposition. It reduces Fehling's solution; dilute nitric acid converts it into nitrophloroglucinol, while the concentrated acid oxidizes it to oxalic acid. Aqueous ammonia converts it into phloramine, $C_6H_3(OH)_2NH_2$, which will be subsequently described. This reaction is very remarkable, since the other phenols only exchange a hydroxyl group for an amido-group with difficulty. Its aqueous solution is coloured a bluish violet by ferric chloride. As already stated, it is, in presence of hydrochloric acid, a delicate reagent for woody substances, which produce with it a reddish violet colouration (Vol. III. Pt. II. p. 697). A solution containing 0·01 per cent. of phloroglucinol colours pieces of pine wood, moistened with hydrochloric acid, a distinct red, and even when only 0·001 per cent. is present, the reaction takes place if the wood be allowed to remain in the solution for twenty-four hours. In this way a lignification of the tissue can be recognized in the very earliest stages of vegetable growth. Phloroglucinol is only a weak antiseptic, and, unlike its isomeride pyrogallol, is not poisonous.[1]

By mixing a very dilute aqueous solution of phloroglucinol and aniline nitrate, or toluidine nitrate, with potassium nitrite, a vermilion-coloured precipitate of azobenzene-phloroglucinol or azotoluene-phloroglucinol,[2] separates out after a short time.

Phloroglucinol does not react with phenylhydrazine in the same manner as with hydroxylamine, but appears to behave towards this reagent as a phenol. The first product is a salt $C_6H_6O_3+3C_6H_5.N_2H_3$, which is colourless and odourless, and melts at 78—83°. By the further action of phenylhydrazine in alcoholic solution, a compound is obtained which has the constitution

NH.NH.C_6H_5

H   H

HO   NH.NH.C_6H_5.

H

---

[1] Andreer, *Ber.* **17**, 334 R.

[2] Weselsky, *ibid.* **9**, 216 ; Weselsky and Benedikt, **12**, 226.

as it is converted by ferric chloride into an azo-compound, and also yields a pentabenzoyl derivative. Attempts to introduce a third phenylhydrazine residue were without success.[1]

1040 *Phloroglucinol trimethyl ether,* $C_6H_3(OCH_3)_3$, is prepared by passing hydrogen chloride into a solution of phloroglucinol in methyl alcohol, and treating the dimethyl ether thus obtained with caustic potash and methyl iodide. It forms colourless crystals, which melt at $52°·5$, and are insoluble in water and alkalis, but dissolve readily in alcohol, ether, and benzene.[2]

If greater quantities of methyl iodide and caustic potash are employed, higher methyl derivatives are obtained. These must be regarded as derived from triketohexamethylene.[3] The hex-

methyl ether, $(CH_3)_2$

crystallizes from ether in acute prisms, which melt at $80°$.

Phloroglucinol diethyl ether, $C_6H_3(OC_2H_5)_2OH$, is obtained by saturating an alcoholic solution of phloroglucinol with hydrochloric acid. It crystallizes from hot water in long, lustrous, snow-white crystals, melting at $75°$.

Phloroglucinol triethyl ether, $C_6H_3(OC_2H_5)_3$, is obtained when the preceding compound is heated with caustic potash, ethyl iodide, and alcohol. Water precipitates it from its alcoholic solution in fine crystals melting at $43°$.[4]

Higher ethyl derivatives can also be obtained, corresponding to the higher methyl compounds. Hexethylphloroglucinol $C_6O_3(C_2H_5)_6$, is an oil, insoluble in alkalis.

Triacetophloroglucinol, $C_6H_3(OC_2H_3O)_3$, is formed by the action of acetyl chloride on phloroglucinol, and crystallizes from alcohol in small prisms, insoluble in water.[5]

Trichlorophloroglucinol, $C_6Cl_3(OH)_3 + 3H_2O$, is formed by the action of chlorine on phloroglucinol in acetic acid or carbon tetrachloride solution. Even when only the theoretical quantity of chlorine is used, some hexchlorotriketohexamethylene is always formed, and it is therefore preferable to prepare it by reducing

---

[1] Baeyer and Kochendorfer, *Ber.* **22**, 2189.
[2] Will, *ibid.* **21**, 603.
[3] Margulies, *Monatsh.* **9**, 1045 ; **10**, 459.
[4] Will and Albrecht, *Ber.* **17**, 2107.
[5] Hlasiwetz, *Annalen,* **119**, 201.

this latter with stannous chloride.[1] It crystallizes from warm water or alcohol in fine needles, which lose their water of crystallization over sulphuric acid, and then melt at 136°, and sublime below this temperature.

Hexchlorotriketohexamethylene, $C_6Cl_6O_3$, is the final product of the action of chlorine on phloroglucinol in chloroform solution. It forms a crystalline mass, which has a penetrating smell, melts at 48°, boils at 268—269°, and is readily soluble in ether, benzene, chloroform, and carbon disulphide. Its constitution is represented by the formula $\begin{vmatrix} CCl_2 . CO . CCl_2 \\ CO—CCl_2—CO \end{vmatrix}$ It is decomposed by water with formation of dichloracetic acid, tetrachloracetone, and carbon dioxide. It is probable that intermediate products having the formulæ $CCl_2H . CO . CCl_2 . CO . CCl_2 . COOH$ and $CCl_2H . CO . CCl_2 . CO . CCl_2H$, are first formed, as if the compound be treated simultaneously with chlorine and water, *octochloracetylactone,* $CCl_3 . CO . CCl_2 . CO . CCl_3$, is obtained. The formation of this compound is explained as follows:

$$\begin{vmatrix} CO.CCl_2.CO \\ CCl_2.CO.CCl_2 \end{vmatrix} + \begin{vmatrix} Cl \\ OH \end{vmatrix} = \begin{vmatrix} CO.CCl_2.COOH \\ CCl_2.CO.CCl_3 \end{vmatrix}$$

$$\begin{vmatrix} CO.CCl_2.COOH \\ CCl_2.CO.CCl_3 \end{vmatrix} + ClOH = \begin{vmatrix} CO.CCl_3 \\ CCl_2.CO.CCl_3 \end{vmatrix} + HO.COOH.$$

Hexchlorotriketohexamethylene is converted by methyl alcohol into tetrachloracetone and methyl dichloromalonate, whilst with ammonia it yields three molecules of dichloracetamide. The latter reaction is a further proof of the symmetric formula of the hexchloro-compound, and of the phloroglucinol from which it is prepared.[2]

Tribromophloroglucinol, $C_6Br_3(OH)_3 + 3H_2O$, is obtained by the action of bromine water on phloroglucinol, or by treating phloroglucinol in acetic acid solution with bromine,[3] and crystallizes in prisms.[4] Cold nitric acid converts it into tribromodinitropropionic acid:[5]

$$C_6H_3Br_3O_3 + 6NO_2.OH = C_3HBr_3(NO_2)_2O_2 + 3CO_2 + 2NO + N_2O_3 + 4H_2O.$$

[1] Zincke and Küster, *Ber.* **22**, 1473. [2] Zincke and Kegel, *ibid.* **23**, 230.
[3] Herzig, *Monatsh.* **6**, 885. [4] Hlasiwetz, *Jahresb.* **1855**, 702.
[5] Benedikt, *Annalen,* **184**, 255.

By gradually adding 10 parts of bromine to an aqueous solution of 1 part of phloroglucinol, Benedikt and Hazura [1] obtained a compound which they termed *phlorobromine*, and gave it the formula C_6Br_9HO. It has however been found that it is in reality *perbromacetylacetone*, $C_5Br_8O_2$, and that the first product of the reaction is *hexbromophloroglucinol* $C_6Br_6O_3$, which by the further action of bromine water yields phlorobromine.[2]

Nitrophloroglucinol, $C_6H_2(NO_2)(OH)_3$, is obtained by dissolving phloroglucinol in warm dilute nitric acid, from which it separates in reddish yellow scales, dissolving in hot water and crystallizing in plates.

Trinitrophloroglucinol, $C_6(NO_2)_3(OH)_3$, is obtained by very gradually adding the potassium salt of trinitrosophloroglucinol, which is described later on, to a mixture of equal parts of concentrated sulphuric acid and nitric acid of specific gravity 1·40, the mixture being continuously shaken. Pieces about as large as a pea kindle on the surface of the liquid and swim about as glowing balls with a hissing noise until they are dissolved. If larger pieces are added, violent explosions may occur.

Trinitrophloroglucinol crystallizes from hot water in yellow hexagonal prisms containing a molecule of water which they lose at 100°. It commences to sublime at 130°, melts at 158°, and explodes when more strongly heated. Like picric acid it dyes wool and silk, but produces fuller and more beautiful tones. It gives the isopurpuric acid reaction with potassium cyanide, and decomposes carbonates forming three series of explosive salts; of these the most remarkable are as follows:

(1) $C_6(NO_2)_3(OK)_3$ forms orange-red needles, often an inch long, having a diamond-like lustre.

(2) $C_6(NO_2)_3(OK)_2OH$ occurs in deep yellow crystals, which have less lustre than those of the normal salt.

(3) $C_6(NO_2)_3OK(OH)_2 + H_2O$ crystallizes in long, very slender, sulphur-yellow needles with a silky lustre.

As already stated, phloroglucinol is also obtained ·by fusing phenol with caustic soda. According to Gautier the isomeric phenoglucinol, $C_6H_3(OH)_3 + 2H_2O$, is obtained in this way. This crystallizes from water in large prisms melting at 200°·5, possesses a very sweet taste, but only gives a faint violet colour with ferric chloride. According to him, the trihydroxybenzene obtained by fusion of quercitol, is also different from phloro-

[1] *Monatsh.* **6,** 702.   [2] *Ber.* **23,** 1706.

glucinol, and he has named it "querciglucin," $3C_6H_3(OH)_3 +$ $2H_2O$. It loses its water and melts at 174°, and has a slightly sweet taste. This substance has been found to be simply impure phloroglucinol.[1] Finally he obtained "oenoglucin" by fusing the colouring matter of red wine from Carignane with potash. This substance is very similar to phloroglucinol, but is only coloured a light violet by ferric chloride.[2]

HYDROXYQUINOL, $C_6H_3(OH)_3$.

1041· This compound is obtained, together with diquinol $C_{12}H_6(OH)_4$, and hexhydroxydiphenyl, $C_{12}H_4(OH)_6$, by fusing quinol with ten times its quantity of caustic soda until the evolution of hydrogen ceases. It crystallizes from ether in microscopic monosymmetric plates or tablets, melting at 140°·5. Its aqueous solution rapidly becomes coloured in the air, and even more rapidly when a drop of a solution of an alkali is added; it is in this respect as sensitive as a solution of catechol. When applied to the skin it produces a brown colouration, which after some hours turns black. By adding a little ferric chloride to the dilute solution, a brownish green colouration is produced which rapidly disappears; on addition of a little carbonate of soda it becomes a fine dark-blue, and on addition of more, wine-red. Ferrous sulphate gives no colouration, but after the addition of a little carbonate of soda the solution turns violet, and on a further addition becomes intensely blue. Hydroxyquinol dissolves in sulphuric acid forming a green solution, which gradually turns violet, and on heating cherry-red.

When hydroxyquinol undergoes dry distillation, part is volatilized undecomposed, and the remainder is reduced to quinol, the mass becoming charred.[3]

Hydroxyquinol methyl ether. $C_6H_3(OCH_3)(OH)_2$, is prepared by the reduction of methoxyquinone with sulphurous acid, and purified by recrystallization or sublimation. It forms colourless plates which melt at 84°.

Hydroxyquinol trimethyl ether, $C_6H_3(OCH_3)_3$, is obtained from the foregoing compound by treatment with caustic potash and methyl iodide.[4] It is a colourless oil, which boils at 247° (*corr.*)

[1] *Journ. Chem. Soc.* **1886**, 1232.
[2] *Bull. Soc. Chim.* **33**, 583.
[3] Barth and Schreder, *Monatsh.* **4**, 176; **5**, 489.
[4] Will and Schweizer, *Ber.* **21**, 603.

and is converted by nitric acid into a dinitro-compound C_6H $(NO_2)_2(OCH_3)_3$.

Hydroxyquinol ethyl ether, $C_6H_3(OC_2H_5)(OH)_2$, and *hydroxyquinol triethylether*, $C_6H_3(OC_2H_5)_3$, are obtained in a similar manner to the methyl ethers.[1] The former crystallizes in colourless prisms, melting at 112·5°, and the latter in long white lustrous needles, melting at 34°.

Triacetohydroxyquinol, $C_6H_3(OC_2H_3O)_3$, is obtained by heating hydroxyquinol with acetic anhydride, and anhydrous sodium acetate. It crystallizes from alcohol in fine white bushy needles, melting at 96·5°.

1042 *Hydroxyquinone* $C_6H_3(OH)O_2$, is not known in the free state, but derivatives of it have been obtained.

Methoxyquinone, $C_6H_3(OCH_3)O_2$ is prepared from *o*-anisidine $C_6H_4(OCH_3)NH_2$, by oxidation with chromic acid. It forms slender yellow needles, which melt at 140°, sublime readily, and are easily soluble in alcohol, less in water.[2]

Ethoxyquinone, $C_6H_3(OC_2H_5)O_2$, is formed in a similar manner by the oxidation of amidodiethylresorcinol. It forms long yellow lustrous needles, which melt at 117°, and sublime at 60— 70°. Both of these compounds have the characteristic quinone smell, but it is not so penetrating as in the case of benzoquinone.[3]

Hydroxyquinhydrone, $C_6H_3(OH)_3 + C_6H_3(OH)O_2$, is prepared by the addition of nitric acid to a cold aqueous solution of hydroxyquinol, and forms small dark greyish blue crystals, with a slight surface lustre.[4]

Tribromohydroxyquinone, $C_6Br_3(OH)O_2$, is obtained by triturating a mixture of hydroxyquinol and dry bromine, and separates from chloroform in splendid orange red crystalline granules.[5]

[1] Will and Pukall, *Ber.* **20**, 1133.
[2] Mühlhäuser, *Annalen*, **207**, 251 ; Will and Schweizer, *Ber.* **21**, 605.
[3] Will and Pukall, *ibid.* **20**, 1131.
[4] Barth and Schreder, *Monatsh.* **5**, 595.   [5] *Ibid.* p. 593.

1043 Three compounds of this constitution are theoretically possible, viz.

$$
\begin{array}{ccc}
\text{OH} & \text{OH} & \text{OH} \\
\text{OH} & \text{OH} & \\
\text{OH} & \text{HO} \quad \text{OH} & \text{HO} \quad \text{OH} \\
\text{OH} & \text{OH} & \text{OH}
\end{array}
$$

One only is at present known in the free state, and corresponds to the second of the above formulæ, but derivatives of the third compound have also been obtained.

Symmetric Tetrahydroxybenzene, $C_6H_2(OH)_4$.

This phenol is prepared by the reduction of the corresponding symmetric dihydroxyquinone, with an acid solution of stannous chloride.[1]

It crystallizes in long thin needles, which melt at 215—220° are very soluble in water, alcohol and ether, less so in hydrochloric and acetic acids, and separate from the latter solution in greyish plates. Its aqueous solution becomes brown on standing in the air, and its alkaline solution readily passes into the corresponding dihydroxyquinone salt. It forms a *tetracetyl* compound $C_6H_2(OC_2H_3O)_4$, which crystallizes in colourless rhombic tablets, melting at 217°.

s-Dimethoxydihydroxybenzene, $C_6H_2(OCH_3)_2(OH)_2$ (1.4.2.5) is prepared by the reduction of the s-dimethoxyquinone described below, and forms colourless lustrous needles, melting at 166°.[2]

s-Diethoxydihydroxybenzene, $C_6H_2(OC_2H_5)_2(OH)_2$ (1.4.2.5), is obtained in a similar manner by the reduction of s-diethoxyquinone.[3] It crystallizes from water in beautiful needles melting at 138°, which on treatment with sodium ethylate and ethyl iodide are converted into *s-tetraethoxybenzene*, $C_6H_2(OC_2H_5)_4$. This forms colourless lustrous plates melting at 143°.

---

[1] Nietzki and Schmidt, *Ber.* **21**, 2377 ; Böniger, *ibid.* **22**, 1288.
[2] Nietzki and Rechberg, *ibid.* **23**, 1217.   [3] *Ibid.* p. 1214.

Hydrochloranilic acid or, *s-Dichlorotetrahydroxybenzene*, C_6Cl_2 $(OH)_4$, is obtained when chloranilic acid is heated with a concentrated solution of sulphurous acid to 100° in a sealed tube (Koch, Gräbe). On cooling, it separates out in long needles. In the moist state and in alkaline solution it is oxidized by the air forming chloranilic acid.

s-Dihydroxyquinone, $C_6H_2(OH)_2O_2$. This compound is obtained from Typke's isodiamidoresorcinol (described later) by treatment with oxidizing agents, which convert it into diimidoresorcinol, and heating this compound with 10 parts of 10 per cent. caustic potash or soda solution to 70°. When a portion of the solution yields a light yellow precipitate on acidifying, and the solution assumes a like colour, excess of alkali is added. The alkali salt of dihydroxyquinone is thus precipitated and converted by acids into free dihydroxyquinone.[1] This may also be obtained by the hydrolysis of ethyl dihydroxyquinonedicarboxylate,[2] but the yield is less satisfactory.

Dihydroxyquinone forms dark yellow zig-zag needles which are insoluble in cold water. The impure compound is decomposed on boiling with water, alcohol and acetic acid, but the pure substance may be recrystallized from these liquids without change. It also crystallizes from ethyl acetate in dark yellow needles, which have a bluish surface lustre. It does not melt, but may be readily sublimed, unites with hydroxylamine forming a dioxime, and yields a *mono-* and *dimethyl ether*. The latter melts at 220° and gives no colouration with sulphuric acid, and is therefore not identical with the *as*-dimethoxyquinone described below. The diamidoresorcinol from which it is obtained is different from the *adj*-diamidoresorcinol, so that the dihydroxyquinone obtained from it cannot correspond to *adj*-tetrahydroxybenzene. There only remains the symmetric constitution, shown in the left-hand formula given below, which is confirmed by the fact that it is converted by nitric acid into nitranilic acid, which as will be shown later, has the formula also shown below.

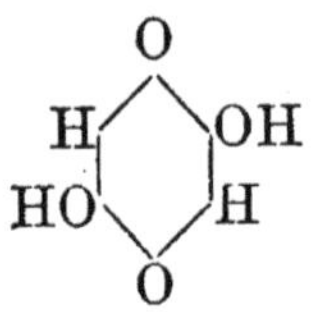

s-Dihydroxyquinone.

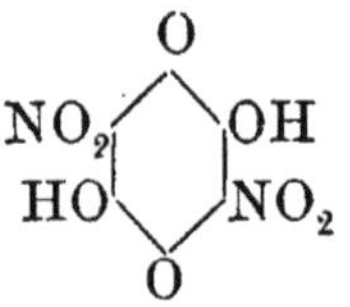

Nitranilic acid.

[1] Nietzki and Schmidt, *Ber.* **21**, 2374.
[2] Loewy, *ibid.* **19**, 2387 ; Böniger, *ibid.* **21**, 1288.

Bromine also converts it into bromanilic acid. It also follows from this that Typke's isodiamidoresorcinol must likewise have the symmetric constitution.

The *potassium* and *sodium* salts resemble one another very closely, and form red stellate groups of needles, which are readily soluble in water, but with difficulty in excess of alkali and in alcohol. The *barium* salt forms blue-black needles, which when dried at 100° have the composition $C_6H_2BaO_4 + H_2O$, and the *silver* salt is an insoluble brown precipitate.

s-Dimethoxyquinone, $C_6H_2(OCH_3)_2O_2$ (1.4.2.5), is obtained by the oxidation of s-diamido-*p*-dimethoxybenzene, and also by the action of methyl iodide on the silver salt of s-dihydroxyquinone. It decomposes with partial carbonization at 220°.

s-Diethoxyquinone, $C_6H_2(OC_2H_5)_2O_2$ (1.4.2.5) is prepared in a similar manner by the oxidation of s-diamido-*p*-diethoxybenzene. It forms sulphur yellow plates, which melt at 183°, and decompose at a higher temperature, a smell somewhat similar to that of vanillin being then observed.[1]

1044 *Chloranilic acid* or *s-Dichlorodihydroxyquinone*, C_6Cl_2 $(OH)_2O_2 + H_2O$. The potassium salt of this acid was first obtained by Erdmann[2] by dissolving chloranil in dilute caustic potash :

$$C_6Cl_4O_2 + 4KOH = C_6Cl_2(OK)_2O_2 + 2KCl + 2H_2O.$$

Gräbe[3] obtained it as follows from trichloroquinone :

$$2C_6HCl_3O_2 + 3KOH = C_6Cl_2(OK)_2O_2 + C_6HCl_3(OH)_2 + KCl + H_2O.$$

In order to prepare it 5 parts of chloranil are moistened with alcohol and a cold solution of 6 parts of caustic potash in 100 parts of water added. When the chloranil is completely dissolved, the potassium salt formed is precipitated by the addition of 10 to 15 parts of common salt. The precipitate is re-dissolved in boiling water, and repeatedly purified by precipitation with common salt. It is then dissolved in 100 parts of boiling water, and 10 parts of hydrochloric acid added to precipitate the free chloranilic acid.[4] It is thus obtained in red crystalline grains, or if it separates slowly in yellowish-red lustrous plates. It loses its

---

[1] Nietzki and Rechberg, *Ber.* **23**, 1213.   [2] *Annalen*, **48**, 315.
[3] *Ibid.* **146**, 24.   [4] Stenhouse, *Journ. Chem. Soc.* **23**, 6.

water of crystallization at 115° and sublimes, undergoing considerable decomposition, at a higher temperature. It dissolves in water, forming a violet-red solution ; on addition of hydrochloric or sulphuric acid the solution is decolourized, most of the acid being precipitated.

When chloranilic acid is treated with only a small quantity of sulphurous acid, the compound, $C_6Cl_2(OH)_2O_2 + C_6Cl_2(OH)_4$, is obtained, which crystallizes in fine, black needles, and belongs to the class of the quinhydrones.

Potassium chloranilate, $C_6Cl_2(OK)_2O_2 + H_2O$, crystallizes in purple prisms or needles, readily soluble in hot water, less so in cold, and still less so in water containing free alkali or common salt. On warming it with phosphorus pentachloride chloranil is formed.[1]

Sodium chloranilate, $C_6Cl_2(ONa)_2O_2 + 4H_2O$, forms dark carmine-red needles. Like the potassium salt it dissolves in water with an intense violet colour, and detonates on heating.

The *ammonium* salt resembles the potassium salt ; the *barium* salt is a rust-coloured, and the *silver* salt a reddish-brown precipitate.

Ethyl chloranilate, $C_6Cl_2(OC_2H_5)_2O_2$, is obtained by the action of ethyl iodide on the silver salt, and crystallizes from alcohol in light red, flat prisms, melting at 107° (Stenhouse).

Bromanilic acid or *s-Dibromodihydroxyquinone*, $C_6Br_2(OH)_2O_2$, is prepared in a similar manner to chloranilic acid from bromanil,[2] and also by the action of concentrated caustic soda solution on dibromo- or tribromoquinone,[3] and by the action of bromine water on a solution of sodium dihydroxyquinone-*p*-dicarboxylate in warm concentrated hydrobromic acid.[4] It forms reddish monosymmetric scales, which have a metallic lustre, form a violet solution in water and alcohol, and sublime on careful heating. Its salts resemble those of chloranilic acid.

The formation of bromanilic acid from dihydroxyquinone-*p*-dicarboxylic acid shows that the bromine atoms must occupy the para-position, and that the constitution of the acid is represented by the formula,

$$
\begin{array}{c}
\text{O} \\
\text{Br}\diagup\diagdown\text{OH} \\
\text{HO}\diagdown\diagup\text{Br} \\
\text{O}
\end{array}\;.
$$

[1] Koch, *Zeitschr. Chem.* **1868**, 202.   [2] Stenhouse, *Journ. Chem. Soc.* **23**, 12.
[3] Saurauw, *Annalen*, **209**, 115.   [4] Hantzsch, *Ber.* **20**, 1303.

Chloranilic acid, which corresponds in all its reactions to this compound must have a similar constitution.[1]

By the action of halogens, chlor- and bromanilic acids are converted into derivatives of pentamethylene, and finally into halogen derivatives of acetone and diacetyl.[2]

1045 *Nitrodihydroxyquinone*, $C_6H(NO_2)(OH)_2O_2$. The potassium salt of this compound is prepared by treating nitrodiimidoresorcinol with aqueous caustic potash, and forms orange-yellow stellate groups of needles. The free nitrodihydroxyquinone is obtained by acting on the barium salt with dilute sulphuric acid.[3] It crystallizes in golden-yellow needles, which are readily soluble in water and have the constitution,

$$\underset{O}{\overset{O}{\underset{HO\diagdown\diagup OH}{H\diagup\diagdown NO_2}}}$$

Nitranilic acid or *s-Dinitrodihydroxyquinone*, $C_6(NO_2)_2(OH)_2O_2$ is obtained by passing nitrogen trioxide into an ethereal solution of quinol kept cool by ice, or by treating diacetylquinol with sulphuric and nitric acids.[4] It is readily prepared by the action of sodium nitrite on chloranil in the following manner: 10 grams of finely powdered chloranil are moistened with a little alcohol, and carefully warmed with a concentrated solution of 20 grams of sodium nitrite. After six hours the mass is extracted with hot water and filtered from unaltered choranil, and the clear solution allowed to cool, when sodium nitranilate separates out, and is converted into the free acid by means of nitric acid.[5] It is further formed by the action of fuming nitric acid on quinoldicarboxylic acid,[6] and by the action of the same reagent on the acid sodium salt of dihydroxyquinone-p-dicarboxylic acid,[7] or by passing nitrogen trioxide into an ethereal solution of the free acid.[8]

Nitranilic acid crystallizes in long golden-yellow prisms, containing water, which is given off at 100°; the anhydrous acid detonates at about 170°. It is readily soluble in water and

[1] Hantzsch, *Ber.* **20**, 1303.

[2] Hantzsch, *ibid.* **21**, 2421 ; **22**, 2841 ; Levy and Jedlicka, *Annalen*, **249**, 66.

[3] Nietzki and Schmidt, *Ber.* **22**, 1661.

[4] Nietzki, *Annalen*, **215**, 138 ; *Ber.* **16**, 2092 ; Nietzki and Benckiser, *ibid.* **18**, 499.

[5] Nef, *Am. Chem. Journ.* **11**, 17.

[6] Herrmann, *Annalen*, **211**, 342.

[7] Hantzsch, *Ber.* **19**, 2399.

[8] Levy, *ibid.* **19**, 2385.

alcohol, insoluble in ether, the aqueous solution decomposing on standing with formation of oxalic and hydrocyanic acids. It has an acid and astringent taste.

The formation of nitranilic acid from dihydroxyquinone p-dicarboxylic acid shows that, as in the case of chlor- and bromanilic acids, the nitro-groups must occupy the para-position, and that its constitutional formula is

$$
\begin{array}{c}
\mathrm{O} \\
\mathrm{NO_2} \diagup \diagdown \mathrm{OH} \\
\mathrm{OH} \diagdown \diagup \mathrm{NO_2} \\
\mathrm{O}
\end{array}
$$

This is further confirmed by the fact that it yields on reduction a diamidotetrahydroxybenzene, which on distillation with zinc dust is converted into p-diamidobenzene.[1]

Potassium nitranilate, $C_6(NO_2)_2(OK)_2O_2$, forms light yellow needles with a bluish surface lustre, which are with difficulty soluble in cold, readily in hot water.

Sodium nitranilate, $C_6(NO_2)_2(ONa)_2O_2$, crystallizes in dark red monosymmetric scaly aggregates, which are strongly dichroic.[2]

Barium nitranilate, $C_6(NO_2)_2(O_2Ba)O_2$, forms small golden-yellow plates, which are insoluble in water. Ethereal salts of this acid do not appear to exist.

β-Dinitrodihydroxyquinone, $C_6(NO_2)_2(OH)_2O_2$, is obtained in small quantity mixed with other bodies, such as dihydroxy-tartaric acid, when nitrogen trioxide is passed through an ice-cold ethereal solution of protocatechuic acid, $C_6H_3(OH)_2COOH$. It is readily soluble in water and slightly in ether, and crystallizes in long greenish needles, which are extremely unstable. Its sodium salt is slightly soluble in cold, readily in hot water, and forms brass-coloured spangles, which have a metallic lustre, and detonate violently on heating.[3]

ASYMMETRIC TETRAHYDROXYBENZENE, $C_6H_2(OH)_4$.

1046 The free phenol is at present unknown. Its *dimethyl ether* was prepared by Hofmann by the reduction of the corresponding quinone, described below. It crystallizes in needles which melt at 160° and readily pass back into the quinone. It appears to form two different acetyl derivatives, one of which

[1] Nietzki, *Ber.* **19**, 2727.   [2] Muthmann, *ibid.* **20**, 2029.
[3] Gruber, *ibid.* **12**, 519.

melts at 132—133°, and gives a green colouration with sulphuric acid, whilst the other melts at 128° and gives no such colouration.[1]

as-Tetramethoxybenzene, $C_6H_2(OCH_3)_4$, is prepared by the action of caustic potash and methyl iodide on the foregoing compound.[2] It crystallizes from ether in lustrous plates, which melt at 47°, and boil at 271°. It is soluble in the common solvents with the exception of water, and readily reacts with bromine forming *Dibromotetramethoxybenzene*, $C_6Br_2(OCH_3)_4$, melting at 76°.

as-Dimethoxyquinone, $C_6H_2(OCH_3)_2O_2$, is prepared by the oxidation of propylpyrogallol dimethyl ether[3] and of pyrogallol trimethyl ether.[4] It is sparingly soluble in alcohol, ether and hot water, readily in acetic acid, and crystallizes from the last-named solution in thick golden prisms, which melt at 249°. As above stated it easily passes into the corresponding quinol on reduction. It unites with bromine, forming *dibromodimethoxyquinone*, $C_6Br_2(OCH_3)_2O_2$, which crystallizes in yellowish red plates melting at 175°.

PENTAHYDROXYBENZENE.

Up to the present time this compound has not been prepared, and only one of its derivatives is known, viz:

1047 *Trihydroxyquinone*, $C_6H(OH)_3O_2$. This is obtained by heating amidodi-imidoresorcinol hydrochloride, $C_6H.NH_2.(NH)_2(OH)_2$, with hydrochloric acid for 2—3 hours at 140—150°. It forms dark scales with a brassy lustre, or an amorphous almost black powder. It is scarcely soluble in the ordinary solvents, but dissolves in alkalis and alkaline carbonates with a brown colour. Its *triacetyl* derivative forms small scales which are very sparingly soluble in alcohol and acetic acid.[5]

1048 *Quercitol*, $C_6H_7(OH)_5$. Although pentahydroxybenzene has not yet been prepared, a substance which is in all probability its hexhydro-derivative, *i.e.* pentahydroxyhexamethylene, or quercitol, has long been known. This compound was

[1] *Ber.* **8**, 67 ; **11**, 332, 333 ; Will, *ibid.* **21**, 608.
[2] *Ibid.* p. 610.   [3] *Ber.* **11**, 332.
[4] Will, *ibid.* **21**, 608, 2020.
[5] Merz and Zelter, *ibid.* **12**, 1040.

discovered by Braconnot in the acorn (the fruit of *Quercus racemosa* and *Quercus sessiliflora*), and considered by him to be milk-sugar.[1] Dessaignes found that this "acorn-sugar" (*sucre de glands*) is a distinct compound, and gave to it the name now generally used.[2] It is also found in the rind of the acorn,[3] and the leaves of the European palm (*Chamærops humilis*).[4] In order to prepare it, the cold concentrated extract of the acorn is decomposed with lead acetate to precipitate tannin, colouring matters, &c., and yeast added to the filtrate to remove glucose. The liquid, freed from lead, is evaporated until crystallization commences, and the crystals, to which inorganic salts still adhere, purified by recrystallization from dilute hydrochloric acid.[5]

Quercitol has a sweet taste, and is more readily soluble in hot water than in cold, but is insoluble in absolute alcohol or ether. It crystallizes in monosymmetric prisms, melting at 225°, and is dextrorotatory.[6] On dry distillation it decomposes with formation of quinol, quinhydrone, and other bodies. These are also formed, together with benzene, phenol, and iodophenol, when it is heated with concentrated hydriodic acid. Volatile iodides are obtained at the same time, which on heating with an excess of hydriodic acid are transformed into hexane (Prunier). Quercitol is also obtained by the oxidation of quinone.

The five hydroxyls of quercitol can be successively replaced by acid radicals. By heating with hydrochloric acid, chlorides are formed, the final product being pentaquercyl chloride, $C_6H_7Cl_5$, which crystallizes in needles melting at 102°.

Pentanitroxyquercitol or *nitroquercitol*, $C_6H_7(NO_3)_5$, is obtained from quercitol by the action of a mixture of concentrated sulphuric and nitric acids, and forms a transparent, resin-like mass, insoluble in water, but readily soluble in alcohol.[7] Various ethereal salts are obtained by heating quercitol with fatty acids or their anhydrides.

Pentacetoquercitol, $C_6H_7(OC_2H_3O)_5$, is obtained by the continued heating of a mixture of quercitol and acetic anhydride, and is

---

[1] *Ann. Chim. Phys.* III. **27**, 392.
[2] *Annalen*, **81**, 103, 251 ; *Compt. Rend.* **33**, 308, 462.
[3] Böttinger, *Annalen*, **202**, 269.
[4] Hugo Müller, private communication.
[5] Prunier, *Ann. Chim. Phys.* V. **15**, 1.
[6] Berthelot, *Chim. Org.* II. **218** ; Prunier, *loc. cit.*
[7] Hofmann, *Annalen*, **190**, 282.

an amorphous mass which is readily soluble in alcohol, slightly in water, and has a very bitter taste.[1]

From these results it would appear that quercitol has the constitution[2]

$$\begin{array}{c}
\text{CH}_2 \\
\text{HO.HC}\diagup\quad\diagdown\text{CH.OH} \\
\text{HO.HC}\diagdown\quad\diagup\text{CH.OH} \\
\text{CHOH}
\end{array}.$$

It has however been recently found by Kiliani and Scheibler[3] that quercitol on oxidation with nitric acid yields mucic and trihydroxyglutaric acids in small quantity, the remainder of the quercitol forming uncrystallizable compounds which readily decompose. The formation of these two acids, although it does not disprove the above constitution, is not in favour of it, as the hexamethylene ring is not usually so readily split by nitric acid.

HEXHYDROXYBENZENE.

1049 In the preparation of potassium according to Brunner's method, an explosive grey or black flocculent mass is formed, which was first observed by Berzelius and Wöhler, and which, as Gmelin found, when exposed to the air soon becomes green and then yellow. On the addition of a little water he obtained a yellow solution, while a cochineal-red powder was left behind. In the solution he found the potassium salt of a peculiar acid, and as this body as well as many of its salts exhibits a yellow colour, he called it *croconic acid* (κρόκον, saffron).[4]

This subject was further examined by Liebig, who said that he had already expressed the opinion that carbonic oxide might be regarded as a radical, of which carbonic and oxalic acids were the oxides, and phosgene the compound with chlorine. He continues: " The pursuit of this idea had led me to most singular and remarkable results, which seem to prove that the resemblance is not confined to these compounds." He found that pure carbonic oxide combines with potassium forming a

---

[1] For other ethereal salts see Prunier, *loc. cit.* ; and Berthelot, *Chim. Org.* and *Ann. Chim. Phys.* III. **54**, 82.

[2] Kanonnikow, *J. Russ. Chem. Ges.* **15**, 460.   [3] *Ber.* **22**, 517.

[4] *Pogg. Ann.* **4**, 31.

black mass, which is violently decomposed by water. On evaporating the solution he obtained the potassium salts of oxalic and croconic acids ; at the same time he observed the cochineal-red body.[1]

At about the same time the latter compound was examined by Heller, and found to be the potassium salt of an acid to which, as its salts exhibit a colour varying from a rose-red to a deep carmine, he gave the name *rhodizonic acid* (ῥοδίζω, I colour rose-red).[2] He also further examined croconic acid.

The compound of potassium with carbonic oxide was later on carefully examined by Brodie, who found that by passing this gas over the warmed metal at about 80°, arborescent growths make their appearance, and finally the whole is converted into a dull grey, crystalline mass. During that time the absorption of the carbonic oxide is but slow ; but now it becomes violent, the grey mass being transformed with evolution of heat, but without change of form, into the dark red compound $(COK)_n$, a body decomposing water with extreme violence, and often exploding spontaneously. As the grey substance changes so readily into the red compound, the former could not be isolated in a pure state ; its formula, however, appears to be $(COK_2)_n$.[3]

Lerch, who made further researches on this subject, used for his experiments the potassium carboxide formed in the preparation of potassium. He found that this body is not affected by dry air, but undergoes oxidation in the presence of moisture, the black mass first turning grey, then again black, and passing through green and red into yellow, containing now only oxalic and croconic acids. But on treating the black substance with hydrochloric acid, air being excluded, colourless *trihydrocarboxylic acid* is formed, according to the equation :

$$C_{10}K_{10}O_{10} + 10HCl = C_{10}H_{10}O_{10} + 10KCl.$$

This compound is readily oxidised to *dihydrocarboxylic acid*, $C_{10}H_8O_{10}$, and *carboxylic acid*, $C_{10}H_4O_{10}$, and finally into *croconic acid*, $C_5H_2O_5$.[4] Lerch further found that the three former compounds are oxidised by nitric acid to *oxycarboxylic acid*.[5]

[1] *Annalen*, **11**, 182. [2] *J. Pr. Chem.* I. **3**, 193.
[3] *Journ. Chem. Soc.* **12**, 269. [4] Will, *Annalen*, **118**, 187.
[5] Lerch, *ibid.* **124**, 20.

No opinion was pronounced as to the constitution of these remarkable compounds until Nietzki and Benckiser found that nitranilic acid (P. 205) is easily reduced to *diamido-tetrahydrcxybenzene*, $C_6(NH_2)_2(OH)_4$, and this can by some reactions which are described below, be converted into *hexhydroxylenzene*, $C_6(OH)_6$. On examining this body they found that it is identical with Lerch's trihydrocarboxylic acid, and consequently the black potassium carboxide is the potassium compound of hexhydroxybenzene.

When Liebig said that the pursuit of the idea that carbonic oxide was a radical had led him to most singular and remarkable results, he could not anticipate that it would lead to a direct synthesis of benzene derivatives from inorganic bodies; a synthesis which in its simplicity cannot be surpassed.

As soon as the constitution of potassium carboxide was clearly made out, that of its derivatives was soon ascertained, oxycarboxylic acid being benzotriquinone, while dihydrocarboxylic acid is identical with tetrahydroxyquinone, and carboxylic acid with rhodizonic acid or dihydroxydiquinoyl.

Hexhydroxybenzene, $C_6(OH)_6$, is prepared by the reduction of triquinoyl (see below), with stannous chloride and hydrochloric acid. It separates after some time in greyish-white needles, which are sparingly soluble in cold, more readily in hot water and are precipitated by hydrochloric acid[1] It is only slightly soluble in alcohol, ether, and benzene, and the solutions turn violet in the air. It reduces silver solution quickly in the cold, and on distillation with zinc dust is converted into benzene and a solid substance which is probably diphenyl. On treatment with acetic anhydride it is converted into the *hexacetyl* derivative, $C_6(OC_2H_3O)_6$, which crystallizes from glacial acetic acid in small well developed, probably rhombic prisms containing acetic acid, which is evolved at 150°. It melts at 203° and on cooling solidifies to a colourless crystalline mass.

Benzotriquinone or *Triquinoyl*, $C_6O_6 + 8H_2O$. Nitranilic acid is converted on reduction into diamidotetrahydroxybenzene, and this compound on treatment with fairly concentrated nitric acid, loses all its nitrogen, and is converted into a substance of the formula $C_6H_{16}O_8$. This must be regarded as the hydrate of a derivative of benzene in which the quinone group is contained

---

[1] Nietzki and Benckiser, *Ber.* **18**, 505.

three times. Its constitution will then be represented by the formula

$$\begin{array}{c} CO \\ CO \diagup \diagdown CO \\ CO \diagdown \diagup CO \\ CO \end{array}$$

or the opposite pairs of oxygen atoms may be linked together as in the first formula of quinone given on p. 49, for it yields on reduction successively *dihydroxydiquinoyl* or *dihydroxybenzodiquinone*, $C_6(OH)_2O_4$, *tetrahydroxyquinone*, $C_6(OH)_4O_2$, and finally *hexhydroxybenzene* $C_6(OH)_6$.

It is almost insoluble in cold water and alcohol, and crystallizes from dilute nitric acid in colourless microscopic needles which melt at 95°, giving off carbon dioxide and water. It does not lose water at 60° *in vacuo*, and it is probable that six of the molecules of water are combined in the same way as in mesoxalic acid, according to which hypothesis triquinoyl would have the formula

$$\begin{array}{c} HO \quad OH \\ \diagdown \diagup \\ HO \diagup \diagdown \diagup \diagdown OH \\ HO \diagup \quad \quad \diagdown OH \\ HO \diagdown \quad \quad \diagup OH \\ HO \diagdown \diagdown \diagup \diagdown OH \\ \diagup \diagdown \\ HO \quad OH \end{array} + 2H_2O.$$

This would also account for the fact that unlike most quinone derivatives, triquinoyl possesses no colour.[1]

Dihydroxydiquinoyl, Dihydroxybenzodiquinone, or *Rhodizonic acid,* $C_6(OH)_2O_4$, is obtained by warming triquinoyl with aqueous sulphurous acid to 40-50°, and saturating the yellow solution with potassium or sodium carbonate. The free substance is best prepared from the potassium or sodium salt by conversion into the barium salt and decomposing this with sulphuric acid. It forms colourless crystals but is extremely unstable.[2] The *potassium* salt forms graphite-like crystals, but was first obtained as a red amorphous powder, from which fact the name of the acid is derived. The *sodium* salt forms long violet needles. The *barium* salt is a bright vermilion red precipitate.

[1] Nietzki and Benckiser, *Ber.* **18**, 504, 1842.   [2] *Ibid.* p. 1838.

Like all diketones containing two carbonyl groups in the adjacent position, rhodizonic acid reacts with *o*-toluylene diamine forming a "quinoxaline" (p. 73) having the constitution

$$HO—C=N\!\!>C_7H_6.$$ On oxidation with nitric acid this compound is converted into the corresponding triquinoyl derivative

$$O—C=N\!\!>C_7H_6,$$ which on further treatment with *o*-toluylene diamine yields a triquinoxaline of the formula:

$$C_7H_6 \cdots C_7H_6 \cdots C_7H_6$$

showing conclusively that triquinoyl contains the group —CO—CO— three times.[1] The four carbonyl groups in rhodizonic acid probably occupy the positions 1.2.3.4.[2]

Tetrahydroxyquinone, $C_6(OH)_4O_2$, is prepared by passing air for a short time through a solution of hexhydroxybenzene in sodium carbonate,[3] and also by the reduction of triquinoyl. It is further formed by the action of nitric acid on inositol.[4] It forms blue black crystals, which do not melt, are difficultly soluble in cold water and ether, but readily in alcohol and hot water. It is oxidized in presence of four molecules of caustic potash to rhodizonic acid. Its *sodium* salt $C_6(OH)_2(ONa)_2O_2$ forms almost black crystals with a greenish surface lustre and the *barium* salt is a dark-red precipitate which on drying shows a green metallic lustre. Its *diacetate* melts at 205°.

1050 *Inositol*, $C_6H_6(OH)_6 + 2H_2O$, was first found by Scherer in the fluid contained in the muscles of the heart of the ox (*is*,

[1] Nietzki and Kehrmann, *Ber.* **22**, 322.
[2] Nietzki, *ibid.* **23**, 3136.
[3] Nietzki and Benckiser, *Ber.* **18**, 507, 1837.
[4] Maquenne, *Ann. Chim. Phys.* VI. **12**, 112.

gen. ἰνός, muscle),[1] and called *inosite* ; according to Sokolow and Panum, it does not occur in any other muscular fluid.[2] It is however, found widely distributed in the animal kingdom, as for instance in the lungs, the liver, the spleen, and the kidneys of the ox,[3] in the brain of the same,[4] in the urine of man in case of Bright's disease (Cloëtta), and even in healthy urine after drinking large quantities of water.[5]

Inositol is also frequently found in the vegetable kingdom. Vohl discovered it first in the young French bean (*Phaseolus vulgaris*), and gave to the new saccharine substance the name of *Phascomannite*, but soon afterwards recognized its identity with inositol.[6] Marmé then found it in the unripe pea, the lentil, and the fruit of the false acacia (*Robinia pseudacacia*), in the common cabbage, foxglove, dandelion, in asparagus, and in the shoots of the potato kept in a cellar during winter.[7] It also occurs in the grape, and hence, as it is not fermented by yeast, it finds its way into wine.[8] It is likewise present in the young leaves of the vine,[9] as well as in the leaves of the ash,[10] and of the walnut tree.[11]

To prepare inositol from the heart or the lungs of the ox, the material is finely chopped up, and exhausted with water, some acetic acid added, and the mixture then heated to the boiling point. The filtrate is treated with neutral lead acetate, filtered off from the precipitate formed, and the filtrate then precipitated by basic lead acetate, the washed precipitate decomposed by sulphuretted hydrogen, and the concentrated solution precipitated by addition of alcohol. It is prepared from vegetable juices by neutralizing the liquid with baryta-water, treating with lead acetate solution, and then proceeding as above described. To precipitate the substance from the concentrated aqueous solution, Hilger recommends a mixture of 10 parts of alcohol and 1 part of ether.

Inositol forms large transparent monosymmetric crystals, which possess a sweet taste, and are soluble at the ordinary temperature in about 6 parts of water, yielding an optically inactive solution.

[1] Dessaignes, *Annalen*, **73**, 322.
[2] *Ibid.* **81**, 375.
[3] Cloëtta, *ibid.* **99**, 289.
[4] Müller, *ibid.* **103**, 140.
[5] Kültz, *Fresenius' Zeitsch.* **16**, 135.
[6] *Annalen*, **99**, 125 ; **101**, 50 ; **105**, 330.
[7] *Ibid.* **129**, 222.
[8] Hilger, *ibid.* **160**, 333.
[9] Caustein and Neubauer, *Ber.* **6**, 1411 ; Neubauer, *Fresenius' Zeitsch.* **12**, 45.
[10] Gintl, *Jahresb.* **1868**, 800.
[11] Tauret and Villiers, *Bull. Soc. Chim.* **29**, 74.

Inositol is but slightly soluble in hot dilute alcohol, and is insoluble in absolute alcohol and in ether. It effloresces in the air and becomes completely dehydrated at 100°. It crystallizes from acetic acid in forms which are free from water of crystallization and melt at 218°. When boiled with alkalis it is not coloured brown, and it has no action on an alkaline copper solution. Its solution yields with lead acetate a gelatinous precipitate which, after drying over sulphuric acid, possesses the empirical formula $2C_6H_{12}O_6 + 5PbO$ (Cloëtta).

If inositol be evaporated almost to dryness with addition of a little nitric acid, a few drops of ammoniacal calcium chloride solution then added, and the mixture again evaporated, a rose-red colouration is produced, and in this way the presence of 0·0005 grams of inositol may be recognized. This reaction is not yielded by saccharine bodies or by starch.[1] If to a few drops of inositol solution one drop of a mercuric nitrate solution (such as is used in the estimation of urea) be added, a yellow precipitate is formed which on heating becomes red; this becomes yellow again when cold, but reddens when re-heated.[2]

Hexnitroxyinositol, $C_6H_6(NO_3)_6$. Inositol is dissolved by concentrated nitric acid, and on addition of sulphuric acid to this solution a gritty precipitate is formed which is insoluble in water. If this be dissolved in hot alcohol, the hexnitrate crystallizes out on cooling in rhombic tables and prisms, which explode violently on percussion. The mother-liquor yields on evaporation fine white needles of the *trinitrate*, $C_6H_9O_3(NO_3)_3$.[3]

Hexacetoxyinositol, $C_6H_6(OC_2H_3O)_6$, is prepared by the action of acetyl chloride on anhydrous inositol.[4] It crystallizes from alcohol in small plates which melt at 212°, and commence to sublime at 200°.

Inositol was formerly supposed to belong to the group of sugars, but Maquenne[5] has shown that it yields derivatives of benzene on reduction and oxidation, and is a derivative of hexamethylene. It is converted by hydriodic acid at 170° into benzene, phenol, and tri-iodophenol, and on treatment with nitric acid, yields tetrahydroxyquinone, rhodizonic acid, and

[1] Scherer, *Annalen*, **81**, 375.
[2] Gallois, *Fresenius' Zeitsch.* **4**, 264.
[3] Vohl, *loc. cit.* ; *Ber.* **7**, 106.
[4] Maquenne, *Bull. Soc. Chim.* **48**, 54.
[5] *Compt. Rend* **104**, 225, 297.

triquinoyl. Its constitution must therefore be represented by the formula

$$\text{CH.OH}$$
$$\text{HO.HC} \diagup \backslash \text{CH.OH}$$
$$\text{HO.HC} \backslash \diagup \text{CH.OH}$$
$$\text{CH.OH}$$

The above mentioned colouration with nitric acid is due to the formation of triquinoyl and its decomposition products.

The substance described formerly as *Dambose* has been found to be identical with inositol.[1]

1051 By the action of hydriodic acid on *a*- and *β*-pinitol on the one hand, and quebrachitol on the other, two compounds are obtained which strongly resemble inositol and one another, but both differ from inositol by being optically active.

Dextro-inositol or β-Inositol, and Laevo-inositol, $C_6H_{12}O_6$. The first of these compounds is obtained by the action of hydriodic acid on *a*- and *β*-pinitol, and the second in a similar manner from quebrachitol. Both of them crystallize from water in hemibedral prisms, and are equally soluble. They become plastic at 210°, and melt at 242°, and are both converted into rhodizonic acid by the action of nitric acid. Dextro-inositol has also been obtained in anhydrous crystals, but not the laevo-compound. Their hexacetyl derivatives are amorphous, become plastic in the hand, and melt at 247°. They differ however, as the name indicates, in their action on the plane of polarized light, having a sp. rotation of + 65° and — 65° respectively. When their concentrated aqueous solutions are mixed, a new optically inactive compound separates out, which is much less soluble in water, and melts at 253° without previously becoming plastic. Its hexacetyl derivative is likewise inactive, and melts at 111°. This compound must from its mode of formation have a constitution similar to that of racemic acid, and the name *racemo-inositol* has therefore been given to it. As it is not identical with the inactive inositol previously known, the latter must correspond in constitution to mesotartaric acid. We have therefore the interesting result that inositol exists in four different forms corresponding exactly to those of tartaric acid.[2]

[1] Maquenne, *Bull. Soc. Chim.* **48**, 58, 162.
[2] Maquenne, Tanret, *Compt. Rend.* **109**, 882, 908 ; **110**, 86.

1052 *Scyllitol*, $C_6H_{12}O_6$ This compound occurs in largest quantity in the kidney of the skate, (*Raja batis* and *clavata*), the dogfish (*Scyllium canicula*), and also of *Spinax acanthias*, and it also occurs in the liver, milt, &c., of these cartilaginous fishes. It may be prepared from these sources by the method described for the extraction of inositol.   It is less soluble in water than the last-named substance, and crystallizes in anhydrous monosymmetric prisms which have a slightly sweet taste.   Like inositol it is neither coloured brown by alkalis nor does it reduce alkaline copper solution.   It does not give Scherer's inositol reaction.

1053 *a-Pinitol*, $C_7H_{14}O_6$, was discovered by Berthelot in a manna-like exudation from a Californian pine,[1] (*Pinus lambertiana*).   It forms hard crystals which melt above 150°.   Its solution rotates the plane of polarization to the right ($[a]j = 58\cdot6°$).

β-Pinitol closely resembles the above, and is obtained from the same source.   It has however a considerably higher sp. rotatory power, viz. $[a]_D = 65\cdot51$.   Both of these compounds are, as stated above, converted into dextro-inositol and methyl iodide, and are therefore methyl ethers of this substance.[2]

The compounds termed *matezite* and *matezodambose* obtained by Geraud [3] from caoutchouc, are identical with β-pinitol and dextro-inositol.[4]   *Sennitol* is also probably identical with β-pinitol.

Quebrachitol, $C_7H_{14}O_6$, is prepared from quebracho bark, (*Aspidosperma quebracho*), and crystallizes from alcohol in anhydrous prisms, which melt at 186—187° and have a very sweet taste.   It does not ferment with yeast, nor reduce Fehling's solution.   Its sp. rotatory power is — 80°.   It is converted by hydriodic acid into methyl iodide and laevo-inositol, of which latter it is therefore the methyl ether.[5]

CROCONIC AND LEUCONIC ACIDS.

1054 It has already been mentioned (p. 209) that all the quinone derivatives of hexhydroxybenzene are converted by the action

[1] *Ann. Chim. Phys.* III. **46**, 76.   [2] Maquenne, *Compt. Rend.* **109**, 882.
[3] Giraud, *ibid.* **73**, 774.
[4] Combes, *ibid.* **110**, 46 ; Giraud, *ibid.* **110**, 84.
[5] Tanret, *Compt. Rend.* **109**, 908.

of caustic potash into the potassium salt of an acid having the composition $C_5H_2O_5$, termed *croconic acid.*

Croconic acid is best prepared by boiling together 1 part of diamidotetrahydroxybenzene hydrochloride, 4 parts of potassium carbonate, 60 parts of water, and 3 parts of precipitated manganese dioxide. To the acidified and filtered solution, barium chloride is then added, which precipitates barium croconate in golden yellow plates.[1] The free acid crystallizes from alcohol in pale sulphur yellow grains or plates, having a bitter taste, and containing three molecules of water which are evolved at 100° *in vacuo.*

Potassium Croconate, $C_5K_2O_5$, crystallizes from hot water in reddish-yellow needles which contain water. These are but little soluble in cold water, and do not dissolve in alcohol. In taste it resembles saltpetre, and it loses its water of crystallization below 100°, assuming a light lemon-yellow colour.

Potassium acid croconate, C_5HKO_5, forms brownish-yellow anhydrous needles, which are distinguished from the normal salt by their darker colour, and their violet surface lustre.

The *barium* and *calcium* salts are lemon-yellow precipitates, whilst the *strontium* salt crystallizes in plates which are readily soluble in water and alcohol.

Copper Croconate, $C_5CuO_5 + 3H_2O$. This characteristic salt is produced when hot solutions of potassium croconate and copper sulphate are mixed together. On cooling it separates out in small rhombic prisms, the faces of which have a semi-metallic lustre and exhibit a deep blue reflection, and by transmitted light have an orange-brown colour. Aggregates of small crystals appear as a dark violet powder, whilst the powdered substance has a lemon-yellow colour, which shows the peculiar surface lustre more strongly the more finely it is divided.[2]

Silver croconate, $C_5Ag_2O_5$, is an orange-red powder, which decomposes on heating with evolution of sparks.

In its properties croconic acid strongly resembles the quinones, and on treatment with concentrated sulphurous acid yields a colourless *hydrocroconic acid,* $C_5H_4O_5$, resembling the quinols. If however the acid be reduced with phosphorus and hydriodic acid, a second reduction product of the same empirical formula is obtained, which yields a potassium salt crystallizing in red needles, and is termed *croconic acid hydride* to distinguish it

---

[1] Nietzki and Benckiser, *Ber.* **19**, 294.
[2] Gmelin, *Annalen,* **37**, 58.

from the foregoing. By the action of sulphuretted hydrogen, croconic acid is converted into *thiocronic acid*, $C_5H_2SO_4$, the *barium* salt of which forms reddish-brown crystals with a violet surface lustre.

1055 *Leuconic acid*, $C_5O_5 + 5H_2O$. This substance was first closely examined by Will and Lerch, the former using the name here given, whilst the latter described it as "oxycroconic acid." It is obtained by the oxidation of croconic acid, preferably in the following manner : pure finely powdered croconic acid is gradually added to 6—8 parts of nitric acid of sp. gr. 1·36 and well cooled by ice, when, after a time the liquid becomes filled with a mass of small colourless crystals, which are filtered off, washed first with a mixture of ether and alcohol, and finally with ether.[1] The air-dried substance has the composition $C_5H_{10}O_{10}$, and loses $\frac{1}{2}$ mol. H_2O over sulphuric acid, and 1 mol. at 100°. It does not melt but becomes brown at 160°, some croconic acid being re-formed. It was supposed by Will and Lerch to form salts, but Nietzki and Benckiser have shown that these are in reality decomposition products of the compound.

Leuconic acid reacts with hydroxylamine, forming a pentoxime of the formula $C_5(NOH)_5$, which is a voluminous yellow precipitate almost insoluble in all the ordinary reagents but soluble in solutions of alkaline carbonates, from which it is precipitated by carbonic acid. It is also formed by acting with hydroxylamine on croconic acid, this being first oxidized by the reagent to leuconic acid. A small quantity of *leuconic acid tetroxime* is always formed at the same time, which like the pentoxime explodes at 166°. The formation of the pentoxime shows that leuconic acid contains the carbonyl group five times, and is probably a pentaketo-derivative of pentamethylene possessing the formula

which also shows its close relation to triquinoyl. This formula has been further confirmed by acting on leuconic acid with o-toluylene diamine, which, as already mentioned in the intro-

[1] *Ber.* **19**, 301.

duction, combines with all *o*-diketones, forming quinoxalines. Leuconic acid combines with two molecules of the diamine forming a compound of the constitution

$$C_7H_6 \diagdown \begin{array}{c} N-C \\ | \\ N-C \end{array} \underset{\displaystyle CO}{\triangle} \begin{array}{c} C-N \\ | \\ C-N \end{array} \diagup C_7H_6$$

This compound still contains an unaltered carbonyl group, and therefore combines with phenylhydrazine in the usual manner, forming a hydrazone.[1]

Croconic acid also unites with *o*-toluylene diamine, and must therefore contain two carbonyl groups in the ortho-position. As it forms salts of the formula $C_5K_2O_5$, and C_5HKO_5 it must contain two hydroxyl groups and has probably the formula

$$\begin{array}{c} CO \\ OC \diagup \triangle \diagdown CO \\ HO.C \underline{\quad\quad} C.OH \end{array}$$

Its formation from rhodizonic acid probably takes place in the following stages[2]

$$\begin{array}{c} CO \\ OC \diagup\!\!\diagdown CO \\ OC \diagdown\!\!\diagup C.OH \\ C.OH \end{array} + H_2O = \begin{array}{c} HO\ COOH \\ \diagdown\!\!\diagup \\ C \\ OC \diagup\triangle\diagdown CO \\ HO.C \underline{\quad} C.OH \end{array} =$$

$$\begin{array}{c} CHOH \\ OC \diagup\triangle\diagdown CO \\ HO.C \underline{\quad} C.OH \end{array} + CO_2$$

$$\begin{array}{c} CHOH \\ OC \diagup\triangle\diagdown CO \\ HO.C \underline{\quad} C.OH \end{array} + O = \begin{array}{c} CO \\ OC \diagup\triangle\diagdown CO \\ HO.C \underline{\quad} C.OH \end{array} + H_2O.$$

Only the second of these intermediate products has been isolated, and is identical with the croconic acid hydride obtained by the action of phosphorus and hydriodic acid on croconic acid (p. 218).

[1] Nietzki and Benckiser, *Ber.* **19**, 293, 772.
[2] Nietzki, *ibid.* **20**, 1619 ; **23**, 3138.

1056 When leuconic acid pentoxime is reduced with stannous chloride and concentrated hydrochloric acid, it is converted into a compound having the formula $C_5H_{11}N_5 + 4HCl + H_2O$, which crystallizes in small tablets united in stellate groups, and decomposes at 80—100°. A second hydrochloride of the formula $C_5H_{11}N_5 + 3HCl + H_2O$, and a sulphate of the formula $(C_5H_{11}N_5)_2 (H_2SO_4)_5 + 2H_2O$ have also been obtained.

From its mode of formation and composition, this base must be regarded as a pentamido-derivative of the hypothetical hydrocarbon *Pentene* (Part II., p. 730), its constitution being represented by the formula

It may be regarded as a double *o*-diamine, and as such reacts with two molecules of diacetyl or of croconic acid, the latter giving rise to a compound of the formula

Leuconic acid tetroxime yields on reduction *tetramidohydroxypentene*, which forms a crystalline *sulphate* of the formula $C_5H(OH)(NH_2)_4(H_2SO_4)_2 + H_2O$. Its *hydrochloride* $C_5H(OH).(NH_2)_4$. 3HCl. crystallizes in octohedra.[1]

AMIDO-DERIVATIVES OF BENZENE.

AMIDOBENZENE, OR ANILINE, $C_6H_5.NH_2$.

1057 In 1826 Unverdorben discovered in the products of the dry distillation of indigo, a peculiar liquid body which combined with acids to form salts distinguished by their power of crystallization, and which he therefore named " crystalline."[2]

[1] Nietzki and Rosemann, *Ber.* **22**, 917. [2] *Pogg. Ann.* **8**, 397.

Runge, in 1834, found a volatile basic compound in coal-tar oil, which gave an azure-blue colouration with bleaching powder solution, while the solutions of its salts were coloured violet-blue. He accordingly named the new compound "blue-oil" or "kyanol." [1] He also found that it coloured pine-wood and the pith of the elder-tree yellow, and that when a solution of cupric chloride is spread on a porcelain plate heated to 100°, allowed to evaporate, and then a drop of "kyanol" nitrate solution added a greenish-black spot is obtained. [2]

Fritzsche, in 1840, examined the products which are obtained when indigo is distilled with caustic soda or caustic potash, and found among them a basic oil, which has the formula, C_6H_7N, and forms salts which crystallize well. This he named aniline, [3] the word being derived from anil (nila, Indian blue; anil, Arabic, the blue), under which designation the Portuguese introduced indigo (ινδικόν, indicum, the Indian colour), which was known to the Greeks and Romans, into Europe.

In a postscript to this memoir, Erdmann says that aniline is undoubtedly identical with kyanol.

Two years later Zinin found that when an alcoholic solution of nitrobenzide (nitrobenzene) is saturated with ammonia and then treated with sulphuretted hydrogen, an oily, basic liquid, which he named benzidam, is obtained, which forms crystallizable salts and has the formula C_6H_7N. [4] Fritzsche then observed that this substance was undoubtedly identical with aniline.

The identity of the three bodies was then experimentally proved by Hofmann, [5] who proposed to name the substance phenamide, but when aniline was recognized as a compound ammonia this term was altered to phenylamine. Since, however the aromatic amines differ in many points from those of the fatty bodies, they were called amido-compounds, according to the proposal of Griess, and hence aniline obtained the name amido-benzene.

Aniline is not only found in the distillation products of coal, [6] but also in those of bone [7] and peat. It is however invariably prepared by the reduction of nitrobenzene. Zinin's method is

---

[1] Pogg. Ann. 31, 65, 513 ; 32, 331.
[2] Runge was therefore the first to observe the formation of aniline black.
[3] J. Pr. Chem. 20, 453.
[4] Ibid. 27, 149.   [5] Annalen, 47, 37.
[6] Anderson, Annalen, 70, 32.   [7] Vohl, ibid. 109, 200.

not generally employed, although it is convenient for the reduc‑
tion of other nitro-compounds. In the case of nitrobenzene the
reaction proceeds as follows :

$$C_6H_5.NO_2 + 3H_2S = C_6H_5.NH_2 + 2H_2O + 3S.$$

Hofmann found that zinc and hydrochloric acid can also be used
for the reduction,[1] and Wöhler showed that aniline is obtained
when nitrobenzene is treated with caustic soda and arsenic
trioxide.[2] In place of the latter grape-sugar may be employed.[3]
In the preparation of small quantities it is best to use tin and
concentrated hydrochloric acid, or a solution of stannous chloride
in hydrochloric acid (Beilstein) :

$$C_6H_5.NO_2 + 3SnCl_2 + 6HCl = C_6H_5.NH_2 + 3SnCl_4 + 2H_2O.$$

When all the nitrobenzene has gone into solution, the tin is
removed by sulphuretted hydrogen, and the solution of the
hydrochloride evaporated to dryness or decomposed by an
alkali.

1058 In the manufacture of aniline, iron filings and acetic
acid were formerly employed as reducing agents according to
Béchamp's method. Since 1864 the cheaper hydrochloric acid
has been generally used :

$$C_6H_5.NO_2 + 3Fe + 6HCl = C_6H_5.NH_2 + 3FeCl_2 + 2H_2O.$$

The quantity of acetic acid or of hydrochloric acid which is
employed in much less than this equation represents. Ferrous
chloride in presence of hydrochloric acid acts as a further reducing
agent :

$$C_6H_5.NO_2 + 6FeCl_2 + 6HCl = C_6H_5.NH_2 + 6FeCl_3 + 2H_2O.$$

This reaction goes on when free acid is present, which is
not long the case, as the aniline produced combines with it to
form the hydrochloride. This, however, in presence of iron and
nitrobenzene, behaves exactly like hydrochloric acid, free aniline
being formed, which then reacts with the ferric chloride in
presence of water, ferric oxide and aniline hydrochloride being
produced. According to the theory, therefore, a small amount
of acid is sufficient, when enough iron is present, to convert an
unlimited amount of nitrobenzene into aniline.

[1] Vohl, *Annalen*, **55**, 200.   [2] *Ibid.* **102**, 127.
[3] Vohl, *Jahresb.* **1863**, 410.

The cast-iron apparatus which is generally employed is shown in Fig. 11. The lower half is constructed in two pieces, in order to facilitate its renewal, as this portion is more quickly destroyed by the acid. In the cover are the necessary openings for the agitator and the condenser, and a hole for the introduction

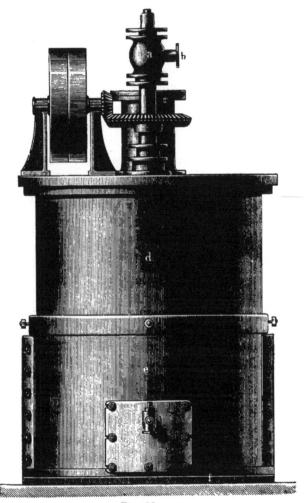

Fig. 11.

of the materials, which can be closed by a wooden plug. In the bottom is an opening through which the iron residues can be removed.

40 parts of water, 25 parts of finely-divided cast-iron filings, and 8 to 10 parts of hydrochloric acid are placed in the vessel,

and 100 parts of nitrobenzene (generally 500 kilos) allowed to flow in, the agitator being kept in motion. In order to start the reaction, steam is blown through, and this is kept up during the gradual addition of 75 parts of damp iron filings which are added as soon as the first violent reaction has ceased, the mixture not being allowed to get cool. If the filings are added too quickly the reaction becomes so violent that some of the aniline is reduced to benzene and ammonia. Finally, from 10 to 20 parts of dry filings are added. Any nitrobenzene which is volatilized during the operation is condensed in a vertical cooler and flows back into the vessel. If all the nitrobenzene has been reduced, the residue consists of aniline, its hydrochloride, and ferric oxide. Milk of lime, or sifted slaked lime, is then added, and the aniline distilled off in a current of high-pressure steam blown through the hollow agitator. The distillate separates into two layers; the lower is aniline, the upper is an aqueous solution of aniline containing 2 to 3 per cent. of the oil, and is used to provide the steam for the next operation. The aniline is then purified by distillation from iron vessels. The iron residues are sent to the blast furnaces.

According to Laurent, small quantities of aniline are formed by heating ammonium phenate to 300° for some time.[1] Berthelot, who repeated this experiment, did not obtain a trace even at 360°;[2] it is obtained, however, together with diphenylamine, by heated phenol with sal-ammoniac and fuming hydrochloric acid to 310°,[3] and also by the continued heating of phenol with ammoniacal zinc chloride to 280—300°,[4] diphenylamine being formed at the same time. It is likewise formed by the action of sodamide $NaNH_2$ on potassium benzene sulphonate.[5]

When pure benzene is used in the preparation of aniline, the aniline obtained is likewise pure, but this is not the case with the "pure" aniline of commerce. In order to obtain perfectly pure aniline from this, it is converted by continued boiling with glacial acetic acid into acetanilide, $C_6H_5.NH(C_2H_3O)$ (p. 237), which is purified by distillation, washing with carbon disulphide and re-crystallization from hot water until its melting-point reaches 112°, and is then decomposed by caustic soda (Beilstein).

1059 *Properties.*—Aniline is a colourless liquid possessing a

[1] *Compt. Rend.* **17**, 1366.   [2] *Bull. Soc. Chim.* **13**, 314.
[3] Bardy and Dusart, *Compt. Rend.* **74**, 188.
[4] Merz and Müller, *Ber.* **19**, 2901.   [5] Jackson and Wing, *Ber.* **19**, 902.

peculiar and characteristic odour, and boiling under the normal pressure at 183°·7.[1] It has at 0° a specific gravity of 1·038, and solidifies at a low temperature to a crystalline mass, melting at − 8°. Aniline which is not perfectly pure remains liquid even at − 20°.[2] On exposure to light and air it becomes brown, and the more impure it is the more rapidly does this colouration take place. It is tolerably soluble in water, 100 parts of the solution containing at the ordinary temperature about 3 parts of aniline; the solubility increases with the temperature. Water also dissolves in aniline, 100 parts of aniline dissolving 5 parts of water at the ordinary temperature, and somewhat more at higher temperatures.[3]

It is miscible with alcohol, ether, and benzene in every proportion. The aqueous solution neither colours red litmus paper blue, nor turns turmeric paper brown, but changes the violet colour of the dahlia to green. Although it has such a weak alkaline reaction it precipitates the salts of zinc, aluminium, and iron, and decomposes ammonium salts on heating.

Runge long ago pointed out that an aqueous solution of aniline is coloured blue by a hypochlorite, or if the aniline be not quite pure, bluish violet, the colour of the solution quickly changing to brown. Very dilute solutions either give no colouration or only a very faint one, but on the addition of a few drops of a very dilute solution of ammonium hydrosulphide the liquid becomes rose-red coloured. The limit of this reaction is 1 part of aniline in 250,000 parts of water.[4]

When aniline, or one of its salts, is mixed with concentrated sulphuric acid and a drop of a solution of potassium bichromate added, a pure blue colouration is obtained which soon disappears.[5]

Aniline coagulates albumen and acts as a powerful poison, its action, according to Letheby and Turnbull, being chiefly on the nervous system. According to Grandhomme the first symptom observed in slight cases of poisoning, caused by inhaling the vapour, is a blue colour on the edge of the lips, accompanied by a state of mild inebriation; the gait becomes reeling, the speech thick, the head affected, and the face pale, while the appetite fails completely. In these cases aperients, such as Epsom salt, Carlsbad salt, &c. are administered; the use of alcohol aggravates

---

[1] Thorpe, *Journ. Chem. Soc.* 1880, **1**, 221.
[2] Lucius and Hofmann, *Ber.* **5**, 154.   [3] Alexejew, *ibid.* **10**, 709.
[4] Jacquemin, *ibid.* **9**, 1433.   [5] Beissenhirtz, *Annalen*, **87**, 376.

the symptoms. In more severe cases, such as arise; for example, from saturation of the clothes with aniline, the lips become dark blue or even black, and the vertigo so violent that the patient falls. In such cases a stimulus in the form of a cold shower, internal application of ether, &c. must be made use of.[1]

Aniline is chiefly used in the colour industry. In the laboratory it is frequently employed as a solvent, since it dissolves many substances—*e.g.* indigo-blue—which are insoluble in the ordinary reagents.

SALTS OF ANILINE.

1060 Although its basic properties are feeble, aniline combines with acids, forming salts which crystallize well. The salts of aniline have been carefully examined, but we shall here only refer to those which have received technical applications, or are specially characteristic.

Aniline hydrochloride, $C_6H_7N.ClH$, is readily soluble in water and crystallizes in needles or large plates, which melt at 192°, and sublime unchanged. It is prepared on the large scale, and generally called "aniline salt"; the platinochloride, $(C_6H_7N.HCl)_2PtCl_4$, crystallizes from hot water in yellow needles.

Aniline sulphate, $(C_6H_7N)_2SO_4H_2$. forms a crystalline powder, readily soluble in water, slightly so in alcohol, and insoluble in ether. This last property is taken advantage of in the separation of aniline from methylaniline.

Aniline nitrate, $C_6H_7N.NO_3H$, crystallizes in large rhombic prisms or needles, which decompose at 190° with formation of nitraniline and other products (Béchamp).

Aniline oxalate, $(C_6H_7N)_2C_2O_4H_2$, is very slightly soluble in cold water, and crystallizes from a hot solution in fine asymmetric prisms; it is slightly soluble in alcohol, and insoluble in ether.

Aniline phenate, $C_6H_5.NH_2.C_6H_6O$, is formed by heating a mixture of aniline and phenol to boiling for some time, or more slowly when the mixture is allowed to stand in the cold. It forms lustrous tablets which, after recrystallization from light petroleum, melt at 31° and boil at 181°. Its smell resembles

[1] Grandhomme, *loc. cit.*

that of phenol, but is weaker; it does not attack the skin like the latter.[1]

Aniline, like ammonia, combines with many salts, *e.g.* metallic chlorides, and also with trinitrobenzene. By mixing the warm alcoholic solutions of these, the compound $C_6H_7N.C_6H_3(NO_2)_3$, separates out in orange-red needles, crystallizing from benzene in splendid lustrous plates, which melt at 123—124° and rapidly give off aniline in the air.[2]

SECONDARY AND TERTIARY ANILINES.

1061 *Methylaniline*, $C_6H_5.NH.CH_3$ is obtained, together with dimethylaniline, by the action of methyl iodide, methyl bromide, or methyl chloride, on aniline,[3] as well as when the latter is heated with methyl nitrate,[4] or when its hydrochloride or hydriodide is heated with methyl alcohol : [5]

$$C_6H_5.NH_2.HCl + CH_3.OH = C_6H_5.N(CH_3)H.HCl + H_2O.$$
$$C_6H_5.NH_2.HCl + 2CH_3.OH = C_6H_5.N(CH_3)_2HCl + 2H_2O.$$

In all cases dimethylaniline is preferably formed, and therefore a portion of the aniline remains unattacked. In order to separate the three bases, dilute sulphuric acid is added as long as aniline sulphate separates out. After the sulphuric acid solution and the free bases have been separated by pressing through a linen cloth, caustic soda is added, and the mixture which separates out dried and treated with acetyl chloride until no further heating takes place; it is then poured into hot water. On cooling, methylacetanilide, $C_6H_5.N(CH_3)C_2H_3O$, separates out in fine, long needles, dimethylaniline hydrochloride remaining in solution. The former is then decomposed by boiling with hydrochloric acid (Hofmann). The moderately concentrated solution of the hydrochlorides of the three bases may also be treated with sodium nitrite in the cold, when diazobenzene chloride and nitrosodimethylaniline hydrochloride remain in solution, while methylphenylnitrosamine, $C_6H_5.N(CH_3)NO$, separates

[1] Dale and Schorlemmer, *Annalen*, **217**, 387 ; G. Dyson, *Journ. Chem. Soc.* 1883, **1**, 466.   [2] Hepp, *Annalen*, **215**, 356.

[3] Hofmann, *ibid.* **74**, 150 ; *Ber.* **10**, 588, 591.

[4] Bardy, *Dingler's Polyt. Journ.* **234**, 233.

[5] Poisrier and Chappot, *Bull. Soc. Chim.* II. **6**, 502 ; Reinhardt and Staedel, *Ber.* **16**, 29.

out as an oil, which is extracted with ether and converted into methylaniline by treatment with tin and hydrochloric acid.[1] Pure methylaniline is also obtained by the action of methyl iodide on sodium acetanilide, $C_6H_5N(C_2H_3O)Na$, and the decomposition of the compound thus obtained by caustic potash.[2]

It may also be prepared from the corresponding methylformanilide, which is obtained by adding methyl bromide to an alcoholic solution of formanilide, and then an alcoholic solution of caustic potash.[3]

Methylaniline is a liquid resembling aniline, boiling at 191° (712 mm.) and giving no colouration with bleaching powder solution.

Methylphenylnitrosamine, $C_6H_5N(CH_3)NO$, is a light yellow oil possessing an aromatic odour, which solidifies on cooling to 2°, and melts on reheating at 12—15°. It is also volatile with steam.[4]

Ethylaniline, $C_6H_5.NH.C_2H_5$, is obtained together with diethylaniline by the action of ethyl bromide on aniline, and also by the action of caustic potash on ethylformanilide and ethylacetanilide. It boils at 203°·5 (712 mm.) and yields a liquid nitrosamine.

Other secondary anilines have been obtained from formanilide in a similar manner to methylaniline. Their boiling points are as follows.

		Boiling-point.	
Propylaniline,	$C_6H_5.NH.C_3H_7$ . .	219·5°	(716 mm.)
Isopropylaniline,	$C_6H_5.NH.C_3H_7$ . .	209—210°	(712 mm.)
Isobutylaniline,	$C_6H_5.NH.C_4H_9$ . .	229—230°	(716 mm.)
Isoamylaniline,	$C_6H_5.NH.C_5H_7$ . .	252·5°	(730 mm.)

1062 *Dimethylaniline*, $C_6H_5.N(CH_3)_2$, is easily obtained pure by heating trimethylphenylammonium iodide[5] in a current of hydrochloric acid, as well as by the dry distillation of its hydroxide. It forms with trinitrobenzene the compound $C_6H_5.N(CH_3)_2,C_6H_3(NO_2)_3$, which crystallizes in fine, lustrous, dark violet needles (Hepp). It is a colourless liquid, boiling at 192°, and solidifying at 0·5°.

It is manufactured on the large scale, and employed for making colours, but the material used for this purpose, as already

[1] Nölting and Boasson, *Ber.* **10**, 795.
[2] Hepp, *ibid.* 327.
[3] Pictet and Crépieux, *Ber.* **21**, 1107.
[4] Fischer, *Annalen*, **190**, 151.
[5] Merril, *J. Pr. Chem.* II. **17**, 286.

explained, is not the pure compound, but contains more or less methylaniline, and sometimes aniline. It was first prepared with methyl iodide, which is still used in the colour industry for other purposes, but when the price of iodine rose methyl nitrate was employed as a substitute. By the use of this very explosive substance, however, many fatal accidents were caused, and commercial "methylaniline" was then prepared from aniline hydrochloride and wood spirit—a method which is still employed.

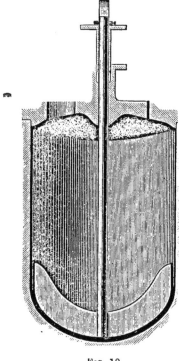

FIG. 12   FIG. 13.

The compounds are heated for some hours in an autoclave (Figs. 12 and 13) to 250—280°, a mixture of 100 parts of aniline hydrochloride with 58 to 80 parts of methyl alcohol, or 40 parts of aniline hydrochloride, 30 parts of aniline, and 45 parts of methyl alcohol being used. The latter must be pure and quite free from acetone, because this not only diminishes the yield of dimethylaniline, but gives a product which is of little value for the preparation of methyl-violet.[1]

[1] Krämer and Grodzky, Ber. **13,** 1006.

In addition to methylaniline and dimethylaniline, the product contains higher boiling bases, such as dimethyltoluidine, $C_6H_4(CH_3)N(CH_3)_2$, dimethylxylidine, $C_6H_3(CH_3)_2N(CH_3)_2$, etc., which are obtained by the entrance of methyl into the benzene nucleus.[1] The portion boiling between 198—205° forms the technical methylaniline.

As methyl chloride is now prepared on the large scale it is used in the manufacture of "methylaniline." A mixture of caustic soda, or milk of lime, and aniline is heated to 100°, and a stream of methyl chloride passed in, the mixture being well agitated (Fig. 14). The proportions of the materials correspond to the following equation :

$$C_6H_5.NH_2 + 2NaOH + 2CH_3Cl = 2NaCl + 2H_2O + C_6H_5.N(CH_3)_2.$$

The pressure must not exceed six atmospheres. The dimethylaniline is then driven over with steam, and contains about 5 per cent. of monomethylaniline, but no other by-product.

Trimethylphenylammonium iodide, $(CH_3)_3C_6H_5.N.I$, is a crystalline body which is formed when methyl iodide and dimethylaniline are brought together, the reaction being very violent. The hydroxide, obtained by means of silver oxide, is crystalline and very corrosive, and has a very bitter taste. It is very deliquescent, but its salts crystallize well, the picrate being very slightly soluble in water.[2]

Methylethylaniline, $C_6H_5.N(CH_3)(C_2H_5)$ is a liquid boiling at 201°, whilst *diethylaniline*, $C_6H_5.N(C_2H_5)_2$ boils at 213°·5.

Hofmann has also prepared several aniline derivatives containing other alcohol radicals.[3] Thus by the action of ethylene bromide on aniline he obtained the following compound :[4]

Melting-point.

Ethylene-aniline, $C_2H_4\!\!\begin{array}{l} \diagup NH.C_6H_5 \\ \diagdown NH.C_6H_5 \end{array}$ small lustrous plates 63°

Hydroxyethylaniline, $C_6H_5.NH.C_2H_4.OH$, is obtained by warming aniline with ethylene oxide, and is an oily liquid

---

[1] Hofmann and Martius, *Ber.* **4**, 742 ; **6**, 345.
[2] Lauth, *Bull. Soc. Chim.* **7**, 448.   [3] *Annalen*, **74**.
[4] *Jahresb.* **1858**, 352 ; **1859**, 388 ; Gretillat, *ibid.* **1873**, 698 ; Morley, *Ber.* **12**, 1794.

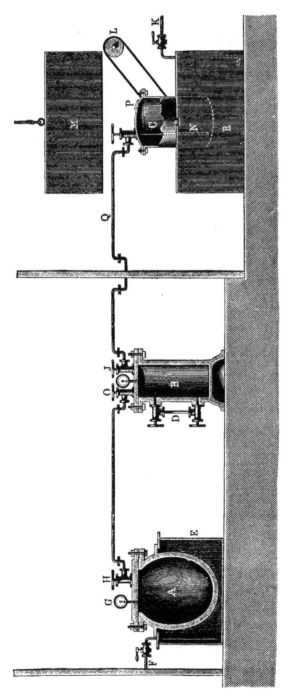

FIG. 14.

A. Autoclave for the preparation of the methyl chloride, with manometer G, and tap H ; it is placed in a steam chest E, which is heated by the pipe F. B. Receiver, in which the methyl chloride is condensed by its own pressure, with manometer, gauge, D, and taps O and J. C. Autoclave in which the methylation takes place ; it is placed in a water bath, R, which is heated by steam from K. Q N. Delivery tube. P. Screw valve by which the pressure can be lessened. L. Shaft for moving the agitator. M. Lid to cover the autoclave during the operation.

which boils at about 280°, and gives a green colouration with bleaching-powder solution. Its salts are very soluble and crystallize with difficulty.[1]

Hydroxymethylethylaniline or *methylphenylethylalkine*, C_6H_5N $(CH_3)C_2H_4OH$, is obtained by heating methylaniline for some time with ethylene chlorhydrin. It is a liquid possessing a weak smell similar to that of aniline, and distilling under diminished pressure without decomposition. On continued standing in the air it is converted into a fine, blue, thick, syrupy liquid, which dissolves in water and still more readily in alcohol. Its salts are readily soluble and very deliquescent.[2]

Phenyltaurine, $C_6H_5.NH.C_2H_4.SO_3H$, is obtained when the aniline salt of chlorisethionic acid, $C_2H_4Cl.SO_3H$ (Vol. III. Part II., p. 62), is heated with aniline. It crystallizes from hot water in fine thin plates which have a silky lustre and an acid reaction and taste. It gives an intense violet colouration with bleaching-powder solution, and on warming its solution with ferric chloride it is coloured green and then gradually indigo-blue.[3]

1063 *Phenylaniline* or *diphenylamine*, $(C_6H_5)_2NH$.—Hofmann first obtained this compound by the dry distillation of aniline blue,[4] and then of aniline and related bodies. It is more simply obtained by heating aniline hydrochloride with aniline :[5]

$$C_6H_5.NH_2 + C_6H_5.NH_2.ClH = (C_6H_5)_2NH.ClH + NH_3.$$

It is prepared in this way on the large scale, by heating 6 parts aniline and 7 parts aniline hydrochloride under a pressure of four to five atmospheres, and at a temperature of 250° for twenty-four hours. From time to time the ammonia formed must be allowed to escape to prevent a transformation of the diphenylamine into aniline. The product is treated with warm hydrochloric acid, and then a large quantity of water added to dissolve the aniline hydrochloride. Free diphenylamine separates out, and may be purified by distillation.

It is almost insoluble in water, readily soluble in alcohol and ether, and crystallizes in small monosymmetric plates, which have an agreeable odour of flowers, and melt at 54°. It boils at 302° (Gräbe), and is a weak base, its salts being decomposed by water.

[1] Demole, *Annalen*, **173**, 126.   [2] Laun, *Ber.* **17**, 675.

[3] James, *Journ. Chem. Soc.* 1885, **1**, 367.

[4] *Annalen*, **132**, 160.

[5] De Laire, Girard and Chapoteaut, *ibid.* **140**, 344.

A deep blue colouration is produced by dissolving it in hydrochloric acid and adding nitric acid drop by drop; this also takes place when its solution in pure sulphuric acid is mixed with sulphuric acid containing nitrous acid, and for this reason it is employed as a test for nitrous acid.[1]

Of its numerous substitution products only the following will be mentioned here:

Hexnitrodiphenylamine, or *dipicrylamine*, $(C_6H_2(NO_2)_3)_2NH$, is obtained by the action of nitric acid on diphenylamine and methyldiphenylamine.[2] It is almost insoluble in water, but dissolves somewhat more readily in alcohol, and crystallizes from hot glacial acetic acid in light yellow prisms, melting at 238°. The hydrogen of the imido-group can be replaced by metals; the ammonium salt, $C_{12}H_4(NO_2)_6N(NH_4)$, occurs in commerce under the name *Aurantia;* it forms reddish-brown crystals, and dyes wool and silk a beautiful orange. Goods dyed with this colouring matter produce an irritation of the skin and exanthema on certain individuals, while on others even concentrated solutions produce no effect. Martius has never observed these poisonous effects, and considers that they arise from an impurity.[3]

Methyldiphenylamine, $(C_6H_5)_2NCH_3$, is obtained when diphenylamine is heated with methyl iodide, as well as by the action of wood-spirit on diphenylamine at 250—300°.[4] On the large scale it is prepared by heating a mixture of 100 parts of diphenylamine, 68 parts of hydrochloric acid of specific gravity 1·17, and 24 parts of methyl alcohol for about ten hours, under a pressure of fifteen atmospheres, to 200—250°.

The base is separated from the product by caustic soda, distilled, and shaken with double its volume of concentrated hydrochloric acid, when the diphenylamine hydrochloride separates out in the solid form, while the salt of the methyl compound remains liquid, and by treatment with a large quantity of water is decomposed.[5]

Methyldiphenylamine is a liquid boiling at 282°. Oxidizing agents give various colour reactions with it; with dilute sulphuric acid a solution is obtained which is very similar to that of potassium permanganate. It is employed in the preparation of blue colouring matters.

[1] E. Kopp, *Ber.* **5**, 284.
[2] Gnehm, *ibid.* **7**, 1399; **9**, 1245; Mertens, **11**, 845.
[3] *Ibid.* **9**, 1247.   [4] Bardy, *ibid.* **3**, 838.
[5] Girard, *Bull. Soc. Chim.* **23**, 2.

Triphenylamine, $(C_6H_5)_3N$.—Potassium dissolves in aniline with elimination of hydrogen, and formation of the compounds, $C_6H_5.NHK$ and $C_6H_5.NK_2$, inasmuch as by the action of bromobenzene a mixture of diphenylamine and triphenylamine is obtained. Fused diphenylamine likewise dissolves potassium, the product thus obtained yielding triphenylamine when treated with bromobenzene. Sodium may be substituted for potassium, but the temperature employed must then be higher.[1] It crystallizes from ether in splendid, vitreous, monosymmetric pyramids melting at 127°. Its solution in glacial acetic acid is coloured green by a little nitric acid, but violet, and then blue by sulphuric acid.[2] It does not combine with acids, and forms no compound with picric acid, but yields a trinitro- and triamidoderivative.[3]

p-Hydroxydiphenylamine, $C_6H_5.NH.C_6H_4OH$, is obtained by heating quinol with aniline and calcium chloride to 250° for ten hours. It is scarcely soluble in cold, slightly soluble in hot water and light petroleum, but readily dissolves in alcohol and benzene, from which it crystallizes in small plates, melting at 70°, and solidifying to a scaly mass ; it boils above 340° without decomposition. As it is at the same time a phenol and an amido-base it forms compounds with bases as well as with acids.

By heating it with methyliodide and caustic potash, the methyl ether of methylhydroxydiphenylamine, $C_6H_5(NCH_3)C_6H_4.OCH_3$. is obtained as a yellow oily liquid, boiling at 313°, and smelling like the violet or geranium.

m-Hydroxydiphenylamine is obtained from resorcinol in an analogous manner to the para-compound, and crystallizes in small plates having a pearly lustre, and melting at 81°·5.

Both compounds are reduced to diphenylamine by heating with zinc dust, while by continued heating with aniline, calcium chloride, and zinc chloride, they are converted into diphenyldiamidobenzenes, $C_6H_4(NHC_6H_5)_2$, which will be subsequently described.[4]

---

[1] Kleber, *Ber.* **18**, 2156.   [2] Merz and Weith, *ibid.* **6**, 1514.
[3] Heydrich, *ibid.* **18**, 2157.
[4] Calm, *ibid.* **16**, 2786 ; Calm and Philip, *ibid.* **17**, 2431.

ANILIDES.

1064 These bodies are obtained by replacing the hydrogen of the amido-group by acid radicals. They are formed by the action of aniline on the ethers, anhydrides, and chlorides of the acids, and also by heating salts of aniline. If more than one hydrogen atom in aniline be replaced by halogens or nitroxyls, and its basicity thus weakened, or entirely destroyed, no anilide is formed on heating with a strong acid, such as acetic acid, and even acetic anhydride has often no action. An anilide is, however, nearly always obtained by heating with an acid chloride in a sealed tube.

When aniline is treated with an acid chloride, aniline hydrochloride is formed, and, being a solid, readily incloses some free aniline, thus rendering the action of the chloride incomplete. It is therefore better, especially in the preparation of the acetyl derivatives, to dissolve the aniline, or substituted aniline, in a molecule of the acid, and then to add a molecule of the chloride.

The anilides are decomposed by boiling with caustic soda into the acid and aniline. In the case of the higher substituted anilides it is often better to effect the decomposition by heating with alcoholic ammonia, or hydrochloric acid in a sealed tube, or to heat with concentrated sulphuric acid to 100°, almost all anilides being in this way readily decomposed. The product is then diluted with water and neutralized with an alkali, the substituted aniline being extracted with ether, chloroform, &c., or obtained by distillation.

The anilides are generally slightly soluble in cold water, crystallize well, and are, by reason of their chemical indifference and stability, peculiarly adapted for the preparation of substituted anilines, which cannot be obtained by the direct action of halogens or nitric acid, since these act too violently upon aniline. Acetanilide is generally employed in the preparation of substitution products of aniline.[1]

Formanilide, $C_6H_5.N(CHO)H$, was obtained by Gerhardt, together with oxanilide, aniline, carbon dioxide and water, by heating normal aniline oxalate.[2] According to Hofmann, the

---

[1] Beilstein, *Org. Chem.* 2nd Edition, **2**, 230.
[2] *Annalen*, **60**, 310.

chief product is oxanilide, but when acid aniline oxalate is rapidly distilled, formanilide is almost exclusively formed :

$$C_6H_5.NH_2 + C_2H_2O_4 = C_6H_5.N(CHO)H + CO_2 + H_2O.$$

At the same time by-products are obtained in small quantity.[1] Pure formanilide is also easily obtained when ethyl formate is digested with aniline.[2]

Formanilide is tolerably soluble in cold, and somewhat more readily in hot water, and crystallizes on spontaneous evaporation of its solution in long, flat, four-sided prisms, melting at 46°.

Alkyl derivatives of formanilide are obtained by treating the latter with caustic potash and an alkyl iodide. They form colourless, almost odourless oils, which solidify in a freezing mixture, and boil at the same temperatures as the alkylacetanilides.[3]

Diphenylformamide, $(C_6H_5)_2N(CHO)$, is obtained as a by-product in the preparation of diphenylamine-blue, which is formed by heating diphenylamine with oxalic acid. It is also obtained from formic acid and diphenylamine. It is insoluble in water, and crystallizes from alcohol in splendid rhombic crystals, which melt at 73—74°.[4]

Thioformanilide, $C_6H_5.N(CHS)H$, is obtained by the direct combination of sulphuretted hydrogen with phenyl carbamine,[5] as well as by the action of phosphorus pentasulphide on formanilide.[6] It crystallizes from boiling water in long white needles, which melt at 137°·5, and have an intensely bitter taste.

1065 *Acetanilide* or *phenylacetamide*, $C_6H_5.N(C_2H_3O)H$.—Gerhardt prepared this substance by the action of acetyl chloride or acetic anhydride on aniline.[7] In order to obtain it, aniline is boiled with glacial acetic acid for one or two days, and the acetanilide isolated from the product by fractional distillation.[8] It can be further purified by recrystallization from water or benzene. It is moderately soluble in ether and alcohol, and slightly in water, from which it crystallizes in lustrous tablets, consisting of small, rhombic plates. It melts at 112°, and boils at 295°. On heating it for a long time with sodium ethylate in a sealed tube to 170° —200°, it forms ethylaniline, but when the mixture is distilled from a retort, ethyl alcohol and sodium acetanilide are formed.[9]

[1] *Annalen*, **142**, 121.   [2] Hofmann, *Jahresb.* **1865**, 410.
[3] Pictet and Crépieux, *Ber.* **21**, 1107.   [4] Girard and Willm, *Jahresb.* **8**, 1195.
[5] Hofmann, *ibid.* **10**, 1095.   [6] *Ibid.* **11**, 338.
[7] *Annalen*, **87**, 164.   [8] Williams, *ibid.* **131**, 288.
Seifert, *Ber.* **18**, 1355, 1358.

Sodium acetanilide, $C_6H_5.N(C_2H_3O)Na$, is also obtained by adding sodium to a solution of acetanilide in xylene;[1] it is a crystalline powder, or a radiating crystalline mass.

Mercuric acetanilide, $(C_6H_5.NC_2H_3O)_2Hg$, is obtained by heating acetanilide with mercuric oxide ; it crystallizes from alcohol in small needles.[2]

Acetanilide hydrochloride, $C_6H_5.NH(C_2H_3O)HCl$, is formed by passing hydrochloric acid through a solution of acetanilide in acetone. It crystallizes in pliable needles, which are decomposed by water into their components, and are gradually converted into aniline hydrochloride and acetic acid in the air.[3]

When acetanilide is heated with zinc chloride for several hours to 250—260°, *Flavaniline*, $C_6H_{14}N_2.ClH$, a beautiful yellow colouring matter, is obtained, having a splendid moss-green fluorescence, which shows brilliantly on silk.[4] This body will be more fully described later on.

Di-acetanilide, $C_6H_5.N(C_2H_3O)_2$, is formed by heating glacial acetic acid with phenyl mustard oil to 130—140° :

$$C_6H_5.NCS + 2C_2H_3O.OH = C_6H_5N(C_2H_3O)_2 + H_2S + CO_2.$$

It resembles acetanilide in its properties, and melts at 111°.[5]

Methylacetanilide, $C_6H_5.N(CH_3)C_2H_3O$, crystallizes from hot water in splendid long needles, which melt at 99°·5 and boil at 245°.[6]

Ethylacetanilide, $C_6H_5N(C_2H_5)C_2H_3O$, is prepared from acetanilide by heating it with caustic potash and ethyl bromide,[7] and by the action of acetyl bromide on diethylaniline. It melts at 54·5°, and boils at 248—250°.

Diphenylacetamide, $(C_6H_5)_2N(C_2H_3O)$, crystallizes from light petroleum in large tablets having a pearly lustre, and melting at 99°·5.[8]

Thiacetanilide, $CH_3CS.NH(C_6H_5)$, is obtained by heating acetanilide with phosphorus pentasulphide. It crystallizes from boiling water in splendid yellowish needles, melting at 75°,[9] and is soluble in caustic soda, forming a sodium salt which can be obtained in compact crystals. By the action of methyl iodide a compound is obtained which has the composition of methyl-

[1] Bunge, *Annalen Suppl.* **7**, 122.   [2] Pfaff and Oppenheim, *Ber.* **7**, 624.
[3] Nölting and Weingärtner, *ibid.* **18**, 1340.
[4] Fischer and Rudolph, *ibid.* **15**, 1500.   [5] Hofmann, *ibid.* **3**, 770.
[6] Hofmann, *ibid.* **10**, 599.   [7] Pictet, *ibid.* **20**, 3423.
[8] Merz and Weith, *ibid.* **6**, 1511.   [9] Hofmann, *ibid.* **11**, 338.

thiacetanilide, but which is entirely different from the substance obtained by the action of phosphorus pentasulphide on methyl acetanilide, and is therefore named methylisothiacetanilide. It is decomposed by hydrochloric acid with formation of methyl-thiacetate and aniline, from which it follows that the methyl is combined with the sulphur, and that in the formation of the isothiacetanilide a molecular change has taken place. The constitution of these isomeric compounds is shown by the following formulæ:

Methylthiacetanilide.

$$CH_3.CS-N\begin{cases}CH_3 \\ C_6H_5\end{cases}$$

Methylisothiacetanilide.

$$CH_3.C\begin{cases}SCH_3 \\ NC_6H_5.\end{cases}$$

Methylthiacetanilide, $CH_3.CS.N(CH_3)C_6H_5$, crystallizes from chloroform in monosymmetric tablets, melting at 58—59°. It is insoluble in water and alkalis, and boils, with slight decomposition, at 290°.[1]

Methylisothiacetanilide, $CH_3.C(SCH_3)NC_6H_5$, is readily obtained by the action of methyl iodide on an alkaline solution of thiacetanilide.[2] It is an oily liquid, boiling at 244—245°.[3] On heating it with methyl iodide to 100°, a crystalline mass is obtained, which is decomposed by water with formation of methyl thiacetate and methylaniline hydriodide, the following reactions probably taking place:[4]

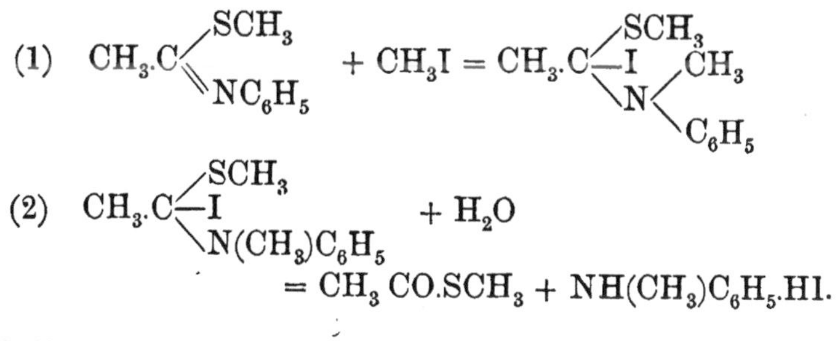

$$(1) \quad CH_3.C\begin{cases}SCH_3 \\ NC_6H_5\end{cases} + CH_3I = CH_3.C{-}I\begin{cases}SCH_3 \\ N\begin{cases}CH_3 \\ C_6H_5\end{cases}\end{cases}$$

$$(2) \quad CH_3.C{-}I\begin{cases}SCH_3 \\ N(CH_3)C_6H_5\end{cases} + H_2O$$
$$= CH_3 CO.SCH_3 + NH(CH_3)C_6H_5.HI.$$

Ethylisothiacetanilide, $CH_3.C(SC_2H_5)NC_6H_5$, is formed when alcoholic solutions of sodium ethylate and thiacetanilide are mixed (Wallach). It is a liquid which is heavier than water, and boils at 255—257° (Wallach and Bleibtreu). On passing hydrochloric acid through its ethereal solution, a white mass separates out, which is very soluble in water, and forms with

---

[1] Wallach, *Ber.* **13**, 528.　　　　[2] *Ibid.* **11**, 1595.
[3] Wallach and Bleibtreu, *ibid.* **12**, 1061.　　[4] Wallach, *ibid.* **13**, 529.

platinum chloride the slightly soluble double salt $(C_{10}H_{13}N.S.HCl)_2$ $PtCl_4$.

Wallach and Bleibtreu have also prepared several other isothiacetanilides.

1066 *Glycolylanilide*, $C_6H_5.NH(CO.CH_2.OH)$, is obtained by heating glycolide with aniline to 130°. It is moderately soluble in cold and very readily in hot water, from which it crystallizes in long prismatic needles, melting at 108°.[1]

Phenylglycocoll or *phenylamido-acetic acid*,$(C_6H_5)NH.CH_2.CO_2H$, is isomeric with the preceding compound. It is best obtained by heating an aqueous solution of equal molecules of aniline, chloracetic acid, and sodium carbonate. It is tolerably soluble in hot water, and separates on cooling in imperfect crystals, which melt at 126—127°. Its solution does not dissolve freshly precipitated mercuric oxide, or silver oxide, but, on the other hand it forms a salt with copper hydroxide, which crystallizes in small dark green plates. In other respects it behaves very similarly to amido-acetic acid.[2]

Several homologues of phenylamido-acetic acid are known.

Phenyl betaïne.—The chloride of this body is obtained by heating dimethylaniline with chloracetic acid:

$$\begin{array}{l} CH_3 \\ \diagdown \\ N{-}C_6H_5 + CH_2Cl.CO_2H = \\ \diagup \\ CH_3 \end{array} \quad \begin{array}{l} CH_3 Cl \\ \diagdown \diagup \\ N{-}C_6H_5 \\ \diagup \diagdown \\ CH_3 CH_2.CO_2H. \end{array}$$

Ether precipitates it from its concentrated aqueous solution in long white needles. Its ethyl salt is obtained in a similar manner from ethyl chloracetate. With moist silver oxide both compounds give the corresponding strongly alkaline, and deliquescent hydroxides.[3]

By heating the ethyl salt for a long time to 130°, and warming the residue with concentrated hydrochloric acid, phenylmethyl-glycocoll hydrochloride is obtained:

$$C_6H_5N(CH_3)_2Cl.CH_2.CO_2.C_2H_5$$
$$= C_6H_5N(CH_3)CH_2.CO_2.C_2H_5 + CH_3Cl.$$

$$C_6H_5N(CH_3)CH_2.CO_2.C_2H_5 + 2ClH$$
$$= C_6H_5N(CH_3)CH_2.CO_2H,ClH + C_2H_5Cl.$$

[1] Norton and Tscherniak, *Bull. Soc. Chim.* **30**, 104.
[2] *Ber.* **10**, 2046. [3] Zimmermann, *Ber.* **12**, 2206.

It crystallizes in prisms, which are readily soluble in water, and slightly in alcohol; on continued heating with water it decomposes into carbon dioxide and dimethylaniline hydrochloride.

Phenyl betaïnamide.—The chloride of this body is formed by heating an alcoholic solution of chloracetamide and dimethylaniline. It forms colourless crystals which on heating to 110— 120° decompose into methyl chloride and methylphenylglycolamide :

$$\begin{array}{c} CH_3 \\ C_6H_5 \\ CH_3 \end{array}\!\!\!> \!<\!\!\!\begin{array}{c} Cl \\ \\ CH_2.CO.NH_2 \end{array} = \begin{array}{c} CH_3 \\ \\ C_6H_5 \end{array}\!\!\!>\!\!N\!-\!CH_2.CO.NH_2 + CH_3Cl.$$

Methylphenylglycolamide is also obtained when methylaniline is heated with chloracetamide. It is slightly soluble in cold, more readily in hot water and alcohol, and crystallizes in small plates or prisms, having a satin-like lustre, and melting at 163°.[1]

Phenylimido-acids are obtained by the action of aniline on ketonic acids, the oxygen of the ketone being replaced by the divalent group C_6H_5N.

Anilpyroracemic acid or *phenyl - a - imidopropionic acid,* $CH_3.C(NC_6H_5)CO_2H$, is obtained by mixing ethereal solutions of pyroracemic acid and aniline, the mixture being kept cool.[2] It forms small crystals, which melt at 122° with evolution of carbon dioxide. On boiling it with water, carbon dioxide, aniline, and aniluvitonic acid, $C_{11}H_9NO_2$,[3] which will be described later on, are formed.

Anilacetoacetic acid or *phenyl - β - imidobutyric acid,* $CH_3.C(NC_6H_5)CH_2.CO_2H$, is obtained by heating aniline with ethyl acetoacetate to 150—160° :

$$C_6H_5.NH_2 + CO\!<\!\!\!\begin{array}{c} CH_3 \\ \\ CH_2.CO_2.C_2H_5 \end{array} = C_6H_5N\!\!=\!\!C\!<\!\!\!\begin{array}{c} CH_3 \\ \\ CH_2.CO_2.C_2H_5 \end{array}$$

$$+ H_2O = C_6H_5N\!\!=\!\!C\!<\!\!\!\begin{array}{c} CH_3 \\ \\ CH_2.CO_2H \end{array} + HO.C_2H_5.$$

It forms crystals which are tolerably soluble in water, and very readily in alcohol and benzene. It is coloured dark violet by ferric chloride, and combines with acids and bases.[4]

[1] Silberstein, *Ber.* **17**, 2660.   [2] Böttinger, *Annalen,* **188**, 336.
[3] *Ibid.* **191**, 321.   [4] Knorr, *Ber.* **17**, 540.

Oxanilide, $C_2O_2(NH.C_6H_5)_2$, is obtained by heating aniline oxalate to 160—180° till no more gas is evolved; the formanilide produced at the same time can be removed by washing with cold alcohol.[1] It is also formed when cyananiline is evaporated with dilute hydrochloric acid.[2] Oxanilide is insoluble in water, ether, and cold alcohol, slightly soluble in boiling alcohol, and more readily in benzene, from which it crystallizes in white pearly scales, melting at 245°, and solidifying on cooling to a radiating mass. It boils at 320°, but sublimes at a lower temperature in small iridescent plates; its vapour has a penetrating odour similar to that of benzoic acid.

Thioxanilide, $C_2S_2(NH.C_6H_5)_2$.—When oxanilide is warmed with phosphorus pentachloride and the product then treated with sulphuretted hydrogen, thioxanilide is obtained, together with an intensely red-coloured body, which is removed by dissolving the product in caustic soda and passing in carbon dioxide; the thio-anilide separates out in splendid plates similar to those of mosaic gold, which melt at 133°.[3]

Oxanilic acid, $NH(C_6H_5)C_2O_2.OH$, is prepared by heating 20 grms. of aniline and 25 of oxalic acid to 130–140°, and dissolving the mass in boiling water. On cooling, the aniline salt of oxanilic acid separates out, and is decomposed by sulphuric acid, the oxanilic acid thus set free being taken up with ether.[4] The residue remaining after distilling off the ether is then recrystallized from boiling water. It is thus obtained in satin-like needles containing a molecule of water; it crystallizes from benzene in long fine, lustrous needles, melting at 150°,[5] and forms crystallizable salts.

Ethyl oxanilate, $NH(C_6H_5)C_2O_2.OC_2H_5$, is obtained by the continued heating of ethyl oxalate and aniline. It crystallizes from alcohol in tablets or prisms, melting at 66—67°. On adding ammonia to its alcoholic solution, *phenyloxamide*, $(C_6H_5)HN.C_2O_2.NH_2$, is precipitated (Klinger), which is also obtained by the action of hydrochloric acid on cyananiline. It separates from hot alcoholic or aqueous solution in snow-white, hair-like flakes with a silky lustre, which melt at 224° and readily sublime; "the sublimate is as light and mobile as precipitated silica" (Hofmann).

Malonanilic acid, $NH(C_6H_5)CO.CH_2.CO_2H$.—By heating

[1] Gerhardt, *Annalen*, **60**, 308.   [2] Hofmann, *ibid.* **73**, 180.
[3] Wallach and Pirath, *Ber.* **12**, 1063 ; Wallach, *ibid.* **13**, 527.
[4] Aschan, *Ber.* **21**, 288c.   [5] Klinger, *Annalen*, **184**, 265.

equal molecules of malonamide and aniline to 200—220°, *phenyl-malonamide*, $NH(C_6H_5)CO.CH_2.CO.NH_2$, is obtained, crystallizing from hot water in fine, matted needles, melting at 163°, and yielding the calcium salt of malonanilic acid on boiling with milk of lime.[1] It is also formed by heating malonic acid with aniline,[2] and by the following remarkable reaction. On passing carbon dioxide through sodium acetanilide, sodium acetylphenylcarbonate is obtained as a crystalline powder, which on heating to 130—140° undergoes an intermolecular change:

$$N \underset{\diagdown CO.CH_3}{\overset{\diagup C_6H_5}{\underset{}{\underline{\quad}CO.ONa}}} = N \underset{\diagdown H}{\overset{\diagup C_6H_5}{\underset{}{\underline{\quad}CO.CH_2.CO.ONa.}}}$$

Malonanilic acid crystallizes from warm water in oblique, lustrous prisms, which melt at 132°, and completely decompose on continued heating, into carbon dioxide and acetanilide.[3]

Anilides of the homologues of malonic, malic, and tartaric acids, &c. are also known.

PHENYLAMIDINES.

1067 Wallach[4] designated by the term amidines a series of bases which are obtained from the acid-amides by replacing the oxygen atom by the group NR, in which R signifies a monad radical or hydrogen. The simplest amidine is acetdiamine, which has already been described as acetamidine $CH_3.C \underset{\diagdown NH.}{\overset{\diagup NH_2}{\underset{}{}}}$

The phenylated amidines are obtained by treating a mixture of aniline and an anilide with phosphorus trichloride, which acts as a dehydrating agent;[5] when acetanilide is employed ethenyldiphenylamidine is obtained:

$$CH_3.\underset{\underset{O}{\|}}{C}.NH.C_6H_5 + NH_2.C_6H_5 = CH_3\underset{\underset{N.C_6H_5.}{\|}}{C}.NH.C_6H_5 + H_2O$$

[1] Freund, *Ber.* **17**, 133.   [2] Rügheimer, *ibid.* **17**, 235.
[3] Seifert, *ibid.* **18**, 1358.   [4] Wallach, *ibid.* **8**, 1575.
[5] Hofmann, *Zeitschr. Chem.* **1866**, 161.

Lippmann obtained this compound by the action of phosphorus pentachloride on acetanilide.[1] According to Wallach and Hofmann [2] the following reactions take place:

(1) $CH_3CO.NH.C_6H_5 + PCl_5 = CH_3.CCl_2.NH.C_6H_5 + POCl_3.$
(2) $CH_3.CCl_2.NH.C_6H_5 + NH_2.C_6H_5 = CH_3.C(NC_6H_5)NH.C_6H_5 + 2HCl.$

The phenylated amidines are monacid bases, which are scarcely soluble in water but take up its elements when boiled with dilute alcohol, decomposing into aniline and an anilide.

On heating those amidines which contain the group NH with carbon disulphide, a thio-anilide is formed, together with thiocyanic acid, which combines with a part of the amidine (Bernthsen):

$$CH_3.C(NH)NH.C_6H_5 + CS_2 = CH_3.CS.NH.C_6H_5 + CNSH.$$

Methenyldiphenylamidine, $CH(NC_6H_5)NH.C_6H_5.$ Hofmann first obtained this compound by heating chloroform with aniline to 180—190°, and named it formyldiphenyldiamine.[3] Ethyl orthoformate can be employed instead of chloroform.[4] It is likewise obtained when formic acid is boiled with aniline or when the latter is heated with phenyl carbamine:[5]

$$NC.C_6H_5 + NH_2.C_6H_5 = HC{\Large\langle}\genfrac{}{}{0pt}{}{NC_6H_5}{NH.C_6H_5}.$$

It crystallizes from hot alcohol or benzene in long needles, melting at 135—136°, and volatilizing at a higher temperature with partial decomposition. On heating it to 140—150° in a stream of sulphuretted hydrogen, it decomposes into aniline and thioformanilide.[6]

Ethenylphenylamidine, $CH_3.C(NH)NH.C_6H_5,$ is prepared by heating aniline hydrochloride and acetonitrile to 170.° It is an oily liquid having an alkaline reaction, and is decomposed on heating.[7]

Ethenyldiphenylamidine, $CH_3.C(NC_6H_5)NH.C_6H_5.$ The formation of this compound has already been explained. It is also

[1] *Ber.* **7**, 541.   [2] *Ibid.* **8**, 1567.   [3] *Jahresb.* **1858**, 354.
[4] Wichelhaus, *Ber.* **2**, 116.   [5] Weith, *ibid.* **9**, 454.
[6] Bernthsen, *Annalen,* **192**, 35.   [7] *Ibid.* **184**, 358.

obtained when acetanilide hydrochloride (p. 238) is heated to 250°, or when aniline hydrochloride is heated with acetonitrile to 230—240° (Bernthsen), as well as by the action of aniline on the isothiacetanilides : [1]

$$CH_3.C\begin{subarray}{l}\diagup SC_2H_5 \\ \diagdown NC_6H_5\end{subarray} + NH_2.C_6H_5 = CH_3.C\begin{subarray}{l}\diagup NH.C_6H_5 \\ \diagdown N.C_6H_5\end{subarray} + HS.C_2H_5.$$

In order to prepare it, 2 parts of phosphorus trichloride are gradually added to a well-cooled mixture of 1 part of acetic acid and 3 parts of aniline, and the whole heated for some hours to 160°. The resinous mass thus obtained is dissolved in boiling water, the cooled filtrate treated with caustic soda, and the precipitate recrystallized from alcohol.[2]

It forms small needles, melting at 131—132°,[3] and having a neutral reaction. The *hydrochloride*, $C_{14}H_{14}N_2.HCl$, crystallizes in small tablets. The *nitrate*, $C_{14}H_{14}N_2.HNO_3$, is very characteristic, and first forms an oily liquid, which soon solidifies to a crystalline mass.

Ethenylisodiphenylamidine, $CH_3.C(NH)N(C_6H_5)_2$, is obtained by heating diphenylamine and acetonitrile to 140—150°. It forms monosymmetric crystals, melting at 62—63°, and is a strong base.[4]

Isopentenyldiphenylamidine, $(CH_3)_2C_2H_3.C(NC_6H_5)NH.C_6H_5$ is a crystalline body, melting at 111°, obtained by Hofmann from valerianic acid.

CYANOGEN COMPOUNDS OF BENZENE.

1068 *Phenyl carbamine*, $C_6H_5.NC$. When chloroform is gradually added to a saturated alcoholic solution of caustic potash mixed with aniline, an energetic reaction is set up, and the carbamine, which is also called phenyl isocyanide, is formed. On distillation it passes over together with aniline; this is removed by oxalic acid, and the residual brown oil dried over potash and rectified.

Phenyl carbamine is a mobile liquid which appears green by transmitted, and blue by reflected light; it possesses the un-

[1] Wallach and Bleibtreu, *Ber.* **12**, 1063.   [2] Hofmann, *Jahresb.* **1865**, 414.
[3] Biedermann, *Ber.* **7**, 540.   [4] Bernthsen, *Annalen*, **192**, 25.

pleasant, penetrating smell of the carbamines, and its vapour produces on the tongue a characteristic bitter taste, and in the throat a sticky sensation. When it is distilled a portion boils constantly at 167°, the thermometer then rises rapidly to 230° and it is converted into an odourless liquid, which solidifies on cooling to a splendid crystalline mass.[1]

On heating phenyl carbamine to 200—220°, a crystalline body, probably identical with the preceding compound, is obtained, while another portion of the carbamine is converted into the isomeric benzonitrile $C_6H_5.CN$.[2] The aqueous mineral acids readily decompose phenyl carbamine, forming aniline and formic acid ; formanilide occurs in this reaction as an intermediate product. It combines with sulphuretted hydrogen to form thioformanilide.

Phenyl isocyanate or *phenylcarbimide*, $CO.N.C_6H_5$. Hofmann obtained this body, which was formerly called anilocyanic acid, or carbanil, by distilling diphenylcarbamide with phosphorus pentoxide ; he then found that it is better to use phenyl urethane :[3]

$$CO \begin{cases} NH.C_6H_5 \\ OC_2H_5 \end{cases} = CO.NC_6H_5 + HO.C_2H_5.$$

It is still more readily prepared by the action of carbonyl chloride on diphenyl carbamide or aniline hydrochloride :[4]

$$C_6H_5.NH_2 + COCl_2 = C_6H_5.NCO + 2HCl.$$

In this manner it can be readily prepared on the large scale.[5]

It is a colourless liquid, the vapour of which has a very penetrating odour and produces a flow of tears. It combines directly with alcohols [6] and phenols,[7] forming additive compounds. Thus with glycerol it reacts in the following manner :

$$C_3H_5(OH)_3 + 3CO.N.C_6H_5 = C_3H_5(O.CO.NH.C_6H_5)_3.$$

Phenylcarbimide hydrochloride, $CON(C_6H_5),HCl$, is obtained by passing hydrochloric acid through pure phenyl isocyanate, and forms a crystalline mass, melting at 45°.[8]

Phenyl di-isocyanate $(CO)_2.(N.C_6H_5)_2$. This compound has been prepared in a very remarkable manner, namely, by adding a

[1] Hofmann, *Annalen*, **144**, 117. [2] Weith, *Ber.* **6**, 213.
[3] Hofmann, *Annalen*, **74**, 9, 33 ; *Jahresb.* **1858**, 348 ; *Ber.* **3**, 654.
[4] Hentschel, *ibid.* **17**, 1284. [5] *Ibid.* **18**, 12 R.
[6] Tessmer, *Ber.* **18**, 968. [7] Snape, *ibid.* **18**, 2428.
[8] *Ber.* **18**, 1178.

drop of triethylphosphine to the foregoing compound. If, however, the isocyanate be present in great excess, crystals appear only after some time and gradually increase in number, but the liquid does not become solid even when allowed to stand for months.[1] It crystallizes from hot ether in thin iridescent plates, which melt at 175°, and on strongly heating are reconverted into the monomolecular compound.[2] By treatment with alcoholic ammonia it is converted into β-diphenylbiuret:

$$C_6H_5N{<}{\overset{CO}{\underset{CO}{}}}{>}NC_6H_5 + NH_3 = (C_6H_5)HN{<}{\overset{CO}{\underset{H_2N.CO}{}}}{>}NC_6H_5.$$

This body crystallizes in prisms, which melt at 164°, and are decomposed by hydrochloric acid with formation of phenyl isocyanate and ammonia.

a-Diphenylbiuret, $NH(CO.NH.C_6H_5)_2$, is obtained by heating biuret or ethyl allophanate (**Vol. III., Part I., p. 373**) with aniline, and by digesting monophenylcarbamide and phenyl isocyanate[3] at 120°. It forms crystals which are slightly soluble in alcohol, melt at 210°, and are converted by dry hydrochloric acid into aniline, cyanic acid, and phenyl isocyanate.[4]

Phenyl isocyanurate, $(CO)_3(NC_6H_5)_3$, is formed when an alcoholic solution of phenyl isocyanuramide is boiled for a short time with hydrochloric acid, or by heating 5 parts of phenyl isocyanate with 3 parts of potassium acetate for three hours to 100°, while if 10 parts of isocyanate are employed the complete polymerization only occurs at 188—200° (Hofmann). Phenyl isocyanurate crystallizes from hot alcohol in prisms, melting at 274—275.°

Phenyl cyanurate, $(CN)_3(OC_6H_5)_3$, is formed on passing cyanogen chloride or cyanuric chloride through an alcoholic solution of sodium phenate. It is insoluble in water and crystallizes from alcohol in long, fine needles, melting at 224°.[5]

1069 *Phenylcyanamide* or *cyananilide*, $C_6H_5NH(CN)$. Cahours and Cloëz obtained this substance by passing cyanogen chloride through an ethereal solution of aniline.[6] It is also obtained by the action of lead oxide on an alcoholic, or better, alkaline

[1] Hofmann, *Ber.* **18**, 764.   [2] *Ibid.* **3**, 755; **4**, 246.
[3] Kühn and Hentschel, *Ber.* **21**, 504.
[4] Hofmann, *Ber.* **4**, 265; Peitzsch and Salomon, *J. Pr. Chem.* II. **7**, 477.
[5] Hofmann and Olshausen, *Ber.* **3**, 257.   [6] *Annalen*, **190**, 91.

solution of phenyl thio-carbamide.[1] On evaporating the alcoholic solution it remains behind as a tenacious mass, which becomes crystalline in moist air or on continued contact with alcohol, and then crystallizes from ether in needles containing $\frac{1}{2}$ mol. of water, and melting at 47°. When water is added to its alcoholic solution, it combines to form phenyl carbamide, and on passing sulphuretted hydrogen through its solution in benzene, phenyl thio-carbamide is formed.[2]

Normal Triphenylmelamine, $C_6H_5.N:C{\Large<}{\substack{NH.C:(N.C_6H_5)\\NH.C:(N.C_6H_5)}}{\Large>}NH$,

is prepared by treating an ethereal solution of cyanuric chloride with aniline, and heating the precipitate formed to 150° with aniline,[3] or by heating trimethyl trithiocyanurate with aniline.[4] It crystallizes in slender needles, melting at 228°, and sublimes without decomposition. It has no basic properties and is decomposed by hydrochloric acid at 200° into cyanuric acid and aniline.

Three other isomeric triphenylmelamines are also known, in which 1, 2, and 3 phenyl groups respectively have passed from the side chain into the ring, on which account Rathke proposes the names, 1 eso-, 2 eso-, 3 eso-Triphenylmelamine for these compounds.

1 *eso-Triphenylmelamine*, $HN:C{\Large<}{\substack{N(C_6H_5).C:(N.C_6H_5)\\NH\text{——}C:(N.C_6H_5)}}{\Large>}NH$,

is prepared from phenylthiammeline by acting on it with ethyl bromide, and heating the mercaptide thus formed with alcoholic ammonia at 100°. It crystallizes from alcohol in small prisms, which melt at 221°. When heated with hydrochloric acid to 100—150° it is converted into normal triphenylmelamine.[5]

2 *eso-Triphenylmelamine*, $HN:C{\Large<}{\substack{N(C_6H_5).C:(NH)\text{—}\\N(C_6H_5).C:N(C_6H_5)}}{\Large>}NH$,

is formed when phenylthiocarbamide is boiled with alcohol and mercuric oxide. It forms small white needles, melting at 217°, which are insoluble in water but readily soluble in chloroform.[6]

3 *eso-Triphenylmelamine*, $HN:C{\Large<}{\substack{N(C_6H_5).C:(NH)\\N.(C_6H_5).C:(NH)}}{\Large>}N.C_6H_5$.

This compound, which is also known as isotriphenylmelamine, is prepared by warming aqueous phenylcyanamide on the water

---

[1] Hofmann, *Ber.* **3**, 266; Rathke, *ibid.* **12**, 733; Feuerlein, *ibid.* **12**, 1602; Berger, *Monatsh.* **5**, 217.   [2] Weith, *Ber.* **9**, 820.

[3] Klason, *J. Pr. Chem.* II. **33**, 294.   [4] Hofmann, *Ber.* **18**, 3218.

[5] Rathke, *Ber.* **21**, 867.   [6] Hofmann, *ibid.* **18**, 3226.

bath, and crystallizes from alcohol in thick needles, which melt at 185°. On boiling with hydrochloric acid the three C(NH) groups are successively converted into CO groups, the final product being triphenylisocyanurate.

A number of higher phenylmelamines, and of phenylammelines have also been prepared.

Cyananiline, $C_{14}H_{14}N_4$, is obtained when cyanogen is passed through an alcoholic solution of aniline :

$$\begin{array}{ccc} C_6H_5.NH_2 & CN & C_6H_5.NH.C{=}NH \\ & + \ | & = \ | \\ C_6H_5.NH_2 & CN & C_6H_5.NH.C{=}NH \end{array}$$

It is insoluble in water, slightly soluble in alcohol, and crystallizes in small lustrous iridescent plates, which melt at 210—220°, and decompose at a higher temperature.[1] As a diacid base it forms salts with acids, but they are rather unstable ; when heated with aqueous mineral acids it is converted into oxanilide and oxamide.

Phenyl thiocyanate, $C_6H_5.S.CN$, is formed by the action of thiocyanic acid on diazobenzene sulphate, or by passing cyanogen chloride through alcoholic lead thiophenate :

$$(C_6H_5S)_2Pb + 2CNCl = 2C_6H_5S.CN + PbCl_2.$$

It is most readily prepared by acting on a solution of diazobenzene sulphate with a paste of cuprous thiocyanate.[2] It is a colourless liquid boiling at 231°, which on standing becomes yellow. An alcoholic solution of potassium hydrosulphide decomposes it into thiophenol and potassium thiocyanate.[3]

Phenyl thiocarbimide or *phenyl mustard oil*, $C_6H_5.N.CS$. Hofmann obtained this body by the distillation of diphenylthiocarbamide (thiocarbanilide) with phosphorus pentoxide, and named it thiocarbanil or phenyl sulphocyanide.[4] It may be also obtained by heating the thiocarbamide with concentrated hydrochloric acid :[5]

$$CS{\Big<}{{}^{NH.C_6H_5}_{NH.C_6H_5}} = CS{=}N.C_6H_5 + NH_2.C_6H_5.$$

[1] Hofmann, *Annalen*, **66**, 129 ; **73**, 180.

[2] Gattermann and Haussknecht, *Ber.* **23**, 738 ; see also Thurnauer, *ibid.* **23**, 769.   [3] Billeter, *Ber.* **7**, 1753.

[4] *Jahresb* **1858**, 349.   [5] Merz and Weith, *Zeitschr. Chem.* **1869**, 589.

It is also formed by the action of thiocarbonyl chloride, $CSCl_2$, on aniline,[1] as well as by the direct combination of sulphur with phenyl carbamine,[2] and when phenyl isocyanate is heated with phosphorus pentasulphide.[3]

It is a liquid which has a smell similar to that of ordinary mustard oil, boils at 218°·5, has a sp. gr. 0·9398 at 20°, and combines with ammonia and the amines to form carbamides. It also combines with alcohols, forming ethers of thiocarbanilic acid (p. 252).

When chlorine is passed through its solution in chloroform, isocyanophenyl chloride, $C_6H_5.N.CCl_2$, is obtained as a heavy, yellow liquid, boiling at 211—212°. It possesses a very unpleasant pungent odour, and its vapour attacks the eyes and mucous membrane. Dry silver oxide acts on it violently with partial carbonization and formation of phenylcarbimide.[4]

Phenyl isoselenocyanate, $CSeNC_6H_5$, is obtained by the action of sodium selenide on isocyanophenyl chloride. It is a yellowish oil, which has a faint odour, and is readily soluble in alcohol and ether, but insoluble in water.[5]

PHENYLCARBAMIDES.

1070 *Phenylcarbamide* or *carbanilamide* $C_6H_5HN.CO.NH_2$. Hofmann obtained this substance by passing the vapour of cyanic acid through cold aniline, and also by the action of ammonia on phenylcarbimide, $C_6H_5N(CO)$. It is, however, most readily prepared by mixing a solution of aniline hydrochloride or sulphate with potassium cyanate, the phenylcarbamide soon separating out.[6] It is also obtained by heating carbamide with the theoretical quantity of aniline in a sealed tube to 150—170°.[7] It crystallizes from hot water in monosymmetric needles, melting at 147°,[8] and does not combine with nitric or oxalic acids.

Symmetric diphenylcarbamide or *carbanilide*, $CO(NH.C_6H_5)_2$, is formed by the combination of aniline with phenylcarbimide, or by the action of water on the latter, or by heating phenyl-

---

[1] Rathke, *Ber.* **3**, 861.   [2] Weith, *ibid.* **6**, 211.

[3] Michael and Palmer, *Am. Chem. Journ.* **6**, 257.

[4] Sell and Zierold, *Ber.* **7**, 1228.   [5] Stolte, *ibid.* **19**, 2351.

[6] *Annalen*, **53**, 57 ; **57**, 265 ; **70**, 130 ; **74**, 14 ; Weith, *Ber.* **9**, 820.

[7] Fleischer, *ibid.* **9**, 820.   [8] Steiner, *ibid.* **8**, 518.

carbamide, or when aniline is treated with carbonyl chloride, &c.[1]

It may also be obtained by heating aniline and phenyl carbonate to 150—180° in a sealed tube.[2]

To prepare it, one part of carbamide is heated with three parts of aniline to 150—170°,[3] or equal molecules of phenylcarbamide and aniline are heated to 180—190°.[4]

It is scarcely soluble in water, readily in alcohol and ether, and crystallizes in prisms, melting at 235° (Weith). On heating it with alcoholic ammonia to 140—150°, it decomposes into aniline and carbamide.[5]

Asymmetric diphenylcarbamide, $(C_6H_5)_2N.CO(NH_2)$. When the vapour of carbonyl chloride is passed through a solution of diphenylamine in chloroform, diphenylamine hydrochloride separates out and the solution contains the chloride of diphenylcarbamide, $(C_6H_5)_2N.COCl$, which crystallizes from alcohol in beautiful small plates, melting at 85°. By the action of alcoholic ammonia it is converted into the carbamide, which crystallizes in long needles, melting at 189°. On distillation with caustic potash, it decomposes into diphenylamine, ammonia and carbon dioxide, and on dry distillation, into cyanic acid and diphenylamine.[6]

Triphenylcarbamide, $(C_6H_5)_2N.CO.N(C_6H_5)H$, is obtained by the action of aniline on the diphenylcarbamide chloride described above, and crystallizes from alcohol in needles, melting at 136° (Michler).

Tetraphenylcarbamide, $CO\{N(C_6H_5)_2\}_2$, is formed when diphenylcarbamide chloride is heated for some minutes with diphenylamine and zinc dust.[7] It is readily soluble in boiling alcohol and forms small crystals, melting at 183°.

Carbanilic acid, $CO \begin{cases} NH(C_6H_5) \\ OH \end{cases}$, is not known in the free state ; its ethers, the phenyl urethanes (Vol. III., Part I., p. 165), are obtained by the action of aniline on the ethers of chloroformic acid.

Ethyl carbanilate, $CO(OC_2H_5)(NHC_6H_5)$, crystallizes from hot water in long needles melting at 51°; it boils at 237—238°, a small quantity decomposing into phenylcarbimide and alcohol,[8]

[1] Hofmann, *Annalen*, **70**, 138 ; **74**, 15.
[2] Eckenroth, *Ber.* **18**, 516.
[3] Baeyer, *Annalen*, **131**, 252.
[4] Weith, *Ber.* **9**, 821.
[5] Claus, *ibid.* **9**, 693.
[6] Michler, *ibid.* **9**, 396 ; 715.
[7] Michler and Zimmerman, *ibid.* **12**, 1166.
[8] Wilm and Wischin, *Annalen*, **147**, 157.

while in the cold these combine to form the original compound.

Phenyl carbanilate, $CO(OC_6H_5)NH(C_6H_5)$, is formed by the action of phenol on phenyl isocyanate and phenyl di-isocyanate.[1] It is more readily obtained, and at a lower temperature, when aluminium chloride is added to the mixture;[2] it is readily soluble in alcohol, slightly in water, and crystallizes in needles melting at 121°.[3]

1071 *Phenyl thiocarbamide*, $C_6H_5NH.CS.NH_2$, is formed by the action of ammonia on phenyl mustard oil,[4] and by heating ammonium thiocyanate with aniline hydrochloride.[5] It is slightly soluble in cold, more readily in boiling water, and still more readily in alcohol; it has a bitter taste, and crystallizes in needles melting at 154°.

s-Ethylphenyl thiocarbamide, $C_6H_5HN.CS.N(C_2H_5)H$, is obtained by the action of ethylamine on phenyl mustard oil, or of aniline on ethyl mustard oil, in ethereal solution, and forms monosymmetric crystals melting at 99.5°.[6]

as-Ethylphenyl thiocarbamide, $(C_6H_5)(C_2H_5)N.CS.NH_2$, is prepared by the action of ethylamine hydrochloride on potassium thiocyanate. It crystallizes from alcohol in long iridescent prisms melting at 113°.

s-Diphenyl thiocarbamide or *thiocarbanilide*, $CS(NH.C_6H_5)_2$, is formed, together with ammonia thiocyanate, when aniline thiocyanate is distilled:[7]

$$2C_6H_5.NH_2.CNSH = CS(NH.C_6H_5)_2 + CNS(NH_4).$$

It is likewise obtained by boiling an alcoholic solution of carbon disulphide with aniline, the sulphuretted hydrogen formed being removed by the addition of caustic potash.[8]

Diphenyl thiocarbamide is scarcely soluble in water, readily in alcohol, and crystallizes in small plates melting at 153°. On heating it with alcoholic ammonia to 100°, it is decomposed into phenyl thiocarbamide and aniline.[9]

Thiocarbanilic acid, $CS \begin{cases} HN(C_6H_5) \\ OH \end{cases}$, is not known in the free state; its ethyl ether, or phenyl thio-urethane, is obtained by

[1] Hofmann, *Ber.* **4**, 249.   [2] Leuckart, *ibid.* **18**, 873.
[3] Gumpert, *J. Pr. Chem.* II. **31**, 119.   [4] Hofmann, *Jahresb.* **1858**, 349.
[5] Schiff, *Annalen*, **148**, 338; Clermont, *Ber.* **9**, 446; **10**, 494.
[6] Weith, *Ber.* **8**, 1524.   [7] Hofmann, *Annalen*, **70**, 142.
[8] Weith, *Ber.* **6**, 967.   [9] Gebhardt, *ibid.* **17**, 3043.

heating phenyl mustard oil with alcohol,[1] as well as by treating it at the ordinary temperature with alcohol potash.[2] Ethyl thiocarbanilate forms asymmetric prisms melting at 71—72°.

It behaves as a mercaptan; by the action of ethyl iodide on its silver salt, phenyl urethane ethyl ether is obtained, which forms beautiful crystals, melting at 29·5—30·5°. On heating with dilute sulphuric acid to 180—200°, it is split up into aniline and ethyl thioxycarbonate, a decomposition which agrees with the following constitution:

$$C\begin{cases} SC_2H_5 \\ N.C_6H_5 \\ OC_2H_5 \end{cases} + H_2O = H_2N.C_6H_5 + CO\begin{cases} SC_2H_5 \\ OC_2H_5. \end{cases}$$

From this we may further conclude that of the following formulæ,

$$C\begin{cases} SH \\ N.C_6H_5 \\ OC_2H_5 \end{cases} \qquad CS\begin{cases} NH(C_6H_5) \\ OC_2H_5, \end{cases}$$

the first, and not the second, which is generally accepted, represents the constitution of phenyl urethane. It is also possible, however, that the second represents the compound in the free state, and that on the formation of a salt, an intermolecular rearrangement takes place.[3]

Diphenylselenocarbamide, $CSe(NH.C_6H_5)_2$, is obtained by the action of aniline on phenyl isoselenocyanate (p. 250).[4] It is a white crystalline substance which melts with decomposition at 186°.

PHENYLGUANIDINES.

1072 *Phenylguanidine*, $C(NH)\begin{cases} NH_2 \\ NH(C_6H_5) \end{cases}$, is obtained by the action of alcoholic ammonia on phenyl thiocarbamide, and is a solid body which on standing decomposes into ammonia and phenylcyanamide.[5]

Diphenylguanidine or *melaniline*, $C(NH)(NH.C_6H_5)_2$. Hofmann obtained this substance by passing cyanogen chloride

[1] Hofmann, *Ber.* **2**, 120; **3**, 772.
[2] R. Schiff, *ibid.* **9**, 1316.
[3] Liebermann, *Annalen*, **207**, 142.
[4] Stolte, *Ber.* **19**, 2351.
[5] Feuerlein, *ibid.* **12**, 1602.

through aniline.[1] Cyananilide and aniline hydrochloride are first formed and then react on each other:

$$CN.NH.C_6H_5 + NH_2.C_6H_5.HCl = C(NH)(NH.C_6H_5)_2HCl.$$

It is, therefore, also obtained by heating the alcoholic solutions of these two compounds,[2] as well as by the action of lead oxide on a solution of thiocarbanilide in alcoholic ammonia.[3] In order to prepare it, the thiocarbanilide is first treated with concentrated caustic potash, concentrated solution of ammonia and washed litharge are then added, and the whole digested on the water bath, the liquid decanted off and the melaniline extracted from the residue by hydrochloric acid.[4] It crystallizes from alcohol in monosymmetric needles melting at 147°, and is a monacid base forming salts which crystallize well. On heating it under pressure with concentrated hydrochloric acid, it forms carbon dioxide, ammonia and aniline.

a-Triphenylguanidine, $C(NC_6H_5)(NH.C_6H_5)_2$, is best obtained by decomposing a boiling alcoholic solution of equal molecules of thiocarbanilide and aniline with lead oxide.[5]

It crystallizes from alcohol in six-sided rhombic prisms, melting at 143°, and combines with acids to form crystalline salts.

β-Triphenylguanidine, $C(NH) \begin{cases} N(C_6H_5)_2 \\ NH(C_6H_5) \end{cases}$, is obtained when cyananilide and diphenylamine hydrochloride are heated to 100—125°. It crystallizes from alcohol in large, strongly refractive tablets, melting at 131°. The hydrochloride, $C_{19}H_{17}N_3HCl + H_2O$, forms thick tablets or prisms.[6] When it is heated with concentrated hydrochloric acid to 260—270°, it takes up water, and splits up into carbon dioxide, ammonia, aniline, and diphenylamine, while the *a*-compound, under the same conditions, only yields carbon dioxide and aniline.

Tetraphenylguanidine, $C(NH)\{N(C_6H_5)_2\}_2$, is formed by passing cyanogen chloride into diphenylamine heated to 150—170°. It crystallizes from light petroleum in rhombic prisms, melting at 130—131°, and is a monacid base, most of its salts being only slightly soluble in water. Concentrated hydrochloric acid decomposes it at 330—340°, forming carbon dioxide, ammonia, and diphenylamine.[7]

[1] *Annalen*, **67**, 129.　　　　　　　[2] Cahours and Clöez, *ibid*. **90**, 93.
[3] Hofmann, *Ber*. **2**, 460 ; Schröder and Weith, *ibid*. **7**, 937.
[4] Rathke, *ibid*. **12**, 772.　　　　　　[5] Hofmann, *ibid*. **2**, 458.
[6] Weith and Schröder, *ibid*. **8**, 294.　　　[7] Weith, *ibid*. **7**, 843.

HALOGEN SUBSTITUTION PRODUCTS OF ANILINE.

1073 In 1842 Fritzsche found that three of the hydrogen atoms in aniline can be replaced by bromine, the tribromaniline which was thus formed being called by him bromaniloïd.[1] This compound was investigated three years later by Hofmann, who observed that, unlike aniline, it does not combine with acids, the basic character of the aniline having been neutralized by the presence of the electronegative element. He therefore endeavoured to obtain compounds in which only one or two of the hydrogen atoms had been replaced by a halogen. The action of chlorine did not yield the desired result, as this substance acts very violently on aniline, forming resinous products, among which, however, he was able to recognize trichloraniline. He was more successful when he started with isatin, $C_8H_5NO_2$, which is an oxidation product of indigo, and yields aniline on distillation with caustic potash. When the substitution products of this substance, which had previously been prepared by Laurent, were heated with caustic potash he obtained the bases monochloraniline and monobromaniline, and also dibromaniline, the basic properties of which are much feebler than those of the monosubstitution products.[2]

It has been stated in the introduction that these important researches were chiefly instrumental in securing universal recognition for the substitution theory. Hofmann himself says that it undoubtedly follows from the facts observed by him, that chlorine or bromine can take the part of hydrogen in organic compounds, that these preserve their electronegative character in the new compounds, and impress it the more strongly upon them the greater the number of hydrogen atoms replaced by halogens.

Liebig adds in a note: "The author seems to me to have definitely proved by this research that the chemical nature of a compound does not in any way depend upon the nature of the elements contained in it, as is assumed by the electrochemical theory, but entirely on their arrangement."

[1] *J. Pr. Chem.* **28**, 204.   [2] *Annalen,* **53**, 1.

Aniline forms a large number of halogen substitution products, which are prepared by the following methods:

1. An anilide, generally acetanilide, is treated with chlorine or bromine, the action proceeding much more quietly and smoothly than with aniline or its salts. The substituted anilides are then decomposed by heating with an alkali or fuming hydrochloric acid. In many cases it is better to decompose the anilide by strong sulphuric acid; the solution is diluted with water, neutralized with caustic soda, and the substituted aniline extracted with ether. Iodine and chloride of iodine act upon aniline or the anilides with formation of iodine substitution products.

2. Substituted anilines are also readily obtained by the reduction of the substituted nitrobenzenes with tin or stannous chloride and hydrochloric acid.

As already mentioned, the replacement of hydrogen by a halogen weakens the basic character of the aniline. In all cases the presence of the amido-group can readily be proved by the action of acetyl chloride, which forms substituted anilides.

CHLORANILINES.

1074 *o-Chloraniline*, $C_6H_4Cl.NH_2$, is obtained by reducing *o*-chloronitrobenzene, and is a liquid which does not solidify at —14°, and boils at 207°. The *hydrochloride*, $C_6H_4Cl.NH_2.ClH$, crystallizes in large rhombic tablets. The *picrate* is almost insoluble in cold water, and only slightly soluble in hot water; it is much less soluble in alcohol than the picrate of the para-compound. This property is made use of in the separation of the two chloranilines. As it is very difficult to obtain *o*-chloronitrobenzene free from *p*-chloronitrobenzene, a mixture of the chloranilines is obtained on reduction, which can readily be separated by means of picric acid. Another method of separation depends upon the fact that *o*-chloraniline is a much weaker base than *p*-chloraniline. When, therefore, the equivalent quantity of sulphuric acid is added to the mixture of the two bases and the whole distilled with water, the sulphate of *o*-chloraniline completely decomposes, and the base passes·over mixed with only a little *p*-chloraniline.[1]

---

[1] Beilstein and Kurbatow, *Annalen*, **176**, 36.

m-Chloraniline may be obtained by reducing *m*-chloronitrobenzene with stannous chloride (Beilstein and Kurbatow). It is a liquid boiling at 230°, and is as strong a base as the ortho-compound, its salts being only partially decomposed on boiling with water.

p-Chloraniline.—Hofmann first obtained this compound by distilling chlorisatin with caustic potash; it is also formed by the action of chlorine on acetanilide,[1] but is best prepared by reducing *p*-chloronitrobenzene with stannous chloride (Beilstein and Kurbatow).

It is very slightly soluble in boiling water, but readily in alcohol, and forms diamond-like crystals, very similar to regular octahedra, but belonging to the rhombic system,[2] which melt at 70—71°. It can be distinguished from *p*-bromaniline, which is very similar to it, as it boils without decomposition at 230—231°.

p-Chloraniline is a powerful base, but its salts have an acid reaction. It does not decompose the salts of aluminium and zinc.

p-Chloraniline hydrochloride, $C_6H_4Cl.NH_2.HCl$, crystallizes in large, monosymmetric prisms. The *platinochloride* forms small plates, soluble in water, alcohol, and even a mixture of the latter with ether (Hofmann).

DICHLORANILINES,[3] $C_6H_3Cl_2.NH_2$.

1075 These may be distinguished according to the position of the chlorine atoms as ortho-, meta-, and para-compounds; in the following tables this is expressed by the numbers, the amido-group being assumed to have position 1.

o-Dichloranilines.

	Melting-point.	Boiling-point.
(a) Asymmetric (3 : 4) broad needles.	71·5°	272°
(b) Adjacent (2 : 3) needles . . .	23—24°	252°

---

[1] Mills, *Jahresber.* **1860**, 349.   [2] Groth, *Ber.* **3**, 453.

[3] Beilstein and Kurbatow, *loc. cit.*

m-Dichloranilines.

	Melting-point.	Boiling-point.
Asymmetric (2 : 4) [1] long needles . . .	63°	245°
Symmetric (3 : 5) needles.	50·5°	259—260°
Adjacent (2 ; 6) needles.	39°	—
p-Dichloraniline (2 : 5) thick needles . .	50°	251°

TRICHLORANILINES, $C_6H_2Cl_3.NH_2$ [2]

Asymmetric (2 : 4 : 5) [3] needles. . . .	95—96°	270°
Symmetric (2 : 4 : 6) long needles . .	77·5°	262°
Adjacent (2 : 3 : 4) needles . . .	67·5°	292°

The second of these was first prepared by Hofmann.

TETRACHLORANILINES, $C_6HCl_4.NH_2$.

Asymmetric (2 : 3 : 4 : 6) [4] needles . .	88°	—
Symmetric (2 : 3 : 5 : 6) [5] needles . .	90°	—
Adjacent (2 : 3 : 4 : 5) [6] broad needles	118°	—

Pentachloraniline, [7] $C_6Cl_5.NH_2$, crystals.

BROMANILINES.

1076 These are best obtained from the three nitrobromo-benzenes.

o-Bromaniline is insoluble in water, but readily dissolves in alcohol. It crystallizes in colourless needles, melts at 31°·5, and boils without decomposition at 229°. [8]

m-Bromaniline forms a colourless, crystalline mass melting at 18—18°·5, and boiling at 251°. [9]

p-Bromaniline was, as already explained, first prepared by Hofmann. It is also obtained by the action of bromine on

[1] Beilstein and Kurbatow, *Annalen,* 182, 95 ; Witt, *Ber.* 7, 1602.
[2] Beilstein and Kurbatow. [3] Lesimple, *Annalen,* 137, 125.
[4] Beilstein and Kurbatow. [5] Lesimple, *Zeitschr. Chem.* 1868, 227.
[6] Beilstein and Kurbatow. [7] Jungfleisch, *Jahresb.* 1868, 354.
[8] Fittig and Mager, *Ber.* 7, 1179. [9] *Ibid.* 8, 364.

aniline,[1] or better, acetanilide,[2] and forms large, well-formed crystals, similar to regular octahedra, but belonging to the rhombic system,[3] which melt at 66·4°.[4] On heating more strongly the liquid suddenly becomes a deep violet colour, and between 190° and 270° a liquid distils over, while a residue is left which forms a splendid blue solution with alcohol. The distillate shows similar phenomena when redistilled, and can be split up by fractionation into aniline, ordinary dibromaniline, and tribromaniline.[5]

p-Bromaniline hydrochloride, $C_6H_4Br.NH_2.HCl$, is isomorphous with chloraniline hydrochloride (Hofmann).

DIBROMANILINES, $C_6H_3Br_2.NH_2$.

	Melting-point.
o-Dibromaniline (3 : 4)[6] white crystals	80·4°

m-Dibromanilines.

(*a*) Asymmetric (2 : 4)[7] long narrow plates . . .	79·5°
(*b*) Symmetric (3 : 5)[8] needles	56·5°
p-Dibromaniline (2 : 5)[9] prisms	51—52°

Asymmetric *m*-dibromaniline was first obtained by Hofmann. It is also obtained by the action of bromine on *o*- and *p*- acetanilide. Its hydrochloride forms arborescent crystals.

TRIBROMANILINES, $C_6H_2Br_3.NH_2$.

	Melting-point.
Symmetric (2 : 4 : 6)[10] long needles	118°
Adjacent (3 : 4 : 5)[11] crystals	—

The first of these is also called ordinary tribromaniline, because it is obtained by the direct bromination of aniline. Unlike

[1] Kekulé, *Zeitschr. Chem.* **1866**, 687.   [2] Mills, *ibid.* **1860**, 834.
[3] Arzruni, *Annalen*, **188**, 23.   [4] Körner, *Jahresb.* **1875**, 342.
[5] Fittig and Büchner, *Annalen*, **188**, 23.   [6] Körner, *Jahresb.* **1875**, 342.
[7] Hofmann, *loc. cit.* ; Griess, *Annalen*, **121**, 266 ; Körner, *loc. cit.* ; Wurster, *Ber.* **6**, 1486 ; Meyer and Stüber, *Annalen*, **165**, 169.
[8] Körner.   [9] Meyer and Stüber.
[10] Körner ; Fittig and Büchner, *Annalen*, **188**, 962 ; Wurster and Nölting, *Ber.* **7**, 1564.   [11] Körner.

the isomeric compound, which is obtained by the reduction of the corresponding nitrobromobenzene, it does not combine with acids. This compound does not melt at 130°, and decomposes when more strongly heated.

Melting-point.

Tetrabromaniline, $C_6HBr_4.NH_2$ (2 : 3 : 4 : 6)[1] needles 115·3°

Pentabromaniline, $C_6Br_5.NH_2$,[2] lustrous needles. . . . 222°

IODANILINES.

1077 *o-Iodaniline*, $C_6H_4I.NH_2$, is obtained by heating *o*-nitroiodobenzene, ferrous sulphate and dilute ammonia on the waterbath. It forms white needles melting at 56·5°, and has an unpleasant pyridine-like smell.[3]

m-Iodaniline, $C_6H_4I.NH_2$.—Griess obtained this substance by reducing *m*-nitroiodobenzene; it crystallizes in small plates melting at 27°.[4]

p-Iodaniline.—While chlorine and bromine readily form substitution products, iodine does not act directly on most carbon compounds. Aniline is among the exceptions to this rule, and is readily converted by iodine into *p*-iodaniline hydriodide.[5] The base may also be obtained by the reduction of *p*-nitroiodobenzene.[6] To prepare it, acetanilide is dissolved in glacial acetic acid and the equivalent quantity of iodine chloride added, *p*-iodacetanilide, which crystallizes in rhombic tablets, being formed. This is then boiled with concentrated hydrochloric acid. and the solution decomposed with ammonia.[7]

p-Iodaniline crystallizes in needles or prisms melting at 63°. It precipitates aluminium salts, but not those of zinc or iron. On heating it with ethyl iodide, ethylaniline, diethylaniline, and free iodine are obtained. Its hydrochloride crystallizes from hot water in plates, or broad thin needles.

m-Di-iodaniline, $C_6H_3I_2.NH_2$(2 : 4).—When hot solutions of aniline and mercuric chloride are mixed, mercuric phenyl-

[1] Wurster and Nölting ; Körner. [2] Körner.
[3] Körner and Wender, *Gazzetta*, **17**, 486.
[4] Griess, *Zeitschr. Chem.* **1866**, 218.
[5] Hofmann, *Annalen*, **67**, 61 ; Mills, *ibid.* **176**, 354.
[6] Griess, *Zeitschr. Chem.* **1866**, 218 ; Kekulé, *ibid.* p. 687.
[7] Michael and Norton, *Ber.* **11**, 108.

ammonium chloride, $N(C_6H_5.HHg)Cl$, separates out as a yellow amorphous precipitate, which is converted into di-iodaniline by an alcoholic solution of iodine.[1] This compound may also be obtained by passing the vapour of two molecules of iodine chloride into a solution of aniline in acetic acid,[2] or by the further action of iodine on o-iodaniline.[3] It crystallizes in needles melting at 95—96°, and forms salts, which are decomposed by cold water. It is converted hy the diazo-reaction into m-di-iodobenzene.

Tri-iodaniline, $C_6H_2I_3.NH_2$, is obtained by passing three molecules of iodine chloride into a solution of aniline in hydrochloric acid. It crystallizes from boiling alcohol in long needles melting at 185·5°.[4]

FLUORINE SUBSTITUTION PRODUCTS OF ANILINE.

1078 *m-Fluoraniline*, $C_6H_4F.NH_2$, is obtained from acetyl m-diamidobenzene by diazotizing and treating with piperidine. The acetamidobenzene-m-diazopiperidide thus obtained is converted by hydrofluoric acid into m-fluoraniline, piperidine, and acetic acid. The former is an oily liquid closely resembling aniline, which sinks in water and is slightly soluble in that liquid.

p-Fluoraniline is prepared from p-nitraniline by converting it into nitrobenzene-p-diazopiperidide, treating this compound with hydrofluoric acid, and reducing the fluonitrobenzene thus obtained with stannous chloride and hydrochloric acid. It forms an oily liquid like aniline, boiling at 187—189°, solidifying in a mixture of ether and solid carbon dioxide, and having a sp. gr. of 1·153 at 25°. It forms well crystallized salts. Acetic anhydride converts it readily into *acetofluoranilide*, $C_6H_4F.NH.CO.CH_3$, a white crystalline substance, almost insoluble in water, but readily soluble in alcohol.[5]

[1] Rudolph, *Ber.* **11**, 78.   [2] Michael and Norton, *loc. cit.*
[3] Körner and Wender, *Gazzetta*, **17**, 486.
[4] Stenhouse, *Journ. Chem. Soc.* **17**, 329 ; Michael and Norton.
[5] Wallach, *Annalen*, **235**, 266 ; Wallach and Heusler, *ibid.* **243**, 222.

NITRO-SUBSTITUTION PRODUCTS OF ANILINE.

1079 Hofmann and Muspratt found that when ordinary dinitrobenzene (*m*-dinitrobenzene) is treated with alcoholic ammonium sulphide, only one nitroxyl is reduced, nitraniline, $C_6H_4(NO_2)NH_2$, being formed.[1] This is also the case with the two isomeric dinitrobenzenes.[2] Stannous chloride and hydrochloric acid may also be employed for the partial reduction of dinitro-compounds, but in this case the nitro group attacked is *not* the one attacked by ammonium sulphide.[3] The nitranilines are also obtained when the monochloronitrobenzenes, or the corresponding bromo- or iodo-compounds are heated with ammonia:

$$C_6H_4Cl(NO_2) + 2NH_3 = C_6H_4(NH_2)NO_2 + NH_4Cl.$$

This reaction takes place most readily with the ortho-compounds. The substituted nitrobenzenes containing more than one halogen are still more readily attacked by ammonia, but only one halogen atom is replaced by the amido-group.

Nitro-substitution products of aniline are also obtained by heating the nitrophenols[4] or their methyl- or ethyl-ethers with ammonia:

$$C_6H_4(NO_2)OCH_3 + NH_3 = C_6H_4(NO_2)NH_2 + HO.CH_3.$$

The basic character of aniline is weakened much more by the introduction of nitroxyl than by that of a halogen atom; hence the mononitranilines are very weak bases, while the dinitranilines do not combine with acids, and trinitraniline behaves as an acid-amide, being decomposed by alkalis with formation of ammonia and trinitrophenol or picric acid, on account of which reaction it was first named picramide.

NITRANILINES, $C_6H_4(NO_2)NH_2$.

1080 *o-Nitraniline* may be obtained by the methods given above from *o*-dinitrobenzene, *o*-nitrobromobenzene,[5] and

---

[1] *Annalen*, **57**, 201. [2] Zincke and Rinne, *Ber.* **7**, 869 and 1372.
[3] Anschütz and Heusler, *ibid.* **19**, 2161.
[4] Merz and Ris, *ibid.* **19**, 1749. [5] Walker and Zincke, *ibid.* **5**, 114.

o-nitranisoïl.[1] It is however best prepared from sulphanilic acid or aniline-*p*-sulphonic acid, $C_6H_4(NH_2)SO_3H$, by converting its sodium salt into sodium acetanilidesulphonate by means of acetic anhydride. This is dissolved in 5 parts of sulphuric acid, well cooled, and heated with the calculated quantity of nitric acid. The solution is poured on to ice, neutralized with lime, and the filtrate evaporated to a small bulk and boiled with dilute sulphuric acid in order to remove the acetyl group. Potassium nitranilinesulphonate is prepared from this, and is then decomposed by heating with concentrated hydrochloric acid to 170—180°; *o*-nitraniline separates out from the solution thus obtained on addition of ammonia.[2] It crystallizes in orange-yellow needles melting at 71°·5, which dissolve in hot water and volatilize when the solution is boiled. Its yellow salts are very unstable.

m-Nitraniline is best obtained by mixing 10 parts of *m*-dinitrobenzene, 30 parts of 90 per cent. alcohol, and 5 parts of concentrated ammonia, and passing sulphuretted hydrogen through the mixture, which is warmed from time to time, until 5 parts by weight have been absorbed. The solution is then precipitated with water, the precipitate boiled with dilute hydrochloric acid, and the filtrate reprecipitated with ammonia, the *m*-nitraniline being finally recrystallized from boiling water.[3] It crystallizes in long yellow needles melting at 114°, and subliming in small plates at 100°.[4] It has a sweet burning taste, boils at 285°, and does not precipitate metallic salts. Its salts are much more stable than those of the ortho-compound, but are decomposed by aniline. It is not attacked by boiling caustic soda.[5]

p-Nitraniline is obtained in a similar manner to the ortho-compound from the corresponding para-compounds, as well as together with the former, by the nitration of the anilides.[6]

To prepare it, 1 part of acetanilide is dissolved in 3 parts of cooled nitric acid of sp. gr. 1·5, the solution precipitated with ice-water, and the *p*-nitracetanilide filtered off. The filtrate contains *o*-nitracetanilide, which can be extracted with chloroform; both anilides are then decomposed with boiling

[1] Salkowski, *Annalen,* **174**, 278.
[2] Nietzki and Benckiser, *Ber.* **18**, 294.
[3] Beilstein and Kurbatow, *Annalen,* **176**, 44.
[4] Hübner, *Ber.* **10**, 1716.
[5] Wagner, *ibid.* **7**, 77.
[6] Arppe, *Annalen,* **93**, 357 ; Hofmann, *Jahresb.* **1860**, 349.

hydrochloric acid.[1] According to Nölting and Collin, acetanilide is dissolved in 4 parts of sulphuric acid, and to this solution, which is kept cold by a freezing mixture of ice and salt, the calculated quantity of 85 per cent. nitric acid is gradually added, the para-compound being formed, together with a little o-nitracetanilide.[2] If aniline be dissolved in concentrated sulphuric acid, well cooled, and then treated with nitric acid largely diluted with sulphuric acid, all three nitranilines are formed. In order to separate them, the solution is precipitated by ice-water, neutralized with carbonate of soda, and distilled, the ortho- and meta-compound being found in the distillate, while the para-compound remains behind.[3]

It crystallizes from hot water in long, yellow, monosymmetric needles, which are almost tasteless, and melt at 147°. It is a very weak base, and on boiling with concentrated caustic soda forms p-nitrophenol (Wagner). Its salts are less stable than those of the meta-compound, but more so than those of o-nitraniline.[4]

DINITRANILINES, $C_6H_3(NO_2)_2NH_2$.

			Melting-point.
Asymmetric	(2 : 4)[5]	yellow monosymmetric needles having a bluish lustre . .	182°
Adjacent	(2 : 6)[6]	long yellow needles	138°

TRINITRANILINE, $C_6H_2(NO_2)_3NH_2$ (2 : 4 : 6).

This compound, which is also named *picramide*, is obtained by the action of ammonia on ethyl picrate and chlorotrinitrobenzene.[7] It crystallizes from glacial acetic acid in dark yellow monosymmetric tablets with a violet lustre, melting at 188°.

Like picric acid, it combines with benzene to form the compound $C_6H_2(NO_2)_3NH_2 + C_6H_6$, crystallizing in broad, light yellow transparent prisms, which rapidly lose benzene in the air. It also combines with other hydrocarbons and with amido-bases.

[1] Grethen, *Ber.* **9**, 775 ; Beilstein and Kurbatow, *Annalen*, **197**, 83 ; Witt, *Ber.* **8**, 144. [2] *Ibid.* **17**, 261.
[3] Hübner, *ibid.* **10**, 1716. [4] Lellmann, *ibid.* **17**, 2719.
[5] Gottlieb, *Annalen*, **85**, 24 ; Rudnew, *Zeitsch. Chem.* **1871**, 202 ; Engelhardt and Latschinow, *ibid.* **1870**, 233 ; Clemm, *J. Pr. Chem.* II. **1**, 145 ; Salkowski, *Annalen*, **174**, 263 ; Schaumann, *Ber.* **12**, 1345.
[6] Salkowski, Körner. [7] Pisani, *Annalen*, **92**, 326.

Picramide-aniline, $C_6H_2(NO_2)_3NH_2 + C_6H_5NH_2$, crystallizes in thick, black, lustrous prisms or very long needles with a dark purple reflection, which form a blood-red powder, and melt at 123—125°.

Picramide-dimethylaniline, $C_6H_2(NO_2)_3NH_2 + C_6H_5.N(CH_3)_2$, forms dark blue, lustrous crystals melting at 139—141°.

Both these compounds decompose in the air, picramide being left behind.[1]

CHLORONITRANILINES, $C_6H_3Cl(NO_2)NH_2$.

1081 These compounds are obtained by the nitration of the monochloracetanilides, as well as by heating the dichloronitrobenzenes with ammonia. The following are known (the first of the subjoined numbers giving the position of the chlorine, and the second that of the nitroxyl, the amido-group occupying position 1):

	Melting-point.
o-Chloro-*m*-nitraniline (2 : 5)[2] yellow needles . . .	117—118°
o-Chloro-*p*-nitraniline (2 : 4)[3] light yellow needles .	104—105°
m-Chloro-*o*-nitraniline (3 : 6)[4] yellow needles . . .	124—125°
m-Chloro-*p*-nitraniline (3 : 4)[5] small yellow plates .	156—157°
p-Chloro-*o*-nitraniline (4 : 2)[6] orange red needles .	115°

BROMONITRANILINES, $C_6H_3Br(NO_2)NH_2$.

	Melting-point.
o-Bromo-*p*-nitraniline (2 : 4)[7] yellow needles . . .	104·5°
m-Bromo-*o*-nitraniline (3 : 6)[8] reddish yellow needles	151·4°
p-Bromo-*o*-nitraniline (4 : 2)[9] orange-yellow needles	111·4°

IODONITRANILINES.

	Melting-point.
o-Iodo-*p*-nitraniline (2 : 4)[10] long needles	105·5°
m-Iodo-*o*-nitraniline (3 : 6)[11] steel-blue plates . .	—
p-Iodo-*o*-nitraniline (4 : 2)[12] orange-yellow needles	122°

Besides these substitution products of aniline, many others can be prepared containing several halogens, and also nitroxyl.

[1] Mertens, *Ber.* **11**, 843; Hepp, *Annalen,* **215**, 358.
[2] Beilstein and Kurbatow, *ibid.* **182**, 98.   [3] *Ibid.*
[4] Körner, *loc. cit.*; Laubenheimer, *Ber.* **9**, 1826.   [5] *Ibid.*
[6] Körner.   [7] Körner; Hübner, *Ber.* **10**, 1709.
[8] Körner; Wurster, *ibid.* **6**, 1542.
[9] Körner; Meyer and Wurster, *Annalen,* **171**, 59.
[10] Michael and Norton, *Ber.* **11**, 113.   [11] Körner.   [12] Michael and Norton.

AMIDOBENZENESULPHONIC ACIDS,
$C_6H_4(NH_2)SO_3H.$

1082 Gerhardt, by heating formanilide, oxanilide, or aniline
with sulphuric acid, obtained an acid which he called sulphanilic
acid.   Laurent considered that he had obtained this compound
by the reduction of his nitrobenzenesulphonic acid, but Schmitt
showed that this was a different substance.[2] Limpricht, who
first prepared the three nitrobenzenesulphonic acids in the pure
state, found that they are readily converted into the amido-
acids by treating their ammonium salts with concentrated ammonia and passing in hydrogen sulphide as long as heat is
evolved.   The solution is then evaporated until all the ammonium sulphide is driven off, filtered from the precipitated
sulphur and acidified with hydrochloric acid.   The amido-acid,
which separates out after long standing, is purified by recrystallization from hot water.[3]

o-Amidobenzenesulphonic acid has been carefully examined
by Limpricht and Berndsen.[4]   It is obtained by converting
nitro-*m*-bromobenzenesulphonic acid into the amido-acid and
heating this with hydriodic acid and phosphorus.[5] It is scarcely
soluble in cold, somewhat more readily in hot water, and crystallizes in anhydrous dull-white rhombohedral forms, or sometimes
in thick rhombic plates, mixed with transparent, four-sided,
pointed prisms, containing half a molecule of water.

m-Amidobenzenesulphonic acid has also been examined by
Berndsen ;[6] it crystallizes in long slender, anhydrous needles, or,
on gradual evaporation of its solution, in transparent monosymmetric prisms, which contain 1·5 molecules of water.   It is
prepared on the large scale from the crude *m*-nitro-acid, and is
employed in the manufacture of colours.

p-Amidobenzenesulphonic acid or *sulphanilic acid* is formed by
the above reactions, and also on heating aniline ethylsulphate,[7]
and aniline *p*-phenolsulphonate, which decomposes into phenol
and sulphanilic acid.[8]

It is prepared by gradually adding 1 part of aniline to 3

[1] *Annalen,* **60**, 308.   [2] *Ibid.* **120**, 164.
[3] *Ibid.* **177**, 79.   [4] *Ibid.* **177**, 18.
[5] *Ibid.* **186**, 128.   [6] *Ibid.* **177**, 82.
[7] Limpricht, *Ber.* **7**, 1349.   [8] Pratesi and Kopp, *ibid.* **4**, 978.

parts of fuming sulphuric acid, warming the mixture to 180° for
3-4 hours, and pouring the product of the reaction into water.
The acid separates as a more or less coloured crystalline mass,
which is purified by dissolving in hot dilute alkali, boiling with
animal charcoal and acidifying the clear filtrate. Ordinary
sulphuric acid may also be employed, but the temperature to
which the mixture is heated must then be higher.

. Sulphanilic acid is slightly soluble in cold, readily in hot water,
and crystallizes in rhombic tablets which contain 1 molecule of
water, and rapidly effloresce in the air. Dilute chromic acid
oxidizes it to quinone.

When treated with hydrochloric acid and sodium hypo-
bromite it is converted into *dibromosulphanilic acid*, $C_6H_2Br_2$.
$NH_2.SO_3H$. If this latter be heated with sulphuric acid to
·170° and a current of superheated steam passed through the
mixture, it is converted into dibromaniline[1] (m.p. 83-84°).

Acetanilide-p-sulphonic acid, $C_6H_4(NH.C_2H_3O).SO_3H$, is prepared by the action of acetic anhydride on sodium sulphanilate
and acidification, or by the sulphonation of acetanilide. It has
not been obtained crystalline and is very unstable.[2]

o-Nitraniline-p-sulphonic acid, $C_6H_3(NH_2)(NO_2)SO_3H$, is prepared by treating 1 part of acetanilide with 3 parts of fuming
sulphuric acid on the water-bath, and adding 2 parts of ordinary sulphuric acid and the calculated quantity of nitric acid
mixed with an equal quantity of sulphuric acid, the temperature
being kept below 0° during the last part of the operation. It is
then poured on to ice, and the yellow needles which separate
filtered off. It is very soluble in water, less so in alcohol and
dilute acids, and is converted by boiling with caustic potash into
the potassium salt of nitrophenolsulphonic acid.[3]

DIAMIDOBENZENES OR PHENYLENEDIA-
MINES, $C_6H_4(NH_2)_2$.

1083 Zinin first obtained the meta-compound of this group in
an impure state by the continued action of ammonium sulphide
on ordinary dinitrobenzene, and called it " semibenzidam,"[4] while

[1] Heinichen, *Annalen,* **253**, 268.
[2] Nietzki and Benckiser, *Ber.* **17**, 707 ; **18**, 294.
[3] Nietzki and Lerch, *ibid.* **21**, 3220.   [4] *J. Pr. Chem.* **33**, 34.

Gerhardt gave to it the name of azophenylamine. Hofmann first obtained it pure by reducing dinitrobenzene with iron and acetic acid, and described it as a-phenylenediamine, to distinguish it from β-phenylenediamine, which he had obtained from para-nitraniline.[1] Besides their preparation from dinitrobenzene and nitraniline, the three diamidobenzenes can be obtained from the six diamidobenzoic acids, $C_6H_3(NH_2)_2CO_3H$, by the elimination of carbon dioxide, as shown by Griess, who first prepared the ortho-compound in this way.[2]

o-DIAMIDOBENZENE, OR o-PHENYLENEDIAMINE.

This body may be obtained by the above-mentioned reactions,[3] and also by the action of sodium amalgam and water on p-bromonitraniline,[4] or by boiling brom-o-phenylenediamine with caustic soda, water and zinc dust, and extracting the solution with alcoholic ether.[5] It crystallizes from hot water in colourless or slightly reddish tablets, and from chloroform in quadratic plates; it melts at 102—103° (Hübner), and boils at 252° (Griess).

On throwing a crystal into water it acquires an extremely rapid rotation in dissolving, which is still more striking than that which is exhibited by camphor. Its isomerides and homologues behave in a similar manner.[6]

o-Diamidobenzene hydrochloride, $C_6H_4(NH_3Cl)_2$, is readily soluble in water, and crystallizes in groups of radiating needles. On adding ferric chloride to its solution, ruby-red needles separate[7] which have the formula $C_{24}H_{18}N_6O(ClH)_2 + 5H_2O$.[8]

On treating an alcoholic solution of the base with a drop of a solution of phenanthrene quinone, $C_{14}H_8O_2$, in hot glacial acetic acid, and boiling for a short time, a bright yellow precipitate of the quinoxaline, $C_{20}H_{12}N_2$, separates out (see p. 73), consisting of small needles, which are coloured deep-red by concentrated hydrochloric acid. In this way 0·5 mgms. of o-diamidobenzene can be detected. o-Diamidotoluene behaves in a similar manner.[9]

[1] Jahresb. **1863**, 421.   [2] Ber. **5**, 201.
[3] Zincke and Sintenis, ibid. **6**, 123 ; Zincke and Rinne, ibid. **7**, 1374.
[4] Meyer and Wurster, Annalen, **171**, 63 ; Hübner, ibid. **209**, 360.
[5] Sandmeyer, Ber. **19**, 2654.   [6] Gattermann, ibid. **18**, 1484.
[7] Griess.   [8] Rudolph, Ber. **12**, 2211.
[9] Hinsberg, ibid. **18**, 1228.

On adding a dilute solution of potassium nitrite to a very dilute solution of *o*-diamidobenzene sulphate, *amido-azophenylene* or *azimidobenzene* separates out as any oily liquid, which soon solidifies :

$$C_6H_4{\Large\langle}^{NH_2}_{NH_2} + NO_2H = C_6H_4{\Large\langle}^{NH}_{N}{\Large\rangle}N + 2H_2O.$$

This body crystallizes from a mixture of benzene and toluene in needles, with a pearly lustre, which melt at $98°·5$ and are not acted on by ferric chloride [1] (see also p. 328).

It is also obtained by the action of *o*-diamidobenzene hydrochloride on *p*-diazobenzenesulphonic acid.[2]

p-Brom-o-diamidobenzene, $C_6H_3Br(NH_2)_2$, is prepared by the reduction of *o*-nitro-*p*-bromaniline [3] or of *o*-nitro-*m*-bromaniline.[4] It forms slender needles melting at $63°$, and readily soluble in water, alcohol and chloroform.

Dicyan-o-diamidobenzene, $C_8H_8N_4$, is obtained by passing cyanogen through a concentrated alcoholic solution of *o*-diamidobenzene :

$$C_6H_4{\Large\langle}^{NH_2}_{NH_2} + {\begin{array}{c}CN \\ | \\ CN\end{array}} = C_6H_4{\Large\langle}^{NH-C\equiv NH}_{NH-C\equiv NH.}$$

It crystallizes from hot water in pale yellow rhombic tablets readily soluble in alcohol, which sublime when cautiously heated, but do not melt even at $280°$.[5] On heating it with hydrochloric acid to $150°$ the following reaction takes place :

$$C_6H_4{\Large\langle}^{NH-C\equiv NH}_{NH-C\equiv NH} + 2H_2O = C_6H_4{\Large\langle}^{N\equiv C.OH}_{N\equiv C.OH} + 2NH_3.$$

The dihydroxyquinoxaline formed will be described in a later volume.

m-DIAMIDOBENZENE, OR *m*-PHENYLENEDIAMINE.

1084 This compound is obtained by reducing *m*-nitraniline, *m*-dinitrobenzene,[6] bromodinitrobenzene,[7] and β-dinitrobenzoic

[1] Ladenburg, *Ber.* **9**, 221.

[2] Griess, *ibid.* **15**, 2195 ; see also Zincke and Campbell, *Annalen,* **255**, 339.

[3] Hübner, *ibid.* **209**, 359. [4] Wurster, *Ber.* **6**, 1544.

[5] Bladin, *ibid.* **18**, 672.

[6] Hofmann, *Jahresb.* **1861**, 512 ; **1863**, 422 ; Gerdemann, *Zeitschr. Chem.* **1865**, 51. [7] Zincke and Sintenis, *Ber.* **5**, 792.

acid, with elimination of carbon dioxide in the latter case.[1] It is also obtained by heating δ-diamidobenzoic acid with baryta,[2] and was found by Hofmann in aniline oil.[3] It is prepared from its hydrochloride by heating with caustic baryta. It generally remains liquid for a long time, but solidifies gradually; if however a crystal of the base be thrown in, the oil at once passes into a crystalline mass, which is scarcely soluble in water but readily in alcohol. It melts at 63°, boils at 287°, and has an alkaline reaction.

m-Diamidobenzene hydrochloride, $C_6H_4(NH_3Cl)_2$, is obtained by the evaporation of its aqueous solution in concentrically arranged crystals. It is employed in the manufacture of colouring matters, for which it is prepared by adding tin to a mixture of hydrochloric acid and dinitrobenzene; zinc is gradually added to the solution to precipitate the metallic tin and so bring it into action again. After the dinitrobenzene has dissolved, the *m*-diamidobenzene hydrochloride is precipitated by concentrated hydrochloric acid. It is also obtained on the manufacturing scale by reducing dinitrobenzene with iron and hydrochloric acid, the solution thus obtained being directly employed for the preparation of colouring matters.

On adding potassium nitrite to the neutral solution of the hydrochloride, a yellow to brown colouration is produced according to the degree of concentration; this reaction is so delicate that even less than 0·1 mgrm. of nitrous acid in 1 litre of water can be detected by means of it.[4] The substance formed is triamido-azobenzene or phenylene brown.

Dimethyl-m-diamidobenzene, or *m-Amidodimethylaniline* is prepared by the reduction of *m*-nitrodimethylaniline.[5] It is a liquid which boils at 268–270°.

Tetramethyl-m-diamidobenzene, $C_6H_4N_2(CH_3)_4$, is obtained by heating *m*-diamidobenzene with methyl alcohol and hydrochloric acid to 180—190°, and decomposing the product with caustic soda.

It is an oily liquid, boiling at 256°, and having a peculiar smell. With methyl iodide it forms the compound $C_6H_4N_2(CH_3)_5I$, which can be obtained in splendid crystals by evaporating the aqueous solution; this body is both an ammonium iodide and an amidobase, and therefore combines with acids.[6]

[1] Wurster, *Ber.* **7**, 149.
[2] Ambühl and Wurster, *ibid.* **7**, 214.
[3] *Ibid.* **7**, 812.
[4] Griess, *ibid.* **11**, 624.
[5] *Ber.* **19**, 200, 1945.
[6] Wurster and Morley, *ibid.* **12**, 1814.

Hexmethyl-m-diphenylammonium iodide, $C_6H_4N_2(CH_3)_6I_2$, was obtained by Hofmann, together with the preceding compound, as a final product of the alternate treatment of *m*-diamido-benzene with methyl iodide and silver oxide; it forms small plates readily soluble in water.[1]

Ethyl-m-diamidobenzene, $C_6H_4(NH_2)(NHC_2H_5)$, and *Diethyl-m-diamidobenzene*, $C_6H_4(NHC_2H_5)_2$, are obtained by the reduction of the corresponding nitro-compounds, and are liquids boiling at 276° and 276–278° respectively.[2]

Diphenyl-m-diamidobenzene, $C_6H_4(NH.C_6H_5)_2$, is obtained by heating resorcinol with aniline, calcium chloride and a little zinc chloride. It readily dissolves in benzene and hot alcohol, and crystallizes in flat needles melting at 95°. Its hydrochloride, $C_6H_4(NH.C_6H_5)_2(ClH)_2$, forms deliquescent needles. Nitric acid added to its solution in concentrated sulphuric acid produces a yellowish green colouration which soon changes to a bluish violet.[3]

m-Phenylene-oxamic acid, $C_6H_4(NH_2)NH.C_2O_2OH$, is formed when *m*-diamidophenol is added to a boiling solution of oxalic acid; it is slightly soluble in water, and crystallizes in bushy needles, which melt at a very high temperature, undergoing considerable carbonization.[4]

p-DIAMIDOBENZENE, OR *p*-PHENYLENEDIAMINE.

1085 This compound may be obtained by the general methods mentioned above.[5] It is most readily prepared by the reduction of *p*-nitracetanilide with tin and hydrochloric acid.[6]

To prepare the free base, the hydrochloride is distilled with anhydrous sodium carbonate.[7] It is slightly soluble in water, more readily in alcohol and ether, crystallizing from the latter in tablets melting at 140°; it sublimes in small plates, and boils at 267°. It is distinguished from its isomerides by the fact that on heating with manganese dioxide and dilute sulphuric acid it is oxidized to quinone.

p-Diamidobenzene hydrochloride, $C_6H_4(NH_3Cl)_2$, is readily

---

[1] Hofmann, *Jahresb.* **1863**, 422.   [2] *Ber.* **19**, 200, 547.

[3] Calm, *ibid.* **16**, 2792.   [4] Klasemann, *ibid.* **7**, 1261.

[5] Hofmann, *loc. cit.*; Zincke and Rinne, *Ber.* **7**, 871; Martius and Griess, *Zeitschr. Chem.* **1866**, 136; Griess, *Ber.* **5**, 201.

[6] Horbrecker, *ibid.* **5**, 920.

[7] Biedermann and Ledoux, *ibid.* **7**, 1531.

soluble in water and crystallizes in tablets which are scarcely soluble in hydrochloric acid. Bleaching powder precipitates quinone-dichlorimide from its solution (p. 302).

-- *Dimethyl-p-diamidobenzene*, or *p-amidodimethylaniline*, C_6H_4 $(NH_2)N(CH_3)_2$, is obtained by the reduction of nitrodimethyl-aniline or nitrosodimethylaniline (p. 310).[1] It dissolves readily in water, melts at 41°, and boils at 257°. It is employed for the manufacture of colouring matters, and as a reagent for the detection of sulphuretted hydrogen. The sulphate, which is employed for this purpose, may be readily prepared from *helianthin*, C_6H_4N $(CH_3)_2.N_2.C_6H_4.SO_3NH_4$, a commercial azo-colour, by heating it with ammonium sulphide on the water-bath until the orange colour disappears. In this way it is decomposed into the ammonium salt of sulphanilic acid and the free base ; the latter is extracted with ether, the solution shaken with lead hydroxide suspended in water, and an ethereal solution of sulphuric acid added to the filtrate. The ether is then poured off from the pulpy mass, and the residue warmed on the water-bath with 4 to 5 parts of absolute alcohol until the salt separates out in fine needles. It is then washed with alcohol on to a filter pump and dried. On treating it with a hydrochloric acid solution of sulphuretted hydrogen, a splendid blue colour is obtained, methylene-blue being formed.[2]

Tetramethyl-p-diamidobenzene, $C_6H_4(N(CH_3)_2)_2$, is obtained by heating the preceding compound with hydrochloric acid and methyl alcohol to 170—180°, and then gradually to 200°. It is slightly soluble in cold, more readily in hot water, and readily in alcohol, crystallizing in small, lustrous, colourless or yellowish tablets melting at 51°, and boiling at 260°. Its aqueous solution becomes deep violet-blue in the air, and its salts are also coloured on addition of oxidizing agents, while its acetic acid solution is converted by sodium nitrite into *nitrosotrimethyl-p-diamidobenzene*, $C_6H_4N_2(CH_3)_3NO$, which crystallizes from water in small greenish yellow plates, melting at 98—99°, and giving Liebermann's reaction. Tin and hydrochloric acid convert it into *trimethyl-p-diamidobenzene*, $C_6H_4N_2(CH_3)_3H$, an oily liquid, the salts of which give a splendid reddish-violet colouration with weak oxidizing agents.

On adding potassium ferricyanide to a solution of the sulphate of the tetramethyl-base, small blue needles are precipitated,

---

[1] Schraube, *Ber.* **8**, 619 ; Weber, *ibid.* **10**, 762 ; Wurster, *ibid.* **12**, 522 and 528. [2] E. Fischer, *ibid.* **16**, 2234.

possessing a cupreous lustre similar to that of indigo; these consist of the ferricyanide, $C_{10}H_{14}N_2.Fe_2(CN)_6H_4$.

The tetramethyl-base combines with methyl iodide to form the compound $C_6H_4N_2(CH_3)_5I$, which crystallizes in small plates.[1]

Hexmethyl-p-diphenylammonium iodide, $C_6H_4N_2(CH_3)_6I_2$, Hofmann obtained this compound by the alternate action of methyl iodide and silver oxide on *p*-diamidobenzene; it crystallizes in small plates (Hofmann).

Ethyl-p-diamidobenzene, $C_6H_4(NH_2)NH(C_2H_5)$, and *Diethyl-p-diamidobenzene*, $C_6H_4(NH_2)N(C_2H_5)_2$, may be obtained by the reduction of the corresponding nitro-compounds. Both are liquids, boiling at about 262°.

Acetyl-p-diamidobenzene or *Amido-acetanilide*, $C_6H_4(NH_2)$ $(NH.C_2H_3O)$, may be obtained by the reduction of nitro-acetanilide with iron and acetic acid, and crystallizes from hot water in clusters of long, thin needles melting at 161°. It is a monacid base, and forms crystalline salts.[2]

p-Amidodiphenylamine, or *Phenyl-p-diamidobenzene*, NH $(C_6H_5)C_6H_4.NH_2$, is obtained by the action of acetic acid and zinc dust on nitrodiphenylamine, phenylamido-azobenzene, $C_6H_5.N_2.C_6H_4.NH(C_6H_5)$, and phenylamido-azobenzenesulphonic acid, the potassium salt of which is known in commerce under the name *Tropaeolin* OO. The product is extracted with water, saturated with caustic potash, and extracted with ether. On shaking the ethereal solution with dilute sulphuric acid, amidodiphenylamine sulphate separates out in small fine plates with a silvery lustre.

On adding ammonia to the hot aqueous solution of the salt, and allowing it to cool, the base crystallizes out in small lustrous plates which melt at 61°, and become coloured green in the air. Ferric chloride added to solutions of its salts, produces a red colouration which soon changes into green, while on strongly concentrating the solution a green precipitate is obtained, which dissolves in concentrated sulphuric acid with a carmine-red colour.[3]

Diphenyl-p-diamidobenzene, $C_6H_4(NH.C_6H_5)_2$, is obtained by heating quinol with aniline, calcium chloride and a little zinc chloride, and crystallizes in lustrous plates melting at 152°. Its solution in concentrated sulphuric acid is coloured cherry red to magenta red by nitric acid.[4]

[1] Wurster, *loc. cit.* ; Wurster and Schobig, *Ber.* **12**, 1807.
[2] Nietzki, *ibid.* **17**, 343.   [3] Nietzki and Witt, *ibid.* **12**, 1399.
[4] Calm, *ibid.* **16**, 2805.

Diamidodiphenylamine, $NH(C_6H_4.NH_2)_2$, is formed, together with *p*-diamidobenzene, when aniline-black is boiled with tin and hydrochloric acid, or with hydriodic acid and phosphorus. It crystallizes from hot water in small, feathery plates which melt at 158°, and yields a diazo-compound which is converted by boiling alcohol into diphenylamine. On oxidation it yields quinone, and ferric chloride colours its solution dark green.

Diphenylamine gives two dinitro-compounds, one of which is yellow and yields diamidodiphenylamine on reduction, while the isomeric one is red and yields an oily base (Nietzki and Witt).

CYANOGEN DERIVATIVES OF PHENYLENE.

1086 *m-Phenylene isocyanate* and *p-Phenylene isocyanate*, C_6H_4 $(N.CO)_2$, are prepared by the action of carbonyl chloride on *m*- and *p*-diamidobenzene. The latter forms white needles melting at 91°. The meta-derivative is very similar, but not so readily obtained.[2]

CARBAMIDE-DERIVATIVES OF THE DIAMIDO-BENZENES.[3]

1087 On treating solutions of the diamidobenzene hydrochlorides with a solution of potassium cyanate, phenylene-carbamides are obtained :

$$C_6H_4 \begin{cases} NH.CO.NH_2 \\ NH.CO.NH_2. \end{cases}$$

o-Phenylene dicarbamide is readily soluble in water and alcohol, and crystallizes in delicate needles, which melt at 290° and sublime in small iridescent plates.

m-Phenylene dicarbamide is only slightly soluble in hot water, and less so in alcohol; it forms crystals which melt above 300° and sublime, with considerable decomposition, in stellate-grouped needles.[4]

[1] Nietzki, *Ber.* **11**, 1097.   [2] Gattermann and Wrampelmeyer, *ibid.* **18**, 2605.
[3] Lellmann, *Annalen*, **221**, 1 ; Lellmann and Würthner, *ibid.* **228**, 199.
[4] Warder, *Ber.* **8**, 1180.

p-Phenylene dicarbamide forms small plates with a silvery lustre, which are extremely difficult to dissolve in the ordinary solvents, and on heating decompose with carbonization.

The diamidobenzenes combine with phenyl isocyanate to form compounds which decompose on heating into aniline and phenylene carbamides:

$$C_6H_4\begin{smallmatrix}\diagup NH.CO.NH.C_6H_5\\ \diagdown NH_2\end{smallmatrix} = C_6H_4\begin{smallmatrix}\diagup NH\diagdown\\ \diagdown NH\diagup\end{smallmatrix}CO + NH_2.C_6H_5.$$

o-Phenylene carbamide is also obtained by heating *o*-nitraniline with ethyl chlorocarbonate and converting the nitro-phenyl urethane formed, by means of tin and hydrochloric acid, into the amido-compound, which crystallizes in long needles, melting at 86° and decomposes when strongly heated.

$$C_6H_4\begin{smallmatrix}\diagup NH.CO.OC_2H_5\\ \diagdown NH_2\end{smallmatrix} = C_6H_4\begin{smallmatrix}\diagup NH\diagdown\\ \diagdown NH\diagup\end{smallmatrix}CO + HO.C_2H_5.$$

o-Phenylene carbamide crystallizes from hot water, in which it is only slightly soluble, and from alcohol in bright, lustrous plates, which become brown and melt at 305°.[1]

m-Phenylene carbamide is also obtained by the action of carbonyl chloride on *m*-diamidobenzene. It is a white amorphous insoluble powder, which commences to char at 300° without melting. On heating with hydrochloric acid to 160—170°, it decomposes into carbon dioxide and *m*-diamidobenzene.[2]

p-Phenylene carbamide is a brown, insoluble powder which carbonizes on heating.

On mixing solutions of the diamidobenzene hydrochlorides with ammonium thiocyanate, thiocyanates of these bases are obtained; those of the meta- and para- compounds are converted on heating into the isomeric phenylene-dithiocarbamides, while that of the ortho-compound is converted into phenylene-mono-thiocarbamide:

$$C_6H_5\begin{smallmatrix}\diagup NH_3SCN\\ \diagdown NH_3SCN\end{smallmatrix} = C_6H_4\begin{smallmatrix}\diagup NH\diagdown\\ \diagdown NH\diagup\end{smallmatrix}CS + CS\begin{smallmatrix}\diagup NH_2\\ \diagdown NH_2.\end{smallmatrix}$$

The diamidobenzenes also combine with phenyl mustard oil forming amidodiphenyl-thiocarbamides, of which the meta-

---

[1] Rudolph, *Ber.* **12,** 1295.   [2] Wichler and Zimmermann, *ibid.* **14,** 2177.

compound melts on heating without decomposition, while the two isomerides lose aniline without melting:

$$C_6H_4\underset{NH_2}{\overset{NH.CS.NH.C_6H_5}{\diagdown}} = C_6H_4\underset{NH}{\overset{NH}{\diagdown}}CS + NH_2.C_6H_5.$$

This is therefore an example of condensation taking place in the para-series.

o-Phenylene thiocarbamide is slightly soluble in water, readily in alcohol, and crystallizes from ammonia in large, colourless plates, which blacken and melt at about 290°, giving off a vapour which condenses forming iridescent plates, and produces on inhalation an intensely bitter taste.

o-Amidodiphenyl thiocarbamide forms colourless, lustrous prisms, which are readily soluble in alcohol and glacial acetic acid, with difficulty in benzene, and insoluble in ether.

m-Phenylene dithiocarbamide is only slightly soluble in the ordinary reagents; it is more readily soluble in alkalis, and is precipitated by acids in microscopic plates, melting at 215°.

m-Amidodiphenyl thiocarbamide forms a yellow, amorphous powder, or colourless prisms, melting at 147—148°, and solidifying to a vitreous mass.

p-Phenylene thiocarbamide forms light brown, almost insoluble microscopic plates, melting at 270—271°.

p-Phenylene dithiocarbamide is insoluble in water, slightly soluble in alcohol, and crystallizes from aqueous ammonia in small needles, melting at 218°.

p-Amidodiphenyl thiocarbamide crystallizes from alcohol in red prisms, which begin to melt at 163° with separation of aniline, this being formed in much larger quantity at 185°.

TRIAMIDOBENZENES, $C_6H_3(NH_2)_3$.

1088 *adj.-Triamidobenzene* (1 : 2 : 3), is obtained by the dry distillation of triamidobenzoic acid, and is a red or brown crystalline mass, melting at 103°, and boiling at 330°. It is a diacid base and forms crystallizable salts. If it be dissolved in concentrated sulphuric acid and a trace of nitric acid added, the liquid is coloured a greenish blue, which subsequently changes to a fine dark blue.[1]

[1] Salkowski, *Annalen*, **163**, 23 ; Hinsberg, *Ber.* **19**, 1253.

Its *hydrochloride*, $C_6H_9N_3.2HCl$, forms needles, which are readily soluble in water, whilst the *sulphate*, $C_6H_9N_3.H_2SO_4 + 2H_2O$, crystallizes in plates.

On boiling with glacial acetic acid, *adj.*-triamido-benzene is converted into *ethenyl-amidophenyleneamidine*,

$$NH_2.C_6H_3 \begin{array}{c} NH \\ \diagdown \diagup \\ N \end{array} C.CH_3.$$ It forms apparently asymmetric

crystals readily soluble in water.

as-Triamidobenzene (1 : 2 : 4) is formed by the reduction of *as*-dinitraniline[1] as well as by the action of tin and hydrochloric acid on dinitroazobenzene-*p*-sulphonic acid, $C_6H_3(NO_2)_2$ $N : N.C_6H_4.SO_3H.$[2]

It crystallizes from chloroform in plates, which melt below 100° and boil at about 340°. Its aqueous solution is as a rule coloured green, but weak oxidizing agents give at once a red colouration. Its *hydrochloride*, $C_6H_9N_3.2HCl$, forms needles which are very soluble in water, very sparingly in alcohol and hydrochloric acid.

s-Triamidobenzene (1 : 3 : 5) is unknown in the free state. The tin double salt is obtained by the action of tin and hydrochloric acid on *s*-trinitrobenzene.[3] If this is decomposed by sulphuretted hydrogen, and the filtrate evaporated over sulphuric acid in vacuo the *hydrochloride*, $C_6H_9N_3.3HCl$, is obtained as a crystalline mass, readily soluble in water, but giving no blue colouration with ferric chloride.

s-Trinitrotriamidobenzene, $C_6(NO_2)_3(NH_2)_3$, is prepared by acting on *s*- tribromotrinitrobenzene with alcoholic ammonia.[4] It forms thin pale yellow tablets, which decompose without melting above 300°.

TETRAMIDOBENZENES, $C_6H_2 (NH_2)_4$.

1089 *adj.-Tetramidobenzene*, $C_6H_4(NH_2)_4$ (1 : 2 : 3 : 4) is obtained by the reduction of *adj.*-diquinoyltetroxime (see p. 300) with stannous chloride and hydrochloric acid. On addition of sulphuric acid and alcohol the *sulphate*, $C_6H_{10}N_4.H_2SO_4$, separates in colourless plates which are scarcely soluble in cold water.

[1] Salkowski, *Annalen*, **164**, 265.
[2] Janovsky, *Monatsh.* **5**, 159.
[3] *Annalen*, **215**, 348.
[4] *Am. Chem. Journ.* **10**, 287.

The free base readily undergoes oxidation in the air, and unites with o-diketones forming diquinoxalines.[1]

When the sulphate is treated with acetic anhydride and sodium acetate it yields an *acetyl* derivative melting at 260° which appears to have the constitutional formula

$$(C_2H_3O.NH)_2.C_6H_2{\Large\langle}{\begin{array}{l} N.C_2H_3O \\ \diagdown \\ N:C.CH_3 \end{array}} + H_2O.$$

On warming this compound with dilute sulphuric acid on the water-bath it yields *diethenyltetramidobenzene*, $C_6H_2(N_2H:C.CH_3)_2$, which crystallizes in needles melting at 145°. An isomeric diethenyltetramidobenzene is also obtained[2] by the reduction of diacetyldinitro-p-diamidobenzene with tin and hydrochloric acid; this forms long needles melting at 210°. The constitution of these two compounds is probably expressed by the formulae

s-*Tetramidobenzene*, $C_6H_2(NH_2)_4$ (1 : 2 : 4 : 5), is prepared by treating dinitro-m-diamidobenzene with tin and hydrochloric acid. Hydrogen chloride is then passed in and the precipitated hydrochloride purified by redissolving in water, and again precipitating with hydrogen chloride. It has the formula $C_6H_{10}N_4.4HCl$ and is extremely soluble in water, but sparingly in concentrated hydrochloric acid. It forms two sulphates $(C_6H_{10}N_4)H_2SO_4$ and $(C_6H_{10}N_4)_2.3H_2SO_4$, both sparingly soluble in water.

The free base cannot be obtained as it at once oxidizes in the air to *tetramidodiphenazine* $C_{12}H_{12}N_6$.

The hydrochloride also combines with o-diketones but only one pair of amido-groups takes part in the reaction.[3]

The *tetracetyl* derivative $C_6H_2(NH.C_2H_3O)_4$ is preparing by acting on the hydrochloride with acetic anhydride and sodium acetate,[4] and forms long needles melting at 285°. The *diethenyl* derivative $C_6H_2{\Large(}{\begin{array}{l} N \\ NH \end{array}}{\Large\rangle}C.CH_3)_2$ is formed by the reduction of

[1] Nietzki and Schmidt, *Ber.* **22**, 1648.
[2] Nietzki and Hagenbach, *ibid.* **20**, 329.   [3] *Ibid.* p. 334.
[4] Nietzki and Müller, *ibid.* **22**, 440.

diacetyldinitro-*m*-diamidobenzene [1] and also crystallizes in needles which do not melt at 360°.

s-Diamidodi-imidobenzene, $C_6H_2.(NH_2)_2(NH)_2$, is prepared by oxidizing an aqueous solution of *s*-tetramidobenzene hydrochloride with ferric chloride. The free base forms small brown needles which are reconverted into tetramidobenzene by reducing agents. The *hydrochloride*, $C_6H_8N_4.2HCl$, forms needles with a brown surface lustre, which are sparingly soluble in water forming a bluish violet solution.[2]

PENTAMIDOBENZENE, $C_6H(NH_2)_5$.

1090 The hydrochloride of this compound is obtained by the reduction of trinitro-*m*-diamidobenzene with tin and hydrochloric acid. It has the composition, $C_6H_{11}N_5.3HCl$, and forms slender needles, readily soluble in water, but insoluble in alcohol and ether. On boiling with acetic anhydride and sodium acetate it yields the *pentacetyl* derivative $C_6H(NH.C_2H_3O)_5$ as an insoluble amorphous powder.[3]

Pentamidobenzene has also been prepared from *s*-tribromodinitrobenzene by treating it with alcoholic ammonia and reducing the triamidodinitrobenzene thus obtained.[4]

AMIDOHYDROXYBENZENES.

AMIDOPHENOLS, $C_6H_4(NH_2)OH$.

1091 These compounds are obtained by the reduction of the nitrophenols and by heating amidosalicylic acid, $C_6H_3(NH_2)$ $(OH)COOH$, and its isomerides with caustic baryta. They are weak bases; their halogen substitution products, on the other hand, behave like phenol, the acid character increasing with the number of hydrogen atoms replaced. The replacement of hydrogen by nitroxyl acts still more strongly in this way.

In these compounds, as in other amido-derivatives, hydrogen can be replaced by acid radicals. It is a characteristic property

---

[1] Nietzki and Hagenbach, *loc. cit.*   [2] *Ibid.*
[3] Barr, *ibid.* **21**, 1547.   [4] Palmer and Jackson, *ibid.* **21**, 1706.

of o-amidophenol that its acid derivatives, like those of many other o-amido-compounds, readily lose water and form anhydro-compounds:

Acetylamidophenol.

$$C_6H_4{\Large\langle}^{NH.CO.CH_3}_{OH.}$$

Ethenylamidophenol.

$$C_6H_4{\Large\langle}^{N}_{O}{\Large\rangle}C.CH_3.$$

o-Amidophenol was discovered by Hofmann, who obtained it by the action of sodium sulphide on o-nitrophenol;[1] the reduction may be more readily effected by tin and hydrochloric acid.[2] It crystallizes in small rhombic plates with a pearly lustre, which melt at 170° but sublime at a lower temperature. Its hydrochloride is readily soluble in water and crystallizes in long needles.

Methyl-o-amidophenate or *o-Anisidine*, $C_6H_4(NH_2)OCH_3$, is formed by the reduction of o-nitranisoïl,[3] $C_6H_4(NO_2)OCH_3$, and is a mobile, strongly refractive liquid, having a peculiar odour, and boiling at 226°·5. It forms salts which crystallize well. Methyl iodide acts violently upon it; if the two substances be mixed and kept cool by a freezing mixture, methyl-o-anisidine, $C_6H_4(NH.CH_3)OCH_3$, is formed as a liquid boiling at 218—220°.[4]

Trimethyl-o-amidophenol or *Trimethylhydroxyphenylammonium hydroxide*, $HO.C_6H_4N(CH_3)_3OH$, is obtained by treating a solution of o-amidophenol hydrochloride in methyl alcohol with 3 parts of methyl iodide, adding a considerable excess of caustic potash and allowing the whole to stand, with repeated additions of potash, until the solution no longer becomes acid.[5] The hydriodide thus obtained is decomposed by freshly precipitated silver oxide. The base is readily soluble in water and alcohol, but not in ether, and crystallizes in white prisms which have an intensely bitter taste. On heating to 105° it forms the anhydride:

$$C_6H_4{\Large\langle}^{N(CH_3)_3}_{\,\,\,|}_{O}$$

On distillation it is converted into the isomeric o-dimethyl-amido-anisoïl, $CH_3O.C_6H_4N(CH_3)_2$, a basic, strongly refractive

[1] *Annalen*, **103**, 351.
[2] Cook and Schmitt, *Kekulé's Lehrb.* **3**, 61 ; *Ber.* **1**, 67.
[3] Mühlhäuser, *Annalen*, **207**, 235.   [4] *Ibid.*
[5] Griess, *Ber.* **13**, 246.

liquid, boiling at 210—212° (Griess, Mühlhäuser).   The corresponding *o*-dimethylamidophenol, $HO.C_6H_4.N(CH_3)_2$, is formed, together with methyl chloride, by the dry distillation of trimethylamidophenol hydrochloride, and crystallizes in rhombic prisms, which melt at 45°, and form an amorphous hydrochloride.

Methenylamidophenol, $C_6H_4 \diagdown{N \atop O} \diagup CH$, is prepared by heating

o-amidophenol with formic acid.   It crystallizes in prisms, melting at 30°·5, and boiling at 182°·5.[1]

Ethenylamidophenol, $C_6H_4 \diagdown{N \atop O} \diagup C.CH_3$, is obtained by heating

o-amidophenol with acetic anhydride, and is a basic liquid boiling at 200—201°.   It is converted by heating with dilute acids into *acetylamidophenol,* $HO.C_6H_4.NH(CO.CH_3)$,[2] which is also formed by the action of tin and acetic acid on *o*-nitrophenol.[3] It crystallizes from dilute alcohol in small plates, melting at 201°.

Carbonyl-o-amidophenol $C_6H_4 \diagdown{NH \atop O} \diagup CO$ is obtained by the

action of carbonyl chloride on *o*-amidophenol,[4] by distilling ethyl-*o*-amidophenylcarbonate[5] and by heating hydroxyphenyl carbamide.[6]   It crystallizes from hydrochloric acid in lustrous needles, which melt at 137—138°.

1092 *m-Amidophenol* is obtained from *m*-amidobenzenesulphonic acid by fusion with caustic alkalis.   It crystallizes from hot water in hard needles, which melt at 121°, are readily soluble in alcohol and ether, sparingly in benzene and cold water, and are stable in the air.[7]   Its *hydrochloride* may also be obtained by reducing *m*-nitrophenol with tin and hydrochloric acid ; it forms hard, colourless granules, an aqueous solution of which soon decomposes and becomes brown.[8]

Methyl-m-amidophenol, $C_6H_4(NH.CH_3)OH$, is obtained by heating methyl-*m*-amidobenzenesulphonic acid with caustic alkalis at 200—220°.   It is a thick oil which is readily soluble in alcohol, ether, and benzene.

Ethyl-m-amidophenol, $C_6H_4(NH.C_2H_5)OH$, is prepared in a

[1] Ladenburg, *Ber.* **10**, 1124.
[2] *Ber.* **9**, 1524.
[3] Morse, *ibid.* **11**, 232.
[4] Chetmicki, *ibid.* **20**, 177.
[5] Bender, *ibid.* **19**, 2269, 2951.
[6] Kalckhoff, *ibid.* **16**, 1828.
[7] D. R. Patent, 44792 ; *Ber.* **21**, 875c.
[8] Bantlin, *ibid.* **11**, 2099.

similar manner from ethyl-*m*-amidobenzenesulphonic acid.[1]   It crystallizes in colourless feathery crystals, melting at 62°.

The corresponding *dimethyl-* and *diethyl-m-amidophenol* are obtained in a similar manner from dimethyl- and diethyl-*m*-amidobenzenesulphonic acid.   The latter compounds are formed by the direct methylation or ethylation of *m*-amidobenzene-sulphonic acid, or by the sulphonation of dimethyl- and diethyl-aniline.[2]

Phenyl-m-amidophenol, or *m-Hydroxydiphenylamine,* C_6H_4 $(NH.C_6H_5)OH$. This compound was first prepared by Merz and Weith by heating resorcinol with aniline and calcium or zinc chloride to 270—280°.[3]   It is more readily obtained by heating *m*-amidophenol hydrochloride with aniline at 210—215° in an autoclave for eight hours. Water and caustic soda are added to the product, and aniline removed by distilling in a current of steam. Addition of acetic acid then precipitates the phenyl-*m*-amidophenol as a reddish-brown mass, which is extracted with hydrochloric acid, precipitated with sodium acetate, distilled with superheated steam and recrystallized.[4] It forms iridiscent plates, melting at 81·5—82°, and boiling at 340°. It is sparingly soluble in hot water, readily in alcohol and ether. Its *hydrochloride*, $C_{12}H_{11}NO.HCl$, forms needles, which readily undergo decomposition. The *sulphate*, $(C_{12}H_{11}NO)_2H_2SO_4$, crystallizes in lustrous needles which are decomposed by water and alcohol.

All these meta-derivatives are largely employed in the pre-paration of colouring matters.

p-Amidophenol. Fritsche obtained this compound by the action of acetic acid and iron filings on *p*-nitrophenol.[5] It is best prepared by reducing the latter compound with tin and hydrochloric acid.[6] It crystallizes in small colourless plates, which rapidly turn brown and melt at 184° with decomposition.[7] Its alkaline solution assumes a violet colour in the air, and it is readily converted into quinone by oxidizing agents.[8] The *hydro-chloride* crystallizes in prisms, which readily dissolve in water,

[1] D. R. P. 48151; *Ber.* **22**, 622c.   [2] D. R. P. 44792; *Ber.* **21**, 875c.
[3] Merz and Weith, *ibid.* **14**, 2345 ; Calm, *ibid.* **16**, 2787.
[4] D. R. P. 46869; *Ber.* **22**, 314c.   [5] *Annalen,* **110**, 166.
[6] Cook and Schmitt, *Kekulé's Lehrb.* **3**, 62.
[7] Lossen, *Annalen,* **175**, 296.
[8] Körner, *Jahresb.* **1867**, 23, 173 ; Schmitt, *J. Pr. Chem.* II. **19**, 317 ; Andresen, *ibid.* II. **23**, 173.

forming a solution which is coloured first violet and then green by bleaching powder (Lossen).

Methyl-p-amidophenate or *p-Anisidine*, $C_6H_4(NH_2)OCH_3$, has been prepared by the reduction of *p*-nitranisoïl,[1] and forms large rhombic tablets, melting at 55·5—56·5°,[2] and boiling at 245—246°.[3] Its *hydrochloride* crystallizes in long needles, and gives a violet colouration with ferric chloride.

Trimethyl-p-amidophenol, $HO.C_6H_3N(CH_3)_3OH$, is obtained in a similar manner to the ortho-compound, and crystallizes in small plates or prisms. On distillation it is converted into *p-dimethylamido-anisoïl*, $CH_3O.C_6H_4N(CH_3)_2$, which crystallizes from alcohol in small rhombic plates (Griess).

1093 *a-Nitramidophenol*, $C_6H_3(NO_2)NH_2(OH)$, is formed by the reduction of *a*-dinitrophenol. It crystallizes with one molecule of water in orange-red prisms, which lose their water of crystallization on heating, and then melt at 142—143°.[4]

β-Nitramidophenol crystallizes in red needles, melting at 110—111° (Stuckenberg).

γ-Nitramidophenol is obtained by boiling nitro-*m*-diamido-benzene with caustic potash, and crystallizes in yellowish red plates, melting at 133—134°.[5]

a-Dinitramidophenol or *Picramic acid*, $C_6H_2(NO_2)_2NH_2(OH)$. By the action of milk of lime or baryta on a mixture of ferrous sulphate and picric acid (carbazotic acid), Wöhler obtained a blood-red solution, and found that it contained the salts of a peculiar crystalline acid, which he called "reduced carbazotic acid,"[6] while Berzelius named it "nitrohaematic acid."[7] Girard, by treating picric acid with ammonium sulphide, obtained a similar compound, and named it picramic acid;[8] Pugh then showed that this is identical with Wöhler's acid, and that it is also formed by the action of stannous chloride on picric acid.[9] It may also be obtained by nitrating *a*-nitramidophenol (Stuckenberg). In order to prepare it, an alcoholic solution of picric acid is evaporated with an excess of ammonium sulphide, the residue extracted with water, and the filtrate decomposed by acetic acid.[10]

Picramic acid ($OH : NO_2 : NH_2 : NO_2 = 1 : 2 : 4 : 6$), crystal-

[1] Cahours, *Annalen*, **74**, 300 ; Brunck, *Zeitsch. Chem.* **1867**, 205.

[2] Lossen, *Annalen*, **175**, 324.   [3] Salkowski, *Ber.* **7**, 1009.

[4] Stuckenberg, *Annalen*, **205**, 66.   [5] Barbaglia, *Ber.* **7**, 1257.

[6] *Pogg. Ann.* **13**, 488.   [7] *Lehrb.* **4**, 675.

[8] *Annalen*, **88**, 281 ; *Jahresb.* **1855**, 535.   [9] *Annalen*, **96**, 83.

[10] Lea, *Chem. News*, **4**, 193 ; Petersen, *Zeitsch. Chem.* **1868**, 377.

lizes in dark red needles or monosymmetric prisms, melting at 165° and forming orange-red solutions in water and alcohol, which are rendered more deeply coloured by alkalis.

Potassium picramate, $C_6H_2(NO_2)_2NH_2(OK)$, crystallizes in long, red, rhombic tablets.

Ammonium picramate, $C_6H_2(NO_2)_2NH_2(ONH_4)$, forms dark, orange-red rhombohedra.

Silver picramate, $C_6H_2(NO_2)_2(NH_2)(OAg)$, is a red, amorphous precipitate.

Methyl picramate or *Dinitranisidine*, $C_6H_2(NO_2)_2NH_2(OCH_3)$. Cabours prepared this compound by the action of alcoholic ammonium sulphide on methyl picrate. It crystallizes from hot alcohol in dark violet needles, and, like picramic acid, combines with acids to form unstable salts.

β-Dinitramidophenol, (OH　$NO_2 : NH_2 : NO_2 = 1 : 4 : 6 : 2$). This compound, which is also known as *isopicramic acid*, is prepared by heating a solution of benzoylamidosalicylic acid, $C_6H_3(NH.C_7H_5O)(OH)CO_2H$, in glacial acetic acid with nitric acid ; carbon dioxide is thus eliminated with formation of benzoylisopicramic acid, which is then decomposed by heating with hydrochloric acid, into benzoic acid and isopicramic acid. This crystallizes from hot water in lustrous yellow plates or needles, which become brown on drying, dissolve in alcohol or water with a cherry-red colour, and melt with decomposition at 170°.

Potassium isopicramate, $C_6H_2(NO_2)_2NH_2(OK)$, is readily soluble in water, and crystallizes in dark blue needles.[1]

a-Diamidophenol, $C_6H_3(NH_2)_2OH$, [OH.$NH_2NH_2 = 1 . 2 . 4$], is formed by the reduction of *a*-dinitrophenol,[2] and is scarcely known in the free state, as it is very readily decomposed. Its salts crystallize well; their aqueous solutions rapidly turn brown in the air, and are coloured an intense dark red by ferric chloride or potassium dichromate.

β-Diamidophenol [OH.$NH_2NH_2 = 1 . 2 . 6$], is also very unstable. Its salts become red in the sunlight.

γ-Diamidophenol [OH.$NH_2NH_2 = 1 . 3 . 4$]. The hydrochloride of this substance is very soluble in water, its solution being coloured blood-red by chloride of lime or ferric chloride.

Triamidophenol, $C_6H_2(NH_2)_3OH$. By the action of iodide of phosphorus and water on picric acid, Lautemann obtained a salt

[1] Dabney, *Am. Chem. Journ.* **5**, 20.

[2] Gauhe, *Annalen*, **147**, 66 ; Hemilian, *Ber.* **8**, 768.

which he named picrammonium iodide, $C_6H_3(NH_3I)_3$.[1] Heintzel, however, found[2] that by the reduction of picric acid with tin and hydrochloric acid, the hydrochloride of triamidophenol, $C_6H_2(OH)(NH_3Cl)_3$, is formed, and that Lautemann's compound is the corresponding hydriodide. Gauhe[3] denied this, but Heintzel then proved[4] that triamidophenol, and not triamidobenzene, is obtained by the reduction of picric acid. It is also formed by the action of tin and hydrochloric acid on picramide,[5] one amido-group being replaced by hydroxyl. It is very unstable in the free state, and its salts also readily oxidize. Its dilute solution is coloured a deep blue by ferric chloride, amidodi-imidophenol, or diamidoquinone-imide, $C_6H_7N_3O$, being formed. This is a monacid base, the hydrochloride of which crystallizes in brown needles with a blue lustre (Heintzel).

a-Triamidophenol, $C_6H_2(OH)(NH_2)_3$ forms needles which are sparingly soluble in water and alcohol. Its *hydrochloride* forms crystalline hexagonal plates.

Tetramidophenol, $C_6H(OH)(NH_2)_4$ is obtained by the reduction of the ethyl ether of trinitroamidophenol. The *hydrochloride*, $C_8H_4N_4O + 2HCl$, crystallizes in plates or long flat prisms, the aqueous solution of which is coloured by chloride of lime or ferric chloride successively dark-green, red, brown, and yellow, finally becoming colourless.[6]

AMIDODIHYDROXYBENZENES.

1094 *Amidocatechol*, $C_6H_3(NH_2)(OH)_2$, is obtained by the reduction of nitrocatechol with tin and hydrochloric acid; its hydrochloride forms long, dark-coloured needles. When the free base is separated by an alkali it immediately oxidizes in the air and dissolves with a dark violet colour.[7]

Diamidocatechol, $C_6H_2(NH_2)_2(OH)_2$, is unknown. Its methyl ether, which is also termed *diamidoguaiacol*, is prepared by the reduction of dinitroguaiacol. It is extremly unstable, and the aqueous solution of its hydrochloride becomes red in the air, and is coloured violet-red by ferric chloride.

o-Amidoresorcinol, $C_6H_3(NH_2)(OH)_2$. The diethyl ether of

[1] *Annalen*, **125**, 1.
[2] *Zeitsch. Chem.* **1867**, 338.
[3] *Ibid.* **1868**, 90.
[4] *Ber.* **1**, 111.
[5] Hepp, *Annalen*, **215**, 350.
[6] Köhler, *J. Pr. Chem.* II. **29**, 81.
[7] Benedikt, *Ber.* **11**, 363.

this compound is obtained by the reduction of o-benzene-azoresorcinol diethyl ether with stannous chloride and hydro-chloric acid. It forms small plates which melt at 124°, and are insoluble in water, readily soluble in alcohol and ether, and quickly undergo oxidation in the air.

Phloramine, $C_6H_3(NH_2)(OH)_2[1:3:5]$, is formed when phloro-glucinol is dissolved in ammonia. This reaction is very re-markable, as in other phenols the hydroxyl can only be replaced by the amido-group with great difficulty. It crystallizes from warm water in small micaceous plates, which in the damp state rapidly become brown, but when dry do not change, wherein it differs from o- and p-amidoresorcinol. Phloramine has a slightly astringent taste, gives no colouration with ferric chloride, and only reduces silver solution when warmed; most of its salts crystallize well.

Phloramine hydrochloride, $C_6H_7NO_2.HCl + H_2O$, forms needles or small plates; the nitrate, $C_6H_7NO_2.HNO_3$, crystallizes in bronze-coloured, and the sulphate, $(C_6H_7NO_2)_2.H_2SO_4 + 2H_2O$, in yellow needles.[1]

p-Amidoresorcinol, $C_6H_3(NH_2)(OH)_2$, is formed by the action of tin and hydrochloric acid on nitroresorcinol[2] and phenylazoresorcinol, $C_6H_5.N_2.C_6H_3(OH)_2$.[3] The *hydrochloride*, $C_6H_7NO_2.HCl + 2H_2O$, crystallizes in large oblique prisms or flat plates, which appear colourless when the light passes through the broad faces, but olive-brown when it passes through the narrow ones. They become coloured green in the air; when caustic soda is added to the solution a deep blue colouration is produced, which then becomes green and finally yellowish brown. Ferric chloride produces a deep brown colouration, and then an almost black precipitate.

Its mono- and diethyl ether are obtained by the reduction of the corresponding derivatives of p-benzene-azoresorcinol. Both readily undergo oxidation in the air.[4]

adj-Diamidoresorcinol, $C_6H_2(NH_2)_2(OH)_2$, may be obtained by the reduction of dinitro- or dinitrosoresorcinol[5] or of *adj*-benzenedisazoresorcinol with tin and hydrochloric acid.[6] The free compound speedily becomes brown in the air, and its

[1] Hlasiwetz, *Annalen*, **119**, 202.
[2] Weselsky, *ibid.* **164**, 6.
[3] R. Meyer and Kreis, *Ber.* **16**, 1330.
[4] Will and Pukall, *ibid.* **20**, 1124, 1135.
[5] Fitz, *ibid.* **8**, 633.
[6] Liebermann and Kostanecki, *ibid.* **17**, 881.

solution is coloured violet-blue by ferric chloride; ammonia likewise colours it violet, but no crystals are deposited on allowing the solution to remain in the air. If the hydrochloride be suspended in chloroform, shaken with a few drops of caustic soda solution, and then largely diluted with water, it is coloured bright-blue. Its *sulphate*, $C_6H_8N_2O_2.H_2SO_4 + 1\frac{1}{2}H_2O$, crystallizes in needles. The constitution of the base is shown by its formation from dinitrosoresorcinol, in which the side-chains occupy the positions 1.2.3.4.

s-Diamidoresorcinol or *Isodiamidoresorcinol*, $C_6H_2(OH)_2(NH_2)_2$, is prepared by the reduction of isodinitroresorcinol[1] or of *s*-benzenedisazoresorcinol.[2] It oxidizes very readily, forming di-imidoresorcinol, and gives a magenta-red colouration with ferric chloride. Its *hydrochloride*, $C_6H_8N_2O_2.2HCl$, forms flat glassy needles; its *sulphate*, $C_6H_8N_2O_2.H_2SO_4 + 2H_2O$, also crystallizes from dilute alcohol in hard needles, which lose their water of crystallization at 105°.

s-Di-imidoresorcinol, $C_6H_2(OH)_2(NH)_2$, is prepared by passing air through a solution of *s*-diamidoresorcinol hydrochloride, to which an excess of ammonia has been added. It forms crystals resembling those of cuprous oxide, and is converted by alkalis into ammonia and *s*-dihydroxyquinone (p. 202).

Triamidoresorcinol, $C_6H(NH_2)_3(OH)_2$, is prepared from trinitro-resorcinol, and is very unstable in the free state. The hydro-chloride, $C_6H_9N_3O_2(HCl)_3 + H_2O$, crystallizes in needles, which dissolve very readily in water and are reprecipitated by hydro-chloric acid. They decompose in the air and become coloured red.

Amidodi-imidoresorcinol, $C_6H(NH_2)(NH)_2(OH)_2 + H_2O$, is obtained by the oxidation of the preceding compound, either by passing air through its concentrated solution or by the addition of ferric chloride. Ammonia precipitates the compound in green needles, having a metallic lustre, which are insoluble in alcohol and dissolve in caustic potash with a deep blue colour. The *hydro-chloride* crystallizes in splendid red needles, is only very slightly soluble in water and is reprecipitated by hydrochloric acid.[3] When it is heated with an acid to 170°, trihydroxyquinone (p. 207) is formed.

Amidodimethylquinol, $C_6H_3(NH_2)(OCH_3)_2$, is formed by acting

---

[1] Typke, *Ber.* **16**, 555.

[2] Kostanecki, *ibid.* **21**, 3113; Nietzki and Schmitt, *ibid.* **22**, 1653.

[3] Schreder, *Annalen*, **158**, 250.

on nitrodimethylquinol with tin and hydrochloric acid;[1] or more readily by gradually adding sodium amalgam to an alcoholic solution to which acetic acid has also been added.[2]

It crystallizes from hot water in small plates with a pearly lustre, but separates from hot benzene in brown crystals and from light petroleum in scales with a silvery lustre, which in thick layers show a brown reflection. It readily decomposes even in vacuo and changes into a black mass. Its *hydrochloride* crystallizes in white, efflorescent needles.

The aqueous solution of the base reduces silver solution, a mirror being formed ; on heating it with ferric chloride, small greenish plates separate out, which dissolve in water with a red colour.

Dimethylquinoltrimethylammonium iodide, $C_6H_3(OCH_3)_2N(CH_3)_3I$, is formed when the foregoing base is heated to 150° with methyl iodide and a little wood spirit. It crystallizes in fine white needles, which melt at 202°, and dissolve readily in water, but with some difficulty in alcohol. Moist silver oxide converts it into the hydroxide, which has a strong alkaline reaction, is exceedingly soluble in water, and crystallizes in transparent needles; the chloride forms white, readily soluble needles, which melt at 172° (Baessler).

a-Diamidoquinol, $C_6H_2(OH)_2(NH_2)_2$, [OH : NH$_2$: OH : NH$_2$ = 1 . 2 . 4 . 6] is obtained from diacetylquinol by treating it with nitric acid and reducing the nitro-compound thus formed with a mixture of tin, stannous chloride and hydrochloric acid, and saturating the products with hydrogen chloride.

a-Diamidoquinol hydrochloride, $C_6H_2(OH)_2(NH_2)_2$, 2HCl, then separates out in colourless needles, readily soluble in water. On treatment with acetic anhydride it yields a tetracetyl compound, melting at 216°, which by the action of an alkali is converted into the corresponding diacetyldiamidoquinone.[3]

β-Diamidoquinol, $C_6H_2(OH)_2(NH_2)_2$, [OH : NH$_2$: OH : NH$_2$ = 1 . 2 . 4 . 5] is obtained by the reduction of dihydroxyquinone dioxime. Its sulphate, $C_6H_2(OH)_2(NH_2)_2 \cdot H_2SO_4$ forms colourless sparingly soluble needles.[4] It also forms a tetracetyl derivative melting at 225°, which however by the action of alkalis does not form the corresponding diacetyldiamidoqui-

[1] Mühlhäuser, *Annalen*, **207**, 254 ; Margatti, *Ber.* **14**, 71.
[2] Baessler, *ibid.* **17**, 2118.
[3] Nietzki and Preusser, *ibid.* **19**, 2247.
[4] Nietzki and Schmidt, *ibid.* **22**, 1656.

none, but an acetylamidohydroxyquinone, $C_6H_2(OH).O_2.NH.$
$C_2H_3O.$

Di-imidoquinol, $C_6H_2(OH)_2(NH)_2$ (?) is formed by the oxidation of the foregoing compound, and crystallizes in red needles
with a violet surface lustre.[1]

β-Diamidoquinol dimethyl ether, $C_6H_2(OCH_3)_2.(NH_2)_2$, is formed
by the reduction of *β*-dinitro-*p*-dimethoxybenzene, obtained,
together with an isomeric compound, by the action of nitric
acid on *p*-dimethoxybenzene. It is isolated in the form of a
sulphate, which on oxidation is converted into *s*-dimethoxy-
quinone (p. 203). The *diethyl ether* is obtained in a similar
way from *p*-diethoxybenzene, which also yields two dinitro-
compounds by the action of nitric acid.

adj-Diamidoquinol, $C_6H_2(OH)_2(NH_2)_2$, [OH : NH_2 : NH_2 : OH
$= 1.2.3.4$], is unknown. As mentioned above, when nitric acid
acts upon *p*-dimethoxybenzene a mixture of dinitro-compounds
is formed. One of these has the symmetrical constitution and
yields the above methyl ether of *β*-diamidoquinol on reduction,
whilst the second yields an isomeric compound, which combines
with *a*-diketones forming quinoxalines, and must therefore be
the dimethyl ether of *adj*-diamidoquinol. The *diethyl ether* is
obtained in a similar manner; its hydrochloride forms long
needles which become grey in the air.[2]

Triamidoquinol, $C_6H(OH)_2(NH_2)_3$, [OH : NH_2:NH_2:OH:NH_2.
$= 1.2.3.4.5$]. By the action of nitric acid on the foregoing
compound or on *β*-diamidoquinol, a nitrodi-imidoquinol of the
formula $C_6H(OH)_2(NH)_2NO_2$ is obtained, which on reduction
yields triamidoquinol.[3] This is best isolated as the sulphate,
which has the formula $[C_6H(NH_2)_3(OH)_2]_2(H_2SO_4)_3$. By oxidation with ferric chloride or in ammoniacal solution with the
oxygen of the air it yields an imido-compound crystallizing in
red needles.

AMIDOTRIHYDROXYBENZENES.

1095 *Amidopyrogallol,* $C_6H_2(NH_2)(OH)_3$, is obtained by the
action of tin and hydrochloric acid on nitropyrogallol. The
hydrochloride forms brown needles, which become a greyish
black on standing; its solution soon becomes dark and deposits

[1] Nietzki and Schmidt, *Ber.* **22**, 1657.
[2] Nietzki and Rechberg, *ibid.* **23**, 1211.
[3] Nietzki and Schmidt, *ibid.* **22**, 1658.

a flocculent blue precipitate; if, however, it be rapidly evaporated, a dark blue mass separates out, and the concentrated liquid then deposits aggregates of short dark prisms which dissolve in water with a deep bluish violet colour. These become black on standing and then only partially dissolve in water with a brown colour. When a freshly prepared solution of the hydrochloride is shaken up with caustic soda, a deep blue colouration which remains for some time, is produced.[1]

AMIDOTETRAHYDROXYBENZENES.

1096 *Amidotetrahydroxybenzene,* $C_6H(OH)_4NH_2$,[OH : OH : NH_2 : OH : OH = 1 . 2 . 3 . 4 . 5]. When nitrodi-imidoquinol, obtained by the action of nitric acid on β-diamidoquinol, (p. 288) is treated with an alkali, it is converted with evolution of ammonia into nitrodihydroxyquinone, which on reduction with stannous chloride is converted into amidotetrahydroxybenzene. Its *hydrochloride* forms flat silvery needles, and on warming with acetic anhydride and sodium acetate yields a *pentacetyl* derivative $C_6H(NH.C_2H_3O)_4O.C_2H_3O$ which melts at $242°$ with decomposition.

Instead of diamidoquinol, *s*-diamidoresorcinol may be employed for the preparation of this compound. On treatment with nitric acid the latter yields a nitro-derivative isomeric with the nitrodi-imidoquinol mentioned above, which on reduction also yields amidotetrahydroxybenzene.[2]

Amidonitrotetrahydroxybenzene, $C_6(NO_2)(NH_2)(OH)_4$, is prepared by reducing the potassium salt of nitranilic acid (p. 205) with a solution of stannous chloride in hydrochloric acid; it separates out in long, brownish violet needles. Its solution in caustic potash or potassium carbonate rapidly blackens in the air, the corresponding quinone being formed.

Diamidotetrahydroxybenzene, $C_6(NH_2)_2(OH)_4$. In order to obtain this substance the method described for the preparation of the preceding compound is followed, and, after this has separated out, zinc is added until the solution becomes colourless. The hydrochloride of the base, $C_6(NH_2.HCl)_2(OH)_4$, is then precipitated in long needles by a stream of hydrogen chloride. The base, which is separated by caustic soda, rapidly

[1] Barth, *Monatsh.* **1**, 882.
[2] Nietzki and Schmidt, *Ber.* **22**, 1658.

decomposes and becomes brown; nitric acid oxidizes it to triquinoyl (p. 211), and it yields p-diamidobenzene on distillation with zinc dust.[1]

AMIDO-DERIVATIVES OF QUINONE.

1097 a-*Dichloranilidoquinone*, $C_6HCl_2O_2.NHC_6H_5$, [$O:NH:Cl:O:Cl = 1.2.3.4.6$], is prepared by dissolving 1 part of a-dichloroquinone in 40 parts of acetic acid, and after cooling adding 0·4—0·5 parts of concentrated hydrochloric acid, and a similar quantity of aniline. Some dichlorodianilidoquinone separates out and is filtered off, the filtrate precipitated with water, and the precipitate recrystallized from dilute alcohol. It forms blue lustrous plates melting at 186°, and readily soluble in alcohol, ether, and chloroform.[2]

β-*Dichloranilidoquinone*, $C_6HCl_2O_2NH.C_6H_5$, [$O:NH:Cl:O:Cl = 1.2.3.4.5$] is prepared by heating m-dichloroquinone with aniline and hydrochloric acid in alcoholic solution.[3] It crystallizes from dilute alcohol in bluish violet needles or plates, which have a metallic lustre and melt at 154°.

Hydroxyanilidoquinone, $C_6H_2O_2(NH.C_6H_5)OH$, is prepared by heating anilidohydroxyquinonephenylimide (p. 303) with dilute caustic potash until the dark red solution becomes brighter. It is set free by acids as a blue crystalline precipitate, which dissolves readily in hot alcohol and glacial acetic acid, and forms a brown solution in sulphuric acid, from which it is reprecipitated by water. It forms soluble alkaline salts.

When dianilidoquinonephenylimide (p. 303) is heated with alcoholic potash, the compound $C_{18}H_{14}N_2O_3$ is formed; this crystallizes in small red needles melting at 191—192°, is readily soluble in alcohol and benzene, dissolves in sulphuric acid with a green colour, and forms metallic salts. Its constitution is expressed by one of the following formulæ:[4]

Dianilidohydroxyquinone.

$$C_6HO_2 \Big\langle\begin{array}{l} NH.C_6H_5 \\ OH \\ NH.C_6H_5 \end{array}$$

Anilidodihydroxyquinonephenylimide.

$$C_6HO(NC_6H_5) \Big\langle\begin{array}{l} OH \\ NH.C_6H_5. \\ OH \end{array}$$

[1] Nietzki, *Ber.* **19**, 2727.
[3] *Ibid.* p. 335. -
[2] Niemeyer, *Annalen*, **228**, 332.
[4] Zincke, *Ber.* **18**, 785.

Dihydroxyanilidoquinone, $C_6H_2(NH.C_6H_4.OH)_2O_2$, is formed when quinone is added to a hot solution of *p*-amidophenol in hydrochloric acid. It crystallizes in small, violet-brown, lustrous plates, which are insoluble in the ordinary solvents, but dissolve readily in alkalis.

o-Amidophenol does not give an analogous compound, but is oxidized by the quinone :

$$4C_6H_7NO + 5C_6H_4O_2 = C_{24}H_{18}N_4O_4 + 5C_6H_4(OH)_2.$$

In order to prepare the condensation product which is formed in this reaction, 4—4·5 parts of quinone are added to a hot, concentrated, alcoholic solution of *o*-amidophenol. It separates out on cooling in small violet needles, melting at 250°, and subliming at a higher temperature in small, red, lustrous needles, which are slightly soluble in alcohol and benzene, and more readily in aniline. It dissolves in acids with a deep red colour.

The *hydrochloride*, $C_{24}H_{18}N_4O_4(HCl)_2$, crystallizes in arborescent needles with a beetle-green lustre, and readily forms double salts with several metallic chlorides. The *sulphate* is a green crystalline powder, the *oxalate* crystallizes in green, and the *picrate* in steel-blue needles with a green reflection.

When the base is boiled with acetic anhydride, the compound $C_{24}H_{16}(C_2H_3O)_2N_4O_4$ is formed; this crystallizes in yellowish brown needles or small plates, melting at 285°.

o-Amidophenol methyl ether, $NH_2.C_6H_4.OCH_3$, reacts with quinone in an analogous manner to aniline and *p*-amidophenol, *o-dimethoxyanilidoquinone*, $C_6H_2(NH.C_6H_4.OCH_3)_2O_2$, being formed; this substance crystallizes in reddish violet needles melting at 230°, and forms a beautiful blue solution in sulphuric acid. Quinone, on the other hand, does not act upon acetyl-*o*-amidophenol, $C_6H_4(OH)NH(C_2H_3O)$, and from this it follows that both the hydroxyl and the amido-group take part in the formation of the base from *o*-amidophenol.[1]

1098 *Dichlorodiamidoquinone* or *Chloranilamide*, $C_6Cl_2(NH_2)_2O_2$, is formed by the action of alcoholic ammonia on chloranil.[2] It forms a dark red crystalline powder with a semi-metallic lustre, sublimes when heated, and is insoluble in water, alcohol, ether and ammonia. It dissolves in alcoholic potash

[1] Zincke and Hedebrand, *Annalen*, 226, 60.

[2] Laurent, *Berz. Jahresb.* 25, 850 ; Knapp and Schulz, *Annalen*, 210, 180.

with a violet-red colour, but is decomposed when the solution is boiled, chloranilic acid and ammonia being formed. Its violet-red solution in sulphuric acid is coloured blue by a little water ; when more is added the colour changes to wine-red and the compound is precipitated.

Diacetyldiamidoquinone, $(NH.C_2H_3O)_2.C_6H_2O_2$, $[O:NH:NH = 1.2.6]$, is obtained by the oxidation of triacetyltriamidophenol [1] and by passing a current of air through tetracetyldiamidoquinol.[2] It forms golden yellow plates, almost insoluble in the ordinary solvents, and melts with decomposition at 205—207°.

Tetramethyldiamidoquinone, $C_6H_2O_2.[N(CH_3)_2]_2$, is obtained by warming quinone with 10 per cent. dimethylamine,[3] and crystallizes from alcohol in red tablets melting at 173—174°.

s-Dianilidoquinone, $C_6H_2O_2(NH.C_6H_5)_2$, $[O : NH : O : NH = 1.2.4.5]$, is formed together with quinol and another compound, by boiling quinone [4] or chloroquinone with an alcoholic aniline solution,[5] and by heating s-dihydroxyquinone with aniline,[6] whence it follows that dianilidoquinone has also a symmetrical constitution. It crystallizes in reddish-brown scales which are almost insoluble in hot alcohol, and form a magenta-red solution in sulphuric acid. It does not melt but sublimes without decomposition.

Chlorodianilidoquinone, $C_6HCl(NH.C_6H_5)_2O_2$, is prepared from trichloroquinone, and forms small brown plates with a metallic lustre, which form a blue solution in sulphuric acid.[7]

Dichlorodianilidoquinone or *Chloranilanilide*, $C_6Cl_2(NH.C_6H_5)_2O_2$, is formed by the action of aniline and alcohol on chloranil,[8] or on trichloroquinone chlorimide,[9] or on a-dichloro-quinone.[10] It forms yellowish-brown needles with a metallic lustre, melts at 285—290°, and sublimes without decomposition. It is insoluble in water, but dissolves readily in alkalis, and forms a bluish-violet solution in sulphuric acid.

By the action of aniline on an alcoholic solution of trichloro-

---

[1] Bamberger, *Ber.* **16**, 2402.

[2] Nietzki and Preusser, *ibid.* **19**, 2247 ; **20**, 797.   [3] Mylius, *ibid.* **18**, 467.

[4] Hofmann, *Jahresb.* **1863**, 415; Wichelhaus, *Ber.* **5**, 851 ; Knapp and Schulz, *Annalen*, **210**, 178.   [5] Niemeyer, *ibid.* **228**, 331.

[6] Nietzki and Schmidt, *Ber.* **22**, 1655.

[7] Neuhofer and Schultz, *ibid.* **10**, 1793 ; Knapp and Schultz, *Annalen*, **210**, 178.

[8] Hesse, *ibid.* **114**, 306 ; Hofmann, *Jahresb.* **1863**, 415; Knapp and Schultz, *Annalen*, **210**, 187.

[9] Schmitt and Andresen, *J. Pr. Chem.* II. **24**, 431 ; **28**, 427.

[10] Niemeyer, *Annalen*, **228**, 333.

quinone chlorimide, $C_6HCl_3O(NCl)$, a compound is obtained which is probably identical with dichlorodianilidoquinone. It crystallizes from hot benzene in small, yellowish-brown plates with a slight metallic lustre, which form a deep blue solution in sulphuric acid.[1]

The three preceding compounds are converted by reduction into the corresponding quinol derivatives, which rapidly oxidize again in the air.

Amidonitrodihydroxyquinone, $C_6(NO_2)(NH_2)O_2(OH)_2$. When the solution of amidonitrotetrahydroxybenzene (p. 290) in potassium carbonate is sufficiently concentrated, the salt, $C_6(NO_2)$ $(NH)_2O_2(OK)_2$, separates out in black needles which have a cupreous lustre. The black solution of this compound is turned yellow by hydrochloric acid, and on standing deposits small orange-red prisms of the compound $C_6(NO_2)(NH_2)O_2(OH)OK$.

Diamido-dihydroxyquinone, $C_6O_2(NH_2)_2(OH)_2$, [$O:NH_2:OH:$ $O:NH_2:OH = 1.2.3.4.5.6$], is prepared by the careful oxidation of the diamidotetrahydroxybenzene, obtained by the reduction of nitranilic acid. Its *hydrochloride* forms reddish-brown needles, which frequently show a green surface lustre. By the action of acetic anhydride and sodium acetate it yields a mixture of acetyl derivatives. The *diacetyl* compound has been obtained by the action of air and alkali on tetracetyldiamidoquinol.

Di-imidodihydroxyquinone, $C_6(OH)_2(NH)_2O_2$, is obtained as the second product of oxidation of diamidotetrahydroxybenzene hydrochloride, and crystallizes in small, almost black plates, with a green lustre. It is scarcely soluble in the ordinary solvents, but imparts a yellowish-red colour to them. It forms a brown solution in sulphuric acid, but is partially decomposed, and is oxidized by nitric acid to triquinoyl.[2]

Di-imidodiquinoyl, $C_6O_4(NH)_2$. [$O:O:NH:O:O:NH = 1.$ $2.3.4.5.6$], is obtained as a by-product in the preparation of triquinoyl by the oxidation of diamidotetrahydroxybenzene with nitric acid, and remains as a colourless residue when the raw triquinoyl is treated with sulphurous acid. It is purified by repeated treatment with the latter and finally washed with water. It has then the composition, $C_6O_4(NH)_2 + 5H_2O$, and like triquinoyl readily undergoes decomposition. It is not soluble in any of the ordinary solvents.[3]

[1] Schmitt and Andresen, *J. Pr. Chem.* II. **24**, 426.
[2] Nietzki, *Ber.* **16**, 2092 ; Nietzki and Benckiser, *ibid.* **18**, 499.
[3] Nietzki and Schmidt, *ibid.* **21**, 1854.

NITROSO- AND ISONITROSO-COMPOUNDS.

1099 Besides the foregoing amido-derivatives of quinone, a very important series of compounds is known, in which one or both of the quinone oxygen atoms are replaced by substituted imido-groups. Many of these are of great importance in the colour industry.

Quinonoxime or *p-Nitrosophenol*, $C_6H_4O(N.OH)$. This compound was first obtained by Baeyer and Caro, together with dimethylamine, by the action of caustic soda on *p*-nitrosodimethylaniline[1] (p. 310). They also found that it may be prepared by treating an aqueous solution of phenol and potassium nitrite with acetic acid,[2] and considered it to be nitrosophenol, $C_6H_4(NO)OH$. Its exact constitution was proved by Goldschmidt, who found that it is formed by the action of hydroxylamine on quinone. In preparing it by this method the free base must not be used, because it reduces the quinone to quinol. The hydrochloride of hydroxylamine does not act in this way; in concentrated solutions, however, it produces such a violent reaction that the mass commences to char, and hence dilute solutions must be employed. If the diketone formula of quinone be correct, the formation of this monoxime is simply the normal reaction of hydroxylamine on a ketone (see also p. 186). If, however, quinone has the peroxide formula, the formation of the monoxime must be represented by the following equation:

$$C_6H_4\!\!<^{O}_{O}\!| + H_2N.OH = C_6H_4\!\!<^{O}_{N.OH}\!| + H_2O.$$

Quinonoxime is best obtained by the action of nitrosulphonic acid on phenol. The reagent, which Groves and Stenhouse call nitrosyl sulphate, is prepared by warming 200 cc. of nitric acid of specific gravity 1·3 with arsenic trioxide to 70°, passing the nitrogen trioxide evolved through an empty flask, and then absorbing it in 250 cc. of concentrated sulphuric acid. Sulphuric acid is then added until the solution contains 15 per cent. of nitrogen trioxide. Somewhat more than the theoretical quantity of this is added to a solution of phenol in 30 parts of

[1] *Ber.* **7** 809   [2] *Ibid.* p. 963.

water. After twenty minutes the separated crystals must be
filtered off, because if allowed to stand they are rendered impure
by a tar-like mass which is formed.[1] Quinonoxime crystallizes
in light brownish-green, thin, rhombic tablets, and is more
readily soluble in hot water than in cold, forming a green
solution. On rapidly cooling it separates out in small, almost
colourless needles, but on slowly cooling crystallizes in brownish-
green plates. It dissolves in alkalis, forming a brown solution,
from which it separates on the addition of acids as an almost
white, amorphous precipitate. It is also readily soluble in alcohol,
ether and acetone, forming green solutions.

In the moist state the crystals rapidly turn brown, and
hence it is best dried on a porous earthenware plate. Quinonoxime is thus obtained as a yellowish powder (Goldschmidt); on
heating to 120—130° it decomposes with a slight detonation;
by dissolving it in phenol, adding a little concentrated sulphuric
acid and then diluting with water, a dark cherry-red solution is
obtained, which on the addition of caustic potash is changed
into a beautiful blue (p. 23).[2] Hydrochloric acid converts it into
chlorophenol mixed with dichloro- and trichloroamidophenol.
It is oxidized by an alkaline solution of potassium ferricyanide,
as well as by nitric acid, to p-nitrophenol. As long as it was
supposed to be nitrosophenol this reaction was very easily
explained, but we must now assume that the quinonoxime
first combines with water, and that the compound thus formed
is then oxidized:

$$(1) \quad HO.N{=}C{<}^{CH=CH}_{CH=CH}{>}CO + HOH.$$

$$= \quad {}^{HO}_{HO}{>}N{-}C{<}^{CH=CH}_{CH-CH}{>}C.OH.$$

$$(2) \quad C_6H_4{<}^{N{<}^{OH}_{OH}}_{OH} + O = C_6H_4{<}^{N{<}^{O}_{O}}_{OH} + H_2O.$$

The salts of quinonoxime, which were formerly named
nitrosophenates, have been examined by ter Meer.[3]

Potassium quinonoximate, $C_6H_4O(NOK)$, is obtained as a

[1] *Journ. Chem. Soc.* 1877, **1**, 544.　　[2] Goldschmidt, *Ber.* **17**, 805.
[3] *Ibid.* **8**, 622.

beautiful amorphous green precipitate, by mixing an ethereal solution of quinonoxime with alcoholic potash. It is very readily soluble in water, and crystallizes from acetone or alcohol in thin bluish-green tablets; sometimes it is also obtained in red crystals. On heating it detonates.

Sodium quinonoximate, $C_6H_4O(NONa) + 2H_2O$, is a vermilion-coloured amorphous precipitate, which crystallizes from alcohol in beautiful short red needles, and is readily soluble in water, forming a reddish-brown solution.

Silver quinonoximate, $C_6H_4(NOAg) + H_2O$, is obtained by the addition of silver nitrate to a hot, dilute solution of the sodium salt, in small, dark violet crystals, which under the microscope appear red in transmitted light, and in sunlight show a green reflection.

Quinonoxime ethylcarbonate, $C_6H_4O(NO.COOC_2H_5)$, is formed by treating the sodium compound with ethyl chlorocarbonate in presence of ether. It crystallizes in golden-coloured needles melting at 109°, slightly soluble in ether, but readily in alcohol and chloroform.[1]

o-Methoxyquinonoxime, $C_6H_3(OCH_3).O.NOH$, is obtained by the action of alkalis on methoxy-*p*-amidonitrosobenzene (p. 308). It crystallizes in slightly yellow needles or prisms, which explode on heating to 140—150°, are scarcely soluble in water, but readily in alcohol and ether.[2]

1100 *Quinonedioxime*, $C_6H_4(N.OH)_2$. It was for some time believed that only one of the oxygen atoms in quinone could be replaced by the oximido-group, a fact which appeared to show that quinone is not a true diketone. By slightly altering the conditions, however, Nietzki and Kehrmann succeeded in obtaining the dioxime from quinol, quinone, or quinonoxime and hydroxylamine,[3] and it may also be obtained by the action of hydroxylamine hydrochloride on *p*-amidonitrosobenzene[4] (p. 307).

For its preparation Nietzki and Guitermann recommend the following process[5]:—Freshly prepared quinonoxime (from phenol) is dissolved in 50 parts of water, and then half a molecule of hydroxylamine hydrochloride and 1 molecule of hydrochloric acid added; after remaining for 6—8 days the precipitated dioxime is filtered off, washed with dilute ammonia, and purified by solution

[1] Walker, *Ber.* **17**, 400.   [2] Best, *Annalen*, **255**, 184.
[3] *Ber.* **20**, 614.   [4] O. Fischer and Hepp, *ibid.* **21**, 685.
[5] *Ibid.* **21**, 429.

in concentrated ammonia and precipitation with carbon dioxide. It crystallizes in two modifications, which form short, colourless needles, and slender yellow needles respectively. The latter fall to a colourless powder on drying.

Quinonedioxime is a fairly stable body, and is less soluble, and less basic than the monoxime. It is converted on reduction into p-diamidobenzene. When heated it becomes brown and decomposes at $240°$. On warming with acetic anhydride it is readily converted into the *diacetyl*-compound, $C_6H_4(NO.C_2H_3O)_2$, which forms colourless needles, very sparingly soluble in water and ether, somewhat more easily in hot alcohol, and readily in acetic acid.

s-Dihydroxyquinonedioxime, $C_6H_2(OH)_2(NOH)_2$, is prepared by the action of hydroxylamine in acid solution on dihydroxyquinone. It forms almost colourless silvery plates, and dissolves in alkalis with a brown colour, but is precipitated unaltered by acids.[1]

o-Methoxyquinonedioxime, $C_6H_3(OCH_3)(NOH)_2$, is obtained by treating either o-methoxyquinonoxime or methoxy-p-amido-nitrosobenzene (p. 308) with hydroxylamine hydrochloride. It crystallizes in greyish white needles which become brown at $235°$ and melt at $250°$ with decomposition.[2]

Phloroglucinoltrioxime, $C_6H_6(NOH)_3$, is obtained by the action of hydroxylamine on phloroglucinol, and forms a sandy powder which is very sparingly soluble in water and alcohol, blackens when heated to $140°$, and explodes with a red flame at $155°$. It has undoubtedly the following constitution:[3]

$$
\begin{array}{ccc}
 & CH_2 & \\
HO.N=C & & C=N.OH \\
| & & | \\
H_2C & & CH_2 \\
 & C & \\
 & \parallel & \\
 & N.OH. &
\end{array}
$$

The halogen substituted quinones vary in their behaviour towards hydroxylamine. According to Kehrmann, the carbonyl group is only attacked when at least one CH group occupies the ortho-position to it.[4]

[1] Nietzki and Schmidt, *Ber.* 21, 2377. [2] Best, *Annalen*, 255, 187.
[3] Baeyer, *Ber.* 19, 159. [4] Kehrmann, *J. Pr. Chem* II. 40, 257.

1101 *Mononitrosoresorcinol*, $C_6H_3(OH)O(NOH)$. Bindschedler and Busch first obtained this compound by the action of amyl nitrite on monosodiumresorcinol,[1] and it was then further examined by Fèvre.[2] In order to prepare it, concentrated alcoholic solutions of equal molecules of sodium ethylate and resorcinol are mixed, and the calculated quantity of amyl nitrite added, the mixture being kept well agitated. The solid mass which separates out is pulverized and freed from amyl alcohol by washing with water.[3] It crystallizes from dilute alcohol in golden needles, which contain one molecule of water, and on heating become black and decompose without melting.

Its salts are not very characteristic; stannous chloride reduces it to amidoresorcinol, concentrated nitric acid converts it into trinitroresorcinol, and by passing nitrogen trioxide through its ethereal solution dinitroresorcinol is obtained.

It gives colour-reactions with all phenols, as well as with amido-compounds. With aniline acetate it forms the compound $C_{18}H_{14}N_2O_2$, which is insoluble in alkalis, and crystallizes from chloroform in small, lustrous steel-blue needles, which dissolve in concentrated hydrochloric acid with a blue, and in sulphuric acid with a green colour (Fèvre).

Dinitrosoresorcinol, $C_6H_2O_2(NOH)_2$, is obtained by the action of sodium nitrite on a very dilute solution of resorcinol in acetic acid. After some time the liquid is poured into dilute sulphuric acid, and the compound which separates out purified by crystallization from alcohol.[4] It is more easily obtained by gradually mixing a dilute solution of resorcinol with nitrosyl-sulphate [5] (p. 295).

Dinitrosoresorcinol crystallizes in yellowish brown or green plates which detonate at 115°; it decomposes carbonates, and, to a slight extent, acetates. With the alkali metals it readily forms soluble normal and acid salts such as, $C_6H_2O_2(NOH)$ $(NONa)$, which are obtained as slightly soluble, green crystalline powders. It is largely used in dyeing under the name of *fast myrtle*.

Nitrous acid, therefore, acts on resorcinol in the same manner as on phenol, the so-called mononitrosoresorcinol corresponding

[1] Krause, *Ber.* **17**, 1850.   [2] *Ibid.* **16**, 1101.
[3] Walker, *ibid.* **17**, 399.   [4] Fitz, *ibid.* **8**, 631.
[5] Stenhouse and Groves, *Journ. Chem. Soc.* **36**, 550.

to quinonoxime, although it is not known whether the isonitroso-group occupies the ortho- or para-position to the carbonyl group.

$$
\begin{array}{ccc}
& O & \\
H & \diagup\!\diagdown N.OH & \\
& or & \\
H & \diagdown\!\diagup OH & \\
& H &
\end{array}
\qquad
\begin{array}{c}
O \\
H \diagup\!\diagdown H \\
H \diagdown\!\diagup OH \\
NOH
\end{array}
$$

In the case of the dinitroso-compound, however, it has been found[1] that the two oxygen atoms and the two isonitroso-groups occupy the positions 1.2.3.4.

$$
\begin{array}{c}
O \\
\diagup\!\diagdown NOH \\
\diagdown\!\diagup O \\
NOH
\end{array}
$$

and the first formula for mononitrosoresorcinol is therefore the more probable.

Diquinoyltetroxime, $C_6H_2(NOH)_4$, is obtained by the action of hydroxylamine hydrochloride on dinitrosoresorcinol.[2] If the product of the reaction is boiled with acetic anhydride, it forms *diquinoyltetroxime anhydride*, $C_6H_2(N_2O)_2$, a compound crystallizing in white needles, melting at 61°.

Trinitrosophloroglucinol, $C_6O_3(NOH)_3$. When 10 grms. of phloroglucinol are dissolved in 300 cc. of water containing 12 grms. of acetic anhydride, and a well-cooled concentrated solution of 16 grms. of potassium nitrite added to the liquid, cooled to 8—9°, an acid potassium salt of this compound separates out in green crusts. If it be allowed to stand for half an hour and treated with excess of caustic potash and then alcohol, the neutral salt, $C_6O_3(NOK)_3$, is precipitated in fine green needles, which are very explosive.

By adding lead acetate to its dilute aqueous solution, a yellow precipitate of the lead salt is obtained, forming, when dry, a cinnnamon-brown powder, which explodes on heating with the greatest violence. By decomposing it with dilute sulphuric acid in presence of alcohol and allowing the filtrate to evaporate, trinitrosophloroglucinol is obtained in nodular

[1] Kostanecki, *Ber.* **20**, 3136.
[2] Goldschmidt and Strauss, *ibid.* **20**, 1610.

aggregates of needles.[1] It has probably the following constitution :

$$\underset{\displaystyle \text{CO}}{\overset{\displaystyle \text{C(NOH)CO}}{\Big<}}\text{CNOH.}$$

$$\underset{\text{C(NOH)CO}}{}$$

1102 *Quinone chlorimide*, $C_6H_4O(NCl)$, was obtained by Schmitt
and Bennewitz by the action of bleaching powder on a solution
of *p*-amidophenol, $C_6H_4(NH_2)OH$ in hydrochloric acid, and was
considered to be dichlorazophenol, $C_{12}H_6Cl_2N_2(OH)_2$.[2] Hirsch[3]
then showed that it was quinone chlorimide, and gave to it the
following constitutional formula :

$$C_6H_4\Big<\overset{\displaystyle O}{\underset{\displaystyle NCl.}{|}}$$

It is thus quinone in which an oxygen atom is replaced by
NCl, and therefore if the more recent views concerning quinone
are adopted, its constitution is represented by the following
formula :

$$\underset{\displaystyle \text{CH}=\text{CH}}{\overset{\displaystyle \text{CH}=\text{CH}}{\Big<}}\text{C}=\text{NCl.}$$

From this it appears that it stands in close relationship to
quinonoxime, from which it is derived by replacement of the
hydroxyl by chlorine ; it has not, however, been hitherto obtained in this way. Schmitt found that it is also formed by
the action of bleaching powder on ethyl-*p*-amidophenate,
$C_6H_4(NH_2)OC_2H_5$.[4]

It is obtained by heating 100 grms. *p*-nitrophenol with 100
grms. of tin and 500—600 grms. of hydrochloric acid, filtering and
diluting to 1 litre. Portions of this solution are then diluted with
four volumes of water cooled to 4–5°, and concentrated chloride
of lime solution added with continuous stirring till the precipitate and liquid assume a pure yellow colour.[5] After washing,
the product is pure enough for employment in further reactions,

[1] Benedikt, *Ber.* **11**, 1374. [2] *J. Pr. Chem.* II. **8**, 1.
[3] *Ber.* **13**, 1903. [4] *J. Pr. Chem.* II. **19**, 315
[5] Fogh. *Ber.* **21**, 890.

or it may be extracted with ether, the latter distilled off and the residue crystallized from glacial acetic acid (Hirsch).

Quinone chlorimide forms golden crystals, which are probably asymmetric, melt at 85°, and detonate at a higher temperature, but are volatile in steam with partial decomposition. It is scarcely soluble in cold water, but readily in hot water, alcohol, and acetic acid. It has a persistent odour, similar to that of quinone, and like this stains the skin a permanent brown. Reducing agents easily reconvert it into p-amidophenol, while sulphur dioxide converts it into the sulphonic acids of this compound. It is converted by strong hydrochloric acid into chlorophenol, and on heating with water to 100° forms ammonium chloride, quinone, and oxidation products of the latter:

$$C_6H_4ONCl + 2H_2O = NH_4Cl + C_6H_4O_2 + O.$$

It dissolves in cold, concentrated sulphuric acid, and in fuming nitric acid without decomposition. On dissolving it in phenol and adding some sulphuric acid, a dark cherry-red coloured solution is obtained, like that produced by quinonoxime, which on addition of water and caustic potash is converted into a splendid blue.[1]

Trichloroquinone chlorimide, $C_6HCl_3O(NCl)$, is obtained in a similar manner to the preceding compound, from trichlorophenol. It is scarcely soluble in cold water, readily in hot, and crystallizes from alcohol in long, yellowish, lustrous prisms, melting at 118°.[2]

Concentrated hydrobromic acid converts it, with separation of bromine, into trichloroquinone.[3]

$$\begin{array}{c} \text{CH=CCl} \\ \diagup \qquad \diagdown \\ \text{CO} \qquad\qquad \text{C=NCl} + H_2O + 2HBr \\ \diagdown \qquad \diagup \\ \text{CCl=CCl} \end{array}$$

$$= \begin{array}{c} \text{CH=CCl} \\ \diagup \qquad \diagdown \\ \text{CO} \qquad\qquad \text{CO} + NH_4Cl + Br_2. \\ \diagdown \qquad \diagup \\ \text{CCl=CCl} \end{array}$$

Quinone dichlorimide, $C_6H_4(NCl)_2$, is obtained by the action of bleaching-powder on a solution of p-diamidobenzene, $C_6H_4(NH_2)_2$, in hydrochloric acid. It is scarcely soluble in cold water,

[1] Schmitt and Andresen, *J. Pr. Chem.* II. **23**, 435.
  Ibid. **24**, 426.   [3] *Ibid.* **28**, 436.

and slightly in boiling water, from which it crystallizes in needles; it is readily dissolved by warm alcohol and acetic acid, forming solutions which stain the skin brown. On reduction it is reconverted into p-diamidobenzene.[1]

1103 *Quinonephenylimide,* $C_6H_4O.NC_6H_5$. This compound is obtained by the action of mercuric oxide on p-phenylamido-phenol in benzene solution. It crystallizes from light petroleum in fiery red crystals, which melt at 97°, and are readily soluble in alcohol, ether, and chloroform, much less in light petroleum. On boiling with acids it forms quinone.[2]

Tetrahydroxyquinonephenylimide, $C_6(OH)_4O(NC_6H_5)$, is formed by the action of aniline on an alcoholic solution of hexhydroxy-benzene in presence of air. It forms lustrous golden-yellow plates, which appear carmine-red in transmitted light.[3]

Anilidoquinonephenylimide, $C_6H_3O(NC_6H_5).NHC_6H_5$, is obtained by reducing azophenine (p. 304) with zinc and hydro-chloric acid in presence of acetic acid, and allowing the diphenylamidophenol thus formed to oxidize in the air.[4]

Dianilidoquinonephenylimide, $C_6H_2O(NC_6H_5).(NH.C_6H_5)_2$, is obtained by heating a solution of 1 part of quinone and 2 parts of aniline in 20 parts of glacial acetic acid for a short time at 100°;[5] or by warming quinonephenylimide with 15 to 20 parts of aniline to the same temperature.[6] It crystallizes in brownish-red needles, which melt at 202—203° according to Zincke and Hagen, and 196—197° according to Bandrowski, and form a blood-red solution in sulphuric acid.

Anilidohydroxyquinonephenylimide, $C_6H_2O.(NC_6H_5)(NHC_6H_5)$ OH. The methyl ether of this substance is formed when the preceding compound is warmed with methyl alcohol and sul-phuric acid:

$$C_6H_2O(NC_6H_5){\Big\langle}\begin{array}{l} NH.C_6H_5 \\ NH.C_6H_5 \end{array} \quad + \ HO.CH_3$$

$$= C_6H_2O(NC_6H_5){\Big\langle}\begin{array}{l} NH.C_6H_5 \\ OCH_3 \end{array} \quad + \ C_6H_5.NH_2.$$

It crystallizes in large brownish-red plates, which melt at 188—189°, and form soluble blue salts. The *platinochloride* is

[1] Krause, *Ber.* **12**, 47 ; Hirsch, *loc. cit.*   [2] Bandrowski, *Monatsh.* **9**, 134.
[3] Nietzki and Schmidt, *Ber.* **21**, 1854.
[4] O. Fischer and Hepp, *Annalen,* **256**, 260.
[5] Zincke and Hagen, *Ber.* **18**, 787.   [6] Bandrowski, *Monatsh.* **9**, 415.

deposited from solution in small dark lustrous plates; the *picrate* forms small brownish-violet crystals.

The *ethyl* ether, obtained in a similar manner, crystallizes in small red plates or prisms, melting at 134°.

When either of the above ethers is decomposed with very dilute alcoholic potash the free *anilidohydroxyquinonephenylimide* is obtained. It crystallizes in scales which have a metallic lustre, and form a green powder slightly soluble in alcohol, readily in acetic acid. It forms very soluble salts, those of potassium and sodium crystallizing in small brown needles which have a silky lustre.[1]

Anilidoquinonediphenylimide, $C_6H_3(NC_6H_5)_2NHC_6H_5,[N:NH: N = 1 . 3 . 4]$, is obtained by the action of hydrochloric acid on azophenine at 160°. It crystallizes in reddish-brown needles, which melt at 229—230°, and dissolve in sulphuric acid with a blue colour, which becomes reddish-violet on dilution.[2]

Dianilidoquinonediphenylimide, or *Azophenine*, $C_6H_2(NC_6H_5)_2 (NHC_6H_5)_2,[N:NH:N:NH = 1 . 2 . 4 . 5]$. This compound was first obtained by Kimich,[3] together with hydroxyazobenzene and a third compound by warming quinonoxime with aniline acetate to 100°. It was further investigated by O. N. Witt,[4] who obtained it by the action of aniline hydrochloride on quinonoxime at 100°, found that it contains no oxygen and gave to it the empirical formula $C_{36}H_{29}N_5$. It was afterwards obtained by the action of aniline on a large number of substances which are related to quinone, *e.g.* the various *p*-nitroso-compounds. O. Fischer and Hepp then pointed out that the analyses of azophenine and its derivatives agree equally well with the formula $C_{30}H_{24}N_4$ as with that given by Witt, and proposed for it the constitutional formula $C_6H_2(NC_6H_5)_2.(NH.C_6H_5)_2$, which they were able to confirm by showing that it is formed in quantity by the action of aniline on dianilidoquinone, and that *vice versâ* it is converted on warming with sulphuric acid and alcohol into dianilidoquinone. As the latter has the symmetrical constitution (p. 293), and moreover azophenine is formed by the action of aniline on *s*-diamidodi-imidobenzene, it must also be regarded as symmetrical.[5]

According to O. Fischer and Hepp, azophenine is best prepared

[1] Zincke, *Ber.* **18**, 788.   [2] O. Fischer and Hepp, *Annalen*, **256**, 261.
[3] *Ber.* **8**, 1028.
[4] *Journ. Chem. Soc.* 1883, **1**, 115 ; *Ber.* **20**, 1539.
[5] O. Fischer and Hepp, *Ber.* **20**, 2479 ; **21**, 676, 2617 ; *Annalen*, **256**, 257.

by heating 100 grams p-phenylamidonitrosobenzene (p. 309) with 100 grams aniline hydrochloride, and 500 grams aniline for eight to ten hours on the water-bath. The product is washed successively with water, dilute alcohol and absolute alcohol, and finally recrystallized from boiling toluene. Azophenine separates in garnet-red plates, which melt at 240°, and are insoluble in alcohol, ether and alkalis, but dissolve in chloroform and also in sulphuric acid, forming a violet solution, which on heating to 300° becomes azure-blue, and shows a carmine-red fluorescence on dilution with water.

Azophenine is important as being the mother-substance of an important class of colouring matters termed Indulines, which are described in a later volume.

Hydrazophenine, $C_{30}H_{26}N_4$, is obtained by heating azophenine with ammonium sulphide and toluene at 140° for two hours. It crystallizes in needles, which melt at 173—174° and are almost insoluble in hydrochloric acid.[1]

Tetrachlorazophenine, $C_{30}H_{20}Cl_4N_4$, is prepared by heating p-chloraniline with quinonoxime, and crystallizes from xylene in red prisms melting at 265°.[2]

Tetrabromazophenine, $C_{30}H_{20}Br_4N_4$, obtained in a similar manner, melts at 243°.

AMIDONITROSOBENZENES OR NITROSOANILINES.

1104 It has already been mentioned (p. 24) that nitrous acid acts on the tertiary anilines, with formation of a nitroso-compound, in which the nitroso-group enters the ring and occupies the para-position to the amido-group. The corresponding nitroso-compounds of the primary and secondary anilines cannot be obtained in a similar manner, as the first are converted by nitrous acid into diazo-compounds, and the second into nitrosamines, such as methylphenylnitrosamine, $C_6H_5.N(NO)CH_3$, in which the nitroso-group is combined with the nitrogen atom.

Otto Fischer and Hepp[3] have found, however, that if the secondary nitrosamines are treated with alcoholic hydrochloric acid, the nitroso-group and the hydrogen atom in the para-position to the amido-group change places. Thus methyl-

[1] Fischer and Hepp, *Ber.* **20**, 2483. [2] *Ibid.* **21**, 678.
[3] *Ber.* **19**, 2991; **20**, 1247, 2471; Mehne, *ibid.* **21**, 729.

phenylnitrosamine is converted into p-methylamidonitroso-benzene : [1]

$$C_6H_5.N\underset{\displaystyle CH_3}{\overset{\displaystyle NO}{<}} = C_6H_4(NO)N\underset{\displaystyle CH_3}{\overset{\displaystyle H}{<}}$$

They found further that the quinonoximes are also converted into amidonitroso-compounds by heating with a mixture of ammonium acetate and chloride : [2]

$$C_6H_4(NOH).O + NH_3 = C_6H_4(NO)NH_2 + H_2O.$$

This reaction can in many cases be performed by ammonia alone, and the substances thus obtained may be likewise classed as imidoximes : [3]

$$C_6H_4\underset{\displaystyle O}{\overset{\displaystyle N.OH}{<}} + NH_3 = C_6H_4\underset{\displaystyle NH}{\overset{\displaystyle NOH}{<}} + H_2O.$$

On oxidation they are converted into amidonitro-compounds, and on reduction yield p-diamido-compounds, whilst on heating with alkalis or acids they are reconverted into quinonoximes. Hydroxylamine converts them into quinonedioximes.

They combine with alkalis and acids at the ordinary temperature without elimination of water. Their constitution, like that of the quinones, may be represented by two different series of formulæ:

---

[1] Fischer and Hepp term this compound p-nitrosomethylaniline, the use of which, and of similar names for the corresponding compounds, may lead to confusion, as methylphenylnitrosamine is also frequently called nitrosomethylaniline. In the names employed above there is no ambiguity, and they will therefore be employed throughout.

[2] *Ber.* **20**, 2471 ; **21**, 684 ; Mehne, *ibid.* **21**, 729.

[3] Ilinski, *ibid.* **19**, 340.

$$C\begin{array}{c}CH-CH\\CH=CH\end{array}C \qquad C\begin{array}{c}CH=CH\\CH=CH\end{array}C$$

N(ONa)———NH$_2$.OH   N(ONa)   NH$_2$.OH.

Fischer and Hepp regard the first series as the more probable. The second series, however, shows more clearly their relation to the quinonoximes, from which they would then be derived by the substitution of an imido-group for oxygen. The imido-quinonoximes, which are simultaneously bases and acids, can exist in the free state as salts, just in the same manner as the free amido-acids (Vol. III. Part II. p. 20).

The following formula must also be considered:

$$O=N-C\begin{array}{c}CH-CH\\CH=CH\end{array}C.NH_2.$$

It is not improbable that we have here another case of tautomerism, and that the azoxy-form, represented in the first series of formulæ, readily passes into the nitroso-form, as many reactions of nitrosodimethylaniline and of nitrosophenol or quinonoxime are more readily understood if the existence of nitrosyl in the molecule is assumed.

1105 *p-Amidonitrosobenzene* or *p-Nitroso-aniline*, $C_6H_4(NO)$. NH$_2$. To obtain this compound 1 part of quinonoxime is mixed with 5 parts of ammonium chloride and 10 parts of dry ammonium acetate, and heated for half an hour on the water-bath, preferably with addition of a little ammonium carbonate. On adding the product to water, *p*-amidonitrosobenzene separates in dark-green crystals, which on recrystallization from benzene form steel-blue needles melting at 173—174°. It forms salts with mineral acids, which dissolve in water with a yellow colour. It likewise forms a sodium salt, obtained by the careful addition of the theoretical quantity of caustic soda in alcoholic solution. It forms a yellow crystalline mass, having the composition $C_6H_6N_2O + NaOH + H_2O$.

On treatment with an excess of aqueous caustic soda, *p*-amido-nitrosobenzene is reconverted into quinonoxime with evolution of ammonia. On reduction it is converted into *p*-diamido-benzene, and on treatment with hydroxylamine yields quinone-dioxime. When heated with aniline and aniline hydrochloride at 80—100° it forms azophenine.[1]

---

[1] O. Fischer and Hepp, *Ber.* **20**, 2474 ; **21**, 684.

Methoxy-p-amidonitrosobenzene, $C_6H_3(NH_2)(OCH_3)(NO)$, (1 . 2 . 4), is prepared in a similar manner from methoxyquinon-oxime.[1] It is a green compound, readily soluble in alcohol, ether, benzene and acids, very sparingly in water. In its properties it closely resembles *p*-amidonitrosobenzene.

1106 *p-Methylamidonitrosobenzene* or *p-Nitrosomethylaniline*, $C_6H_4(NO).NH.CH_3$. To prepare this compound, methylphenyl-nitrosamine, $C_6H_5.N(NO)CH_3$, is dissolved in 2 parts of ether, and 4 parts of absolute alcohol saturated in the cold with hydrogen chloride then added. The mixture immediately assumes a dark-orange colour, and after a time a vigorous reaction takes place with evolution of heat and separation of yellow needles. After some hours the crystals are filtered off and washed with a mixture of ether and alcohol. These crystals, which are the hydrochloride of *p*-methylamidonitrosobenzene, are only stable when quite pure, and are best at once converted into the free base by treating a moderately concentrated aqueous solution of the salt with sodium carbonate or ammonia. The base separates in small yellowish-green plates, and sometimes in steel-blue prisms, which are readily soluble in alcohol, ether, and chloroform, sparingly in benzene and light petroleum. It melts at 118°, and forms a sodium salt, which may be obtained by adding to an alcoholic solution of the base the theoretical quantity of caustic soda or sodium ethylate also dissolved in alcohol. It separates on addition of ether in beautiful yellow needles, which have the composition $C_7H_8N_2O + NaOH$.

By the action of an excess of caustic soda, *p*-methylamidonitroso-benzene is readily converted into methylamine and quinonoxime, and is converted on reduction into methyl *p*-diamidobenzene. On treatment with nitrous acid it yields *p-nitrosobenzene-methylnitrosamine*, $C_6H_4(NO).N(NO).CH_3$, a compound which separates from alcohol in nodular aggregates of crystals melt-ing at 101°, and is more stable than methylphenylnitrosamine. On careful oxidation it is converted into *p*-nitromethylphenyl-nitrosamine, which may be also obtained by the direct nitration of methylphenylnitrosamine.[2]

Methoxy-p-methylamidonitrosobenzene, $C_6H_3(NO)(OCH_3).NH.$ $CH_3 (1 . 3 . 4)$, is prepared from methylanisidine in a similar manner to *p*-methylamidonitrosobenzene.[3] It crystallizes from benzene in grass-green plates, which melt at 110°. Its *hydro-*

---

[1] Best, *Annalen*, **255**, 186.   [2] Fischer and Hepp, *Ber.* **19**, 2991.
[3] Best, *Annalen*, **255**, 178.

chloride crystallizes from water or alcohol in greenish-yellow plates. On reduction it yields the corresponding derivative of *p*-diamidobenzene.

p-Ethylamidonitrosobenzene, $C_6H_4(NO).NH.C_2H_5$. To prepare this compound, ethylphenylnitrosamine is treated with alcoholic hydrochloric acid in the manner already described. It crystallizes in green plates which melt at 78°, dissolves readily in alcohol and ether, but with difficulty in water, and crystallizes best from benzene. Its *hydrochloride* forms stellate groups of needles. Towards alkalis and reducing agents it behaves in a similar manner to the methyl compound.[1]

The corresponding propyl and isobutyl compounds have also been prepared, and exhibit properties corresponding to those already described.[2]

p-Phenylamidonitrosobenzene, or *p-Nitrosodiphenylamine*, C_6H_4 $(NO).NH.C_6H_5$, is prepared by the action of alcoholic hydrochloric acid on diphenylnitrosamine, the hydrochloride first formed being decomposed by water. The free base crystallizes from benzene in green tablets, which have a bluish reflex, and separates from alcoholic solution on the addition of water in steel-blue prisms or flat tablets. It melts at 143°, and is readily soluble in alcohol, chloroform, benzene and acetic acid, forming a brown solution, and in ether forming a green solution. With concentrated sulphuric acid it gives a red colouration which becomes violet on warming. It forms salts with alkalis, which are decomposed by carbon dioxide, a reaction which forms a ready method for its purification. Its *hydrochloride*, $C_{12}H_{10}N_2O$. HCl, forms brown tablets with a bronze surface lustre.

This base is not so readily decomposed by alkalis as those previously described, but on boiling with concentrated caustic soda for several hours it yields aniline and quinonoxime. Caustic potash, on the other hand, acts as a reducing agent, forming *p*-amidodiphenylamine, which is, however, much more readily obtained by treating the base with tin and hydrochloric acid.[3]

By the action of nitrous acid, *p*-phenylamidonitrosobenzene is converted into the *p-nitrosobenzenephenylnitrosamine* $C_6H_4(NO).$ $N(NO)C_6H_5$, which crystallizes in yellowish-green plates, melting at 98° with decomposition. It readily gives Liebermann's re-

[1] Fischer and Hepp, *Ber.* **19**, 2993.
[2] Wacker, *Annalen*, **243**, 290.
[3] Fischer and Hepp, *Ber.* **19**, 2994 ; Ikuta, *Annalen*, **243**, 272.

action with phenol and sulphuric acid, but does not yield a di-p-nitroso-compound on treatment with alcoholic hydrochloric acid.

p-Phenylacetylamidonitrosobenzene, $C_6H_4(NO)N(C_2H_3O).C_6H_5$, is obtained by the action of acetic anhydride on the nitroso-base.[1] It forms red prismatic crystals, which melt at 96—97°.

Diphenylamidonitrosobenzene, $C_6H_3(NO)(NHC_6H_5)_2$.[NH : NH : NO $= 1.3.4$], is obtained from m-diphenylamidobenzene by the addition of amyl nitrite and alcoholic hydrochloric acid. It forms brownish-red prisms with a bluish surface lustre, and yields amido-m-diphenylamidobenzene on reduction. By the action of aniline it readily yields azophenine (p. 304), a further proof of the symmetrical structure of this compound.[2]

1107 *p-Dimethylamidonitrosobenzene,* or *p-Nitrosodimethylaniline,* $C_6H_4(NO)N(CH_3)_2$, is obtained by the action of amyl or sodium nitrite on dimethylaniline hydrochloride.[3] It is prepared on the large scale as follows :—10 parts of dimethylaniline are dissolved in 50 parts of concentrated hydrochloric acid and 200 parts of water, and the cold solution gradually mixed with a solution of 5·7 parts of sodium nitrite in 200 parts of water. The hydrochloride then separates out in small sulphur-yellow needles, by the decomposition of which with alkalis the free base is obtained. It crystallizes from ether in large green plates, which melt at 85°, and are somewhat volatile with steam. On reduction it yields dimethyl-p-diamidobenzene, and on oxidation forms p-nitrodimethylaniline. On boiling with caustic potash it yields quinonoxime and dimethylamine, and its constitution is therefore represented by one of the following formulæ, according as the diketone or peroxide formula of quinone is adopted:

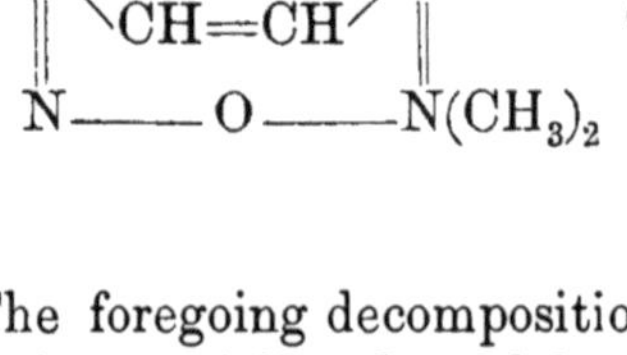

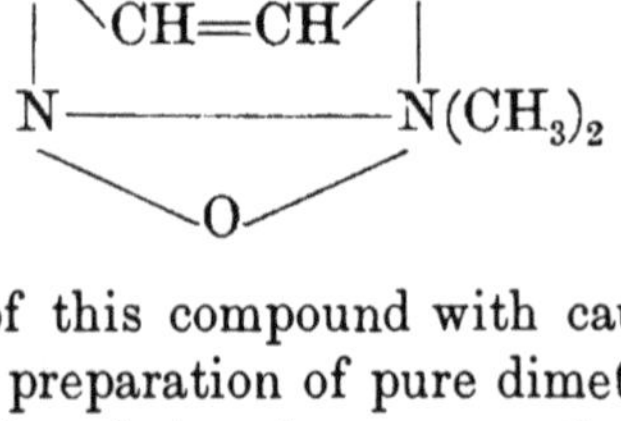

The foregoing decomposition of this compound with caustic soda is especially adapted for the preparation of pure dimethylamine.[4] The base is also employed in the preparation of colouring matters, such as methylene-blue.

[1] Ikuta, *Annalen,* **243**, 276.   [2] Fischer and Hepp, *ibid.* **255**, 144.
[3] Baeyer and Caro, *Ber.* **7**, 963 ; Schraube, *ibid.* **8**, 616 ; Wurster, *ibid.* **12**, 523 ; Meldola, *Journ. Chem. Soc.* 1881, **1**, 37.
[4] Baeyer and Caro, *loc. cit.*

p-Diethylamidonitrosobenzene or *p-Nitrosodiethylaniline*,C_6H_4 $(NO)N(C_2H_5)_2$, is prepared in a similar manner to the methyl compound,[1] and crystallizes from ether in large green prisms which melt at 84°.

m-Tetramethyldiamidonitrosobenzene, $C_6H_3(NO)N_2(CH_3)_4$, is obtained by adding sodium nitrite to a solution of the tetramethyl-base in hydrochloric acid. The *hydrochloride*, C_6H_3 $(NO)N_2(CH_3)_4ClH$, thus obtained crystallizes in lustrous, deep garnet-red, or almost black needles, which dissolve in water with a fine wine-red colour. On adding caustic soda the solution becomes orange-yellow, with formation of the free nitroso-base, which may be extracted with ether, and is thus obtained as a deep brown oil, which gradually solidifies. It forms colouring matters of various shades with phenols and amido-bases.[2]

Trimethyl-p-diamidonitrosobenzene, $C_6H_3(NO).NH(CH_3)N$ $(CH_3)_2$, is obtained by the action of sodium nitrite on an acetic acid solution of tetramethyl-*p*-diamidobenzene, and crystallizes from water in small greenish-yellow plates, which melt at 98—99°, and give Liebermann's reaction. On reduction with tin and hydrochloric acid it yields trimethyl-*p*-diamidobenzene.[3]

The nitroso-compounds react with phenylhydrazine in a somewhat peculiar manner, forming substances which appear to be diazo-compounds. Thus *p*-dimethylamidonitrosobenzene yields a compound to which Fischer and Wacker [4] assign the following constitutional formula :

$$C_6H_4 \left\langle \begin{array}{l} N{=}(CH_3)_2 \\ |\ {>}O \\ N{=}N.NH.C_6H_5 \end{array} \right.$$

DIAZO-DERIVATIVES OF BENZENE.

1108 The formation and constitution of this important class of compounds, and the general decompositions which they undergo, have already been discussed in the Introduction (p 26). It may be of interest to add here the account of their discovery as related by Dr. Griess.[5]

---

[1] Kapp, *Ber.* **8**, 621.   [2] Witt, *ibid.* **18**, 877.
[3] Wurster and Schobig, *ibid.* **12**, 1809.
[4] Fischer and Wacker, *ibid.* **22**, 622.
[5] Private communication to Watson Smith.

" Dr. Gerland, when working in the laboratory of Prof. Kolbe, in Marburg, investigated the action of nitrous acid on amido-benzoic acid at the request of Kolbe. Thus oxybenzoic acid was prepared, indicating a chemical change then considered of much importance. In like manner I investigated a means of converting picramic acid (amidodinitrophenylic acid) into the oxydinitrophenylic acid, $C_6H_2(NO_2)_2(OH)_2$, but I obtained instead of the latter, a compound possessed of such striking and peculiar properties, that I at once concluded it must belong to a completely new class of compounds. Analysis soon showed me that this peculiar compound had the composition, C_6H_2 $(NO_2)_2N_2O$. Naturally I soon submitted many other amido-compounds in like manner to the action of nitrous acid, and obtained thus, in almost every case, the corresponding diazo-compound. But the circumstance to which I was indebted for my success in obtaining the diazo-compounds was that of the treatment of the amido-compounds with nitrous acid in the cold, whereas, in the earlier experiments of Hunt and Gerland a higher temperature was always attained, and con-sequently no diazo-compounds could exist. Having obtained these diazo-compounds, I then tried their action on all possible substances, among which of course are the numerous class of amido-compounds. I found that the diazo-compounds combine directly with these, forming frequently brilliantly coloured substances which dye animal fibres directly. The first colouring matter thus prepared by me, which I obtained in the years 1861-2, was the benzeneazo-a-naphthylamine.[1] It was first prepared on the large scale, to the best of my recollection, in the years 1865-6 by Caro, who was then chemist in the works of Messrs. Roberts, Dale and Co., of Manchester. I first re-commended the oxyazobenzene obtained by me for use as a colouring matter in 1866." [2]

For use in the preparation of substitution-products, azo-colours, &c., it is as a rule unnecessary to prepare the pure diazo-salt, a solution of it being all that is required. A solution of the chloride is obtained by adding a solution of one molecule of sodium nitrite to a well-cooled dilute solution of one molecule of aniline and two molecules of hydrochloric acid, stirring well during the addition ; the end of the reaction is recognized by the evolution of free nitrous acid, which may be detected by its smell, or by holding some paper soaked in potassium iodide and starch over the

[1] *Phil. Trans.* **1864**, III. 679. [2] *Annalen*, **137**.

liquid. In this way either sodium nitrite or aniline in solution can be roughly estimated.[1]

A solution of diazobenzene nitrate or sulphate is often used instead of the chloride, and is obtained in a similar manner. When a solution of the sulphate is boiled, pure phenol is formed, and by means of this reaction many amido-bases can be completely converted into phenols.[2]

Diazobenzene chloride, $C_6H_5.N:NCl$. To prepare this compound in the crystalline condition, a cold saturated alcoholic solution of aniline hydrochloride, to which a few drops of concentrated hydrochloric acid have been added, is cooled to 5° and mixed with the theoretical quantity of amyl nitrite, the liquid being well cooled during the addition. Colourless needles of diazobenzene chloride soon separate and may be filtered off, washed with a little alcohol and with ether, and dried over sulphuric acid *in vacuo*. It decomposes with a slight explosion on heating, and can only be exploded with difficulty by percussion. It is insoluble in ether, light petroleum, and benzene, but dissolves in alcohol and with extreme avidity in water, being deliquescent in the air, undergoing decomposition at the same time.[3] When hydrochloric acid and stannic chloride are added to a concentrated solution, the double salt, $(C_6H_5N_2Cl)_2SnCl_4$, separates out in small, white indistinct plates, which are only very slightly soluble in alcohol and ether, but readily in lukewarm water, from which they are reprecipitated by concentrated hydrochloric acid. The dry compound decomposes on heating with vigorous decrepitation; on boiling with water it yields phenol. When allowed to stand in the air for a considerable time, the compound undergoes a decomposition in which the greater portion of it is apparently converted into p-diphenol, $OH.C_6H_4.C_6H_4.OH$.[4]

Diazobenzene bromide, $C_6H_5N_2Br$, is prepared by the addition of bromine to an ethereal solution of diazo-amidobenzene:

$$C_6H_5N{=}N{-}NH.C_6H_5 + 3Br_2 = C_6H_5N{=}NBr$$
$$+ C_6H_2Br_3.NH_2 + 2HBr.$$

The tribromaniline remains in solution, while the bromide separates out in small plates with a mother-of-pearl lustre, which are very explosive in the dry state. By shaking its

[1] Nietzki, *Ber.* **17**, 1351. [2] V. Meyer, *ibid.* **8**, 1074.
[3] Knoevenagel, *ibid.* **23**, 2996. [4] Griess, *ibid.* **18**, 965.

aqueous solution with silver chloride, a solution of pure diazo-benzene chloride is obtained.

Diazobenzene perbromide, $C_6H_5NBr—NBr_2$, is formed when a solution of the nitrate is treated with hydrobromic acid and bromine. It separates out as a brown oil, which, after being washed with ether, solidifies in large yellow plates. It is tolerably stable in the dry state, but on standing in contact with water or ether, in which it is insoluble, or with alcohol, in which it dissolves to a slight extent, it rapidly decomposes. On heating it detonates violently; when it is heated with anhydrous carbonate of soda, or boiled with absolute alcohol, bromobenzene is formed.

1109 *Diazobenzene nitrate*, $C_6H_5.N_2NO_3$, was first obtained by Griess, and used as a starting-point for other diazobenzene-compounds.[1] In order to prepare it, the nitrous gases evolved from a mixture of arsenic trioxide and nitric acid are led slowly into a well-cooled paste of aniline nitrate and water, until the addition of caustic potash to a small portion ceases to set free aniline. It is then filtered from a brown resinous body which is generally formed, and the diazobenzene nitrate precipitated in long, white needles by the addition of alcohol and ether to the filtrate. It may be more readily obtained by acting on an alcoholic solution of aniline nitrate containing a little nitric acid with amyl nitrite, precipitating if necessary with ether.[2] It is readily soluble in water and slightly in alcohol but is insoluble in ether; in the dry state it explodes when gently heated, or on percussion, even more violently than fulminating mercury or iodide of nitrogen.

If its cold, saturated solution be allowed to drop into concentrated ammonia, diazo-amidobenzene, $C_6H_5.N_2.NH.C_6H_5$, an amorphous brownish red substance, $C_{18}H_{24}N_2O$, and a compound, which has the formula $C_{12}H_{13}N_5O$, are formed. This last compound may be obtained in yellow prisms by the spontaneous evaporation of its deep yellow solution, and explodes even more readily and with greater violence than the nitrate. On boiling with hydrochloric acid it decomposes with formation of phenol and aniline:

$$C_{12}H_{13}N_5O = C_6H_6O + C_6H_7N + 2N_2.$$

Acid diazobenzene sulphate, $C_6H_5.N_2SO_4H$, is obtained by the addition of dilute sulphuric acid to an aqueous solution of the

[1] *Annalen*, **137**, 39.   [2] Knoevenagel, *loc. cit.*

nitrate. The nitric acid and excess of sulphuric acid are removed by washing with alcohol and ether, and the solution then allowed to evaporate over sulphuric acid; a little phenol and free sulphuric acid are thus formed and must be removed by absolute alcohol. The crystals are then dissolved in the smallest possible quantity of water, treated with alcohol and ether, and the aqueous layer again allowed to evaporate. It may also be prepared in almost theoretical quantity by adding amyl nitrite to an alcoholic solution of aniline sulphate.[1]

Acid diazobenzene sulphate crystallizes in white prisms which decompose and deliquesce in the air. It detonates at about 100°.

Diazobenzene platinochloride $C_6H_5N_2PtCl_6$, is formed when an acid solution of platinum chloride is added to the solution of a diazobenzene salt; it crystallizes in slender, yellow prisms, slightly soluble in water and insoluble in alcohol. When it is ignited with carbonate of soda, chlorobenzene is obtained.

This reaction may be used to replace the amido-group by chlorine, but this is more simply effected by Sandmeyer's method.

Diazobenzene aurochloride, $C_6H_5N_2AuCl_4$, crystallizes from warm alcohol in beautiful small plates with a golden lustre, which are insoluble in water.

Diazobenzene cyanide, $C_6H_5N_2CN + HCN$, is obtained by the addition of the sulphate to a cooled solution of potassium cyanide, and crystallizes in orange-yellow prisms, which melt at 69° and are very unstable.[2]

Diazobenzene picrate, $C_6H_5N_2OC_6H_2(NO_2)_3$, is a yellow, crystalline precipitate, insoluble in water, alcohol, ether and benzene.[3]

Potassium diazobenzenesulphonate, $C_6H_5N_2SO_3K$, is prepared by adding diazobenzene nitrate to a cooled, slightly alkaline solution of potassium sulphite and then treating with caustic potash; this precipitates the compound in yellow crystals, which detonate violently when heated. Like all the diazo-compounds, it gives Liebermann's reaction (p. 23) with phenol and sulphuric acid.[4] On reduction it is converted into potassium phenylhydrazinesulphonate, $C_6H_5.N_2H_2SO_3K$.

1109 *Diazobenzenesulphonic acids* are obtained by the action

[1] Knoevenagel, *loc. cit.*
[2] Gabriel, *Ber.* **12**, 1638.
[3] Baeyer and Jäger, *ibid.* **8**, 893.
[4] Fischer, *Annalen*, **190**, 73.

of nitrous acid on the anilinesulphonic acids according to the following equation :

$$C_6H_4{\Large<}{\overset{NH_2}{\underset{SO_3H}{}}} + NO_2H = C_6H_4{\Large<}{\overset{N\diagdown}{\underset{SO_3}{}}}N + 2H_2O.$$

As may be seen from this formula, they are not true acids, but must be considered as salts of diazobenzene.

m-Diazobenzenesulphonic acid forms small, reddish yellow prisms, which readily dissolve in cold water, but are decomposed by water at 60° with formation of *m*-phenolsulphonic acid. It is not attacked by boiling absolute alcohol, and in the dry state detonates violently when heated.[1]

p-Diazobenzenesulphonic acid was first obtained by Schmitt.[2] To prepare it, rather more than the calculated quantity of sodium nitrite is added to a solution of sulphanilic acid (*p*-amidobenzenesulphonic acid) in dilute caustic soda, and the mixture poured into cooled, dilute sulphuric acid. The diazo-compound separates out in white crystals,[3] which are insoluble in cold water, and readily soluble in water at 60—70°, but are decomposed by it at higher temperatures with formation of *p*-phenolsulphonic acid, while boiling alcohol converts them into benzenesulphonic acid. Gaseous ammonia decomposes the diazo-compound with explosive violence; it is not attacked by phosphorus pentachloride even at 100°.[4]

o-Nitrodiazobenzenesulphonic acid, $C_6H_3(NO_2){\Large<}{\overset{N\diagdown}{\underset{SO_3}{}}}N$, is

obtained by treating *o*-nitranilinesulphonic acid (p. 267), with sodium nitrite and hydrochloric acid. It forms yellow needles, which are very sparingly soluble in water, and when dry, explode violently on heating.[5]

Diazobenzenedisulphonic acids are formed by the action of nitrous acid on the amidobenzenedisulphonic acids.[6] They are monobasic acids of the following constitution :

$$HO.SO_2.C_6H_3{\Large<}{\overset{N\diagdown}{\underset{SO_3}{}}}N.$$

---

[1] Berndsen, *Annalen,* **177**, 88.   [2] *Ibid.* **120**, 144.
[3] Fischer, *ibid.* **190**, 76.   [4] Laar, *J. Pr. Chem.* II. **20**, 263.
[5] Nietzki and Lerch, *Ber* **21**, 3221.
[6] Drebes, *ibid.* **9**, 553 ; Heinzelmann, *Annalen,* **188**, 157 ; **190**, 223 ; Zander, *ibid.* **198**, 5 and 24.

Diazobenzene potassoxide, $C_6H_5N_2OK$, (?) is obtained, according to Griess, by gradually adding a cold saturated solution of diazobenzene nitrate to a large excess of concentrated caustic potash and evaporating the yellow liquid, which has a peculiar aromatic odour, until it solidifies as a crystalline mass. It is then well pressed between porous plates, extracted with alcohol, the solution evaporated, and the residue again pressed, dried over sulphuric acid, and washed with ether to remove a brownish red decomposition product. In order to obtain it perfectly pure, ether is added to its solution in a little absolute alcohol; the diazobenzene potassoxide is thus precipitated in small white plates which have an alkaline reaction and absorb carbon dioxide very readily. It detonates feebly at 130°, and decomposes in aqueous solution, slowly in the cold, more rapidly on heating, with separation of a brownish red substance.

Diazobenzene argentoxide, $C_6H_5N_2OAg$, (?) is obtained, as a greyish white precipitate by mixing a solution of the potassium compound with a silver solution. It can be preserved without undergoing any change, and detonates violently on heating.[1]

Diazobenzene hydroxide, $C_6H_5N_2OH$. (?)—Griess prepared this compound by the addition of an equivalent of acetic acid to an aqueous solution of diazobenzene potassoxide. He looked upon it as free diazobenzene, $C_6H_4N_2$, and assumed the existence of this group of atoms in its derivatives, giving, *e.g.*, to the nitrate, the formula $C_6H_4N_2.NO_3H$. It is a thick yellow oil, which has an aromatic odour, begins to evolve nitrogen a few minutes after it has been isolated, and soon changes to a brownish red elastic substance; with larger quantities the spontaneous decomposition proceeds with explosive violence. The formula given for this substance is deduced from the fact that when freshly prepared it combines with caustic potash, nitric acid, &c., to form the diazo-derivatives which have just been described.

The above potassium compound has been again examined by Curtius,[2] who finds that it is formed in the manner described by Griess, and that the amount of potassium it contains corresponds with the formula given above. A determination of the nitrogen however, only gave two-thirds of the quantity required by the above formulæ. The constitution assigned by Griess to these substances cannot therefore be correct, but their real nature has not yet been ascertained.

[1] *Annalen*, **137**, 57.   [2] *Ber.* **23**, 3035.

1110 *p-Amidodiazobenzene chloride*, $C_6H_4\genfrac{<}{>}{0pt}{}{N=NCl\ (1)}{NH_2\quad (4)}$ is

obtained by passing nitrous fumes through a solution of *p*-di-
amidobenzene hydrochloride. It is very unstable in the free
state, but on addition of gold chloride yields an *aurochloride*,

$C_6H_4\genfrac{<}{}{0pt}{}{NH_3AuCl_4}{N=N.AuCl_4}$, from which other salts of *p*-amidodiazoben-

zene may be prepared.[1]

m-Bidiazobenzene chloride, or *m-Tetrazobenzene chloride,*

$C_6H_4\genfrac{<}{}{0pt}{}{N=N.Cl\ (1)}{N=N.Cl\ (3)}$. By the action of nitrous acid on *m*-

diamido compounds they are usually converted into derivatives
of triamidoazobenzene. If, however, the nitrous acid is largely
in excess, and comes in contact only with small quantities of
the hydrochloride, the *m*- and *p*-diamido compounds may also be
converted into diazo-compounds. *m*-Diamidobenzene is thus
converted into *m*-bidiazo- or *m*-tetrazobenzene chloride. This is
very unstable, but forms a *platinochloride*, $C_6H_4N_4PtCl_6$, and
an *aurochloride*, $C_6H_4N_4Cl_2$, $2AuCl_3$, both of which are yellow
crystalline explosive compounds. The bidiazo-compound behaves
in its general reactions exactly like diazobenzene.[2]

p-Bidiazobenzene chloride or *p-Tetrazobenzene chloride,*

$C_6H_4\genfrac{<}{}{0pt}{}{N=NCl\ (1)}{N=N.Cl\ (4)}$, is obtained from *p*-diamidobenzene and

nitrous acid, when the precautions mentioned above are taken.
It yields a perbromide, crystallizing in reddish yellow crystalline
grains, which are converted by ammonia into hexazobenzene[3]
(p. 327).

1111 *Diazohydroxybenzene compounds* are obtained by treating
alcoholic solutions of the salts of the amidophenols with nitrogen
trioxide. They are decomposed by hydriodic acid in the cold with
formation of the iodophenols. Alcoholic solutions of the substituted amido-phenols are not converted by nitrous acid into
diazohydroxybenzene salts, but into anhydrides of the substituted
diazophenols;

$$C_6H_3(NO_2)\genfrac{<}{}{0pt}{}{OH}{NH_2} + NO_2H = C_6H_3(NO_2)\genfrac{<}{>}{0pt}{}{O}{N}N + 2H_2O.$$

[1] Griess, *Ber.* **17**, 603.   [2] Griess, *ibid.* **19**, 317.   [3] *Ibid.* p. 319

When, on the contrary, the ethers of these amidophenols are employed, the formation of such anhydrides is impossible, and they behave like the amidophenols or aniline.

o-Diazophenol chloride, $C_6H_4(OH)N_2Cl$, is obtained by treating *o*-amidophenol hydrochloride with absolute alcohol saturated with nitrogen trioxide, the mixture being cooled by ice-water. An indigo-blue solution is first formed, and soon changes to brown. On addition of ether, the chloride is precipitated and crystallizes in beautiful rhombohedra, which are readily soluble in alcohol and become milk-white in the air, losing their water of crystallization. It forms a *platinochloride* which crystallizes well and yields *o*-chlorophenol on heating.[1]

p-Diazophenol chloride, $C_6H_4(OH)N_2Cl$, is obtained in a similar manner to the preceding compound from *p*-amidophenol hydrochloride, and crystallizes in long needles, slightly soluble in alcohol (Schmitt). The *platinochloride*, $[C_6H_4(OH)N_2]_2PtCl_6$, on ignition yields *p*-chlorophenol, and quinol is obtained by heating the chloride with concentrated hydrochloric acid.[2]

p-Diazophenol nitrate is formed in a singular manner by passing nitrogen trioxide through a well-cooled ethereal solution of phenol :

$$C_6H_5(OH) + 2N_2O_3 = C_6H_4(OH)N_2.NO_3 + NO_3H.$$

The free nitric acid formed converts a portion of the phenol into *o*- and *p*-nitrophenol.[3] The nitrate is also obtained when nitrogen trioxide is passed through an ethereal solution of quinonoxine.[4] It is readily soluble in alcohol, insoluble in ether, and crystallizes in light brown needles which are very explosive and fire like gunpowder, leaving a bulky residue of carbon. By dissolving it in dilute sulphuric acid and adding alcohol and then ether, the sulphate is obtained as a splendid crystalline mass. This is not explosive, and is converted by barium chloride into diazophenol chloride, which is identical with that obtained from *p*-amidophenol. (Weselsky and Schuler).

p-Diazo-anisoïl nitrate, $C_6H_4(OCH_3)N_2.NO_3$, is obtained when nitrogen trioxide is passed through water containing nitrate of anisidine (*p*-methoxyamidobenzene) in suspension. It forms colourless crystals which explode on percussion, and gradually

---

[1] Schmitt, *Ber.* **1**, 67.
[2] Weselsky and Schuler, *ibid.* **9**, 1160.
[3] Weselsky, *ibid.* **8**, 98.
[4] Jäger, *ibid.* **8**, 894.

decompose on boiling with water, more rapidly on heating to 140°, with formation of quinol.[1]

Diazonitrophenol anhydride, $C_6H_3(NO_2)\underset{O}{\overset{N}{<}}N.$ Griess obtained this compound by passing nitrogen trioxide through an ethereal solution of *p*-amido-*o*-nitrophenol. It is a brownish yellow, granular mass, which is slightly soluble in hot water, and explodes at 100° with extreme violence.[2] Its methyl ether, *diazonitranisoïl,* is only known as the nitrate, $C_6H_3(NO_2)(OCH_3)$ $N_2.NO_3$, and may be obtained from nitranisidine nitrate; it crystallizes in small plates.[3]

Diazodinitrophenol anhydride, $C_6H_2(NO_2)_2\underset{O}{\overset{N}{<}}N$, was prepared by Griess from amidodinitrophenol; it crystallizes from alcohol in small plates, which explode violently on heating.[4]

Diazo-amidonitranisoïl, $C_6H_3(NO_2)(OCH_3)N_2.NH.C_6H_3(NO_2)$ OCH_3. Griess obtained this compound by passing nitrogen trioxide through an alcoholic solution of nitranisidine. It forms yellow microscopic needles, insoluble in water, and only slightly soluble in hot alcohol.[5]

Diazonitrodihydroxyquinone. When amidonitrotetrahydroxy-benzene (p. 290) is made into a paste with dilute hydrochloric acid, well cooled with ice, and a solution of sodium nitrite added drop by drop, nitric oxide is evolved and a clear solution is obtained, which after some time solidifies to a pulpy mass, consisting of long, golden-yellow needles, having the composition $C_6H_4NaN_3O_8$. This compound is also obtained by dissolving amidonitrotetrahydroxybenzene in cold dilute nitric acid, and partially neutralizing the golden-yellow solution with carbonate of soda. Its constitution is expressed by one of the following formulæ:

$$\begin{array}{l}HO\\ \\HO.N:N\end{array}\!\!>\!C_6O_2\!<\!\!\begin{array}{l}NO_2\\ \\ONa\end{array} + H_2O, \text{ or } N\!\!\begin{array}{l}\\ \\ \\N\end{array}\!\!<\!\!\begin{array}{l}O\\ \\ \end{array}\!\!>\!C_6O_2\!<\!\!\begin{array}{l}NO_2\\ \\ONa\end{array} + 2H_2O.$$

It is readily soluble in water and is reprecipitated by alcohol. It can be recrystallized from water at 50°, but decomposes on

[1] Salkowski, *Ber.* **7**, 1009.   [2] *Annalen,* **113**, 212.
[3] *Jahresb.* **1866**, 459.   [4] *Annalen,* **113**, 205.
[5] *Ibid.* **121**, 278.

boiling with a violent evolution of gas. In the dry state it explodes with great violence.

When potassium nitrate is used in its preparation, the less soluble potassium salt is obtained, which is much more explosive, while the silver salt, obtained in small yellow plates by precipitating the sodium salt with silver nitrate, explodes by pressure or percussion with almost greater violence than silver fulminate.[1]

DIAZOAMIDO-COMPOUNDS.

1112 As already mentioned in the introduction (p. 29) diazo-amido-compounds are readily obtained by the action of nitrous acid on an excess of an amido-compound, or when an amido-compound is added to the solution of a diazo-salt. It was further stated that any pair of dissimilar amines can only yield one diazoamido-compound, whether the one or the other be first diazotized, whereas one would naturally expect under these circumstances the formation of two compounds having the general formulæ :

$$X.N{=}N.NH.Y \text{ and } X.NH.N{=}N.Y.$$

The theoretical considerations and experimental evidence offered by Goldsmidt and Molinari have also been mentioned.

Meldola and Streatfeild [2] have carefully examined the alkyl derivatives of the "mixed" diazo-amido-compounds (*i.e.*, compounds in which the two radicals, combined with the group N_3H, are different), and find that in all cases three isomeric alkyl derivatives exist. These are formed by the three following general reactions :

I. By the action of $X.N_2Cl$ on $Y.NH.R'$
II. By the action of $Y.N_2Cl$ on $X.NHR'$
III. By the action of KOH and $R'I$ on $X.N_3H.Y.$

X and Y representing the two radicals combined with the group N_3H, and R' a monad alkyl group. They find further that the isomerides obtained by the third method, *i.c.*, direct alkylation

[1] Nietzki and Benckiser, *Ber.* **18**, 499.
[2] *Journ. Chem. Soc.* **1886**, I. 624 ; **1887**, I. 102, 434 ; **1888**, I. 664 ; **1889**, I. 412 ; **1890**, I. 785.

of the mixed diazo-amido-compounds, are also formed when the compounds obtained according to I and II are heated together in equimolecular proportions, and may therefore be regarded as a molecular compound of those two isomerides. These investigators think it most probable that the combination takes place in the following manner:

$$
\begin{array}{c}
\text{X.N=NR'.Y} \\
\text{Y.N=N.R'X}
\end{array}
=
\begin{array}{c}
\text{X.N—N.R'Y} \\
\quad | \qquad | \\
\text{Y.N—N.R'Y}
\end{array}
$$

An attempt was made to determine the molecular weight of one of these compounds by Raoult's method, but the results obtained agreed with the simple formula, whence it appears that the molecular compound splits up into its constituents in solution.

As these molecular compounds are formed by the direct alkylation of the diazoamido-compounds, it is possible that these also have double the molecular weight usually assigned to them. This view is to some extent favoured by the fact that the dry silver salt of p-chlorobenzenediazoamido-p-toluidine, yields the molecular compound on treatment with methyl iodide in anhydrous benzene solution. This is however not incapable of explanation by the simple formula, and until a satisfactory proof of the truth of the double formula for the unalkylated compound is brought forward, the simpler one may be employed.

1113 *Diazoamidobenzene,* $C_6H_5.N : N.NH.C_6H_5$. Griess first obtained this compound by passing nitrous acid into an alcoholic solution of aniline,[1] and also prepared it by the action of aniline on diazobenzene nitrate:[2]

$$
\begin{array}{l}
C_6H_5.N_2NO_3 + 2C_6H_5NH_2 = \\
C_6H_5.N_2.NHC_6H_5 + C_6H_5.NH_2.HNO_3.
\end{array}
$$

In order to prepare it, 10 grams of aniline are dissolved in 100 grams of water containing $2\frac{1}{2}$ molecules of hydrochloric acid, carefully cooled with ice water, and a cold solution of 1 molecule of sodium nitrite in a little water slowly added. To this a cold solution of 10 grams aniline in 50 grams of water mixed with just sufficient hydrochloric acid for complete solution are added, and then a solution of 30 grams of sodium acetate in the

[1] *Annalen,* **121**, 258.   [2] *Ibid.* **137**, 58 ; see also Kekulé, *Lchrb.* **2**, 726.

minimum quantity of water. Diazoamidobenzene separates as a yellow crystalline precipitate, which, after washing and drying between filter paper, may be recrystallized from light petroleum boiling at 70—100°.[1]

Diazoamidobenzene forms well developed dark-yellow crystals which are insoluble in water, slightly soluble in hot alcohol, crystallizing therefrom in golden-yellow plates, or from benzene in large flat prisms, melting at 91° to a reddish brown oil which detonates at a higher temperature. On warming with strong hydrochloric acid it decomposes into nitrogen, aniline, and phenol.

It forms no salts with acids, but chloroplatinic acid produces a precipitate of $(C_{12}H_{11}N_3)_2PtCl_6$, in small red needles or prisms which readily decompose. Diazoamidobenzene, in presence of salts of aniline, is, as already explained, readily converted into amidoazobenzene, $C_6H_5.N_2.C_6H_4.NH_2$.

Diazobenzene-amidobromobenzene, $C_6H_5.N_2.NH.C_6H_4Br$.—Griess obtained this compound by the action of p-bromaniline on diazobenzene nitrate. It crystallizes in yellow needles or small plates[2] which melt at 90—91°.[3] When he treated aniline with diazobromobenzene nitrate he did not obtain the isomeric compound $C_6H_4Br.N_2.NH.C_6H_5$, but the same substance as before.[4]

Diazobenzene-ethylamine, $C_6H_5.N_2.NH(C_2H_5)$, is obtained by the action of ethylamine (1 mol.) on an aqueous solution of diazobenzene nitrate (1 mol.); it is an oily liquid, very similar to the following compound, which has been more carefully examined.

Diazobenzenedimethylamine, $C_6H_5.N_2.N(CH_3)_2$, is a yellowish oil having a peculiar aromatic odour, and can be distilled in small quantities without decomposition, but in large quantities decomposes with a tolerably violent explosion, forming dimethylamine. It is a weak base, and forms salts which are slowly decomposed by water in the cold, more rapidly on heating :

$$C_6H_5.N_2.N(CH_3)_2 + H_2O = C_6H_5.OH + N_2 + NH(CH_3)_2.$$

The alcoholic solutions of both these compounds yield amido-

[1] E. Fischer, *Org. Präparate,* p. 11.
[2] *Annalen,* **137**, 60.
[3] Nölting and Binder, *Ber.* **20**, 3012.
[4] *Ber.* **7**, 1618.

azobenzene when treated with a salt of aniline, ethylamine or dimethylamine being formed at the same time.[1]

Dis-diazobenzenemethylamine, $CH_3.N \begin{subarray}{l} \diagup N_2C_6H_5 \\ \diagdown N_2C_6H_5 \end{subarray}$ is obtained by

the action of 2 molecules of diazobenzene chloride on 3 molecules of methylamine, two of the latter being converted into methylamine hydrochloride. It crystallizes in long pale yellow needles, which melt at 112—113°, are readily soluble in ether and benzene, less easily in alcohol. On treatment with hydrochloric acid it forms nitrogen, methylamine, aniline, methyl alcohol, and a little amidoazobenzene, and on reduction yields methylamine and phenylhydrazine.[2]

Dis-diazobenzene-ethylamine, $C_2H_5.N \begin{subarray}{l} \diagup N_2.C_6H_5 \\ \diagdown N_2.C_6H_5 \end{subarray}$ is prepared in

a similar manner from ethylamine, and crystallizes from ether in yellow prisms melting at 70—71°.

As an example of the isomeric alkyl derivatives of mixed diazoamido-compounds, the following "triplet" may be mentioned, derived from *m*-nitraniline and *p*-bromaniline.

The first isomeride (*m*) $NO_2.C_6H_4.N_2.N(CH_3)C_6H_4Br$ (*p*) is prepared by mixing solutions of *m*-nitrodiazobenzene chloride and methyl-*p*-bromaniline oxalate. The precipitate, after repeated crystallization from alcohol, forms ochreous needles melting at 144°. It is decomposed by hydrochloric acid in the cold, forming *m*-nitrodiazobenzene chloride and methyl-*p*-bromaniline in almost theoretical proportions.

The second isomeride (*p*) $Br.C_6H_4.N_2N(CH_3)C_6H_4.NO_2$ (*m*) is obtained in a corresponding manner by acting on diazotized *p*-bromaniline with *m*-methylnitraniline. The precipitate, after two crystallizations from alcohol, forms long slender yellow needles, which melt at 160—161°, and are converted by hydrochloric acid into *p*-bromodiazobenzene chloride and *m*-methylnitraniline.

The third isomeride $\begin{subarray}{l} (m)\ NO_2.C_6H_4 \diagdown \\ \qquad\qquad\qquad N_4 \\ (p)\ \ Br.C_6H_4 \diagup \end{subarray} \begin{subarray}{l} \diagup N.CH_3.C_6H_4Br(p) \\ \diagdown N.CH_3.C_6H_4NO_2(m) \end{subarray}$

is formed by the direct methylation of *m*-nitro-*p*-bromodiazoamidobenzene, or by warming together the two isomerides already

[1] Baeyer and Jäger, *Ber.* **8**, 148.
[2] Goldschmidt and Badl, *ibid.* **22**, 933.

described in equal molecular proportions. It crystallizes from alcohol in slender yellow needles, which, like all the corresponding compounds, have no distinct melting point, the temperature ranging from 125—137.° On treatment with hydrochloric acid it yields a mixture of m-nitrodiazobenzene chloride, p-bromodiazobenzene chloride, p-bromaniline and m-nitraniline, which is in full agreement with the theory that it is a molecular compound.[1]

Diazoimides or Triazo-compounds.

1114 The first of these compounds was prepared by Griess by the action of ammonia on diazobenzene perbromide. Its empirical formula is $C_6H_5N_3$, and Kekulé proposed for it the constitutional

formula $C_6H_5.N\begin{matrix}N\\ \| \\ N.\end{matrix}$ The above reaction is a general one,

and the compounds may also be obtained by acting on the nitrosohydrazines with potash or by treating diazo-salts with hydroxylamine and soda.[2] They are also formed by the action of diazo-salts on the hydrazines,[3] and may therefore also be prepared direct from a diazo-salt by adding to the latter one half the quantity of stannous chloride necessary for its complete reduction, the hydrazine formed then acting on the unaltered diazo-compound.[4]

$$C_6H_5.N_2H_3.HCl + C_6H_5N_2Cl = C_6H_5N_3 + C_6H_5NH_2.HCl + HCl.$$

They are usually termed "Diazoimides," but Griess has proposed the term triazo-compounds as more suitable.[5]

The triazo-compounds may be regarded as derivatives of

hydrazoic acid, $HN\begin{matrix}N\\ \| \\ N\end{matrix}$ which has recently been obtained by

Curtius. The *phenyl* derivative of this acid is identical in every respect with the triazobenzene described below.[6]

1115 *Triazobenzene or Diazobenzeneimide*, $C_6H_5N_3$, is obtained as mentioned above by the action of ammonia on diazobenzene

---

[1] Meldola and Streatfeild, *Journ. Chem. Soc.* **1889**, I. 425 ; **1890**, I. 786.
[2] E. Fischer, *Annalen*, **190**, 92.   [3] Griess, *Ber.* **9**, 1659.
[4] Culmann and Gasiorowsky, *J. Pr. Chem.* II. **40**, 99.
[5] Griess, *Ber.* **19**, 315.   [6] *Ber.* **23**, 3023.

perbromide, of caustic potash on nitrosophenylhydrazine, and by the treatment of diazobenzene chloride with hydroxylamine and soda, or with $\frac{1}{2}$ mol. of stannous chloride. To prepare it, crude phenylhydrazine hydrochloride is dissolved in 15 parts of water, an excess of sodium nitrite gradually added, and the mixture heated to boiling in an apparatus connected with an inverted condenser until the gentle evolution of gas ceases.[1]

Triazobenzene is a yellowish, oily liquid having a stupefying, ammoniacal and aromatic odour. It is insoluble in water, and can be distilled with steam, or in a vacuum, but explodes on distillation under the ordinary pressure. On reduction it forms ammonia and aniline, and is converted by boiling hydrochloric acid into o- and p-chloraniline, and by sulphuric acid into p-amidophenol.[2]

p-Triazobenzenesulphonic acid, $C_6H_4N_3.SO_3H$, is obtained by the action of an excess of phenylhydrazine on p-diazobenzene-sulphonic acid, triazobenzene, amidobenzoic acid and aniline being formed at the same time. The phenylhydrazine salt of p-triazobenzenesulphonic acid separates out from the aqueous solution, and crystallizes in long yellowish narrow plates, sparingly soluble in cold water and alcohol, readily in the hot liquid. On addition of barium chloride it yields barium p-triazoben-zenesulphonate, which crystallizes from hot water in white six-sided prisms. On decomposition with the requisite quantity of sulphuric acid this yields the free acid, crystallizing in white deliquescent needles, readily soluble in alcohol, which have at first an acid and then an intensely bitter taste. It explodes on heating, leaving a porous mass of charcoal, and is decomposed by mineral acids on warming.[3]

m-Triazobenzenesulphonic acid, $C_6H_4N_3.SO_3H$, is obtained in a similar manner from m-diazobenzenesulphonic acid and m-hydrazinesulphonic acid. It is also a white crystalline deli-quescent compound, which may be boiled with water without decomposition, but evolves nitrogen on treatment with hydro-chloric acid. Zinc and acetic acid convert it into m-amido-benzenesulphonic acid.[4]

m-Amidotriazobenzene, $C_6H_4N_3.NH_2$ (1.3). To obtain this compound m-phenyleneoxamic acid (p. 271) is diazotized, the diazo-compound converted into the perbromide, and treated with ammonia. The oxamic acid of m-amidotriazobenzene

[1] *Annalen,* **190**, 92, 96.   [2] Griess, *Ber.* **19**, 313.
[3] *Ibid.* **20**, 1529   [4] Limpricht and Neumann, *ibid.* **21**, 3416.

$C_6H_4\Big\langle\begin{array}{l}N_3\\ NH.C_2O_2OH\end{array}$ thus obtained, crystallizes in white needles almost insoluble in water, and on boiling with concentrated caustic potash, is converted into m-amidotriazobenzene, which may be driven over with steam. It forms a yellow oil, has a smell resembling bitter almonds, and decomposes explosively when heated alone. Its *hydrochloride*, $C_6H_4N_3.NH_2.HCl$, forms beautiful white pointed rhombic plates. In its general properties it resembles aniline, and may be converted into a diazo-compound from which colouring matters possessing very remarkable properties may be obtained in the usual manner.[1]

p-Amidotriazobenzene, $C_6H_4N_3(NH_2)$ (1.4), is prepared in a similar manner from p-phenyleneoxamic acid, and forms long white four-sided plates, which are readily soluble in alcohol, ether, and chloroform, sparingly in hot water. It melts in a capillary tube at 65°, explodes at a higher temperature in the dry state, and is gradually decomposed by boiling water. Its *hydrochloride*, $C_6H_4N_3.NH_2.HCl$, forms reddish needles or plates which are readily soluble in water, whilst the *platinochloride*, $C_6H_4N_3.NH_2, H_2PtCl_6$, crystallizes in yellow needles.[2]

p-Diazotriazobenzene chloride, $C_6H_4.N_3.N_2Cl$, is obtained by the action of nitrous acid on the foregoing compound. Its *platinochloride*, $C_6H_4N_3.N_2PtCl_6$, crystallizes in dark-yellow plates or pale yellow needles according to the concentration of the solution, and in the dry state explodes violently on heating. The *perbromide*, $C_6H_4N_3.N_2Br_3$, forms reddish-yellow crystals.

p-Ditriazobenzene, or *Hexazobenzene*, $C_6H_4(N_3)_2$ (1.4), is obtained by the action of ammonia on the perbromide. It crystallizes in white well-developed tablets which become brown on exposure to light. It is insoluble in water, sparingly soluble in alcohol, readily in ether and chloroform, and is indifferent towards acids and bases in the cold. It melts at 83°, but on heating even in small quantities to a higher temperature, explodes with exceptional violence. It is however volatile with steam, an overpowering aromatic odour being then observed. It may also be prepared from p-tetrazobenzene (P. 318), but the yield is not so good.[3]

[1] Griess, *Ber.* **18**, 963.   [2] *Ibid.* **21**, 1559.   [3] *Ibid.* p. 1561.

AZIMIDO-COMPOUNDS.

1116 The first of these compounds was obtained by Hofmann [1] in 1860 from common nitrophenylenediamine, and others were obtained by Griess [2] from β- and γ-diamidobenzoic acid, and by Ladenburg [3] from o-phenylene- and o-toluylenediamine. The simplest compound, obtained by the action of nitrous acid on o-phenylenediamine, has the composition $C_6H_4.N_3H$. The name azimido-compounds was first employed by Griess,[4] whilst Kekulé [5] proposed for them the constitutional formula

$$C_6H_4\left\langle\begin{array}{c}N\\\\NH\end{array}\right\rangle N,$$ regarding them as inner anhydrides of o-amido-

diazo-compounds. Griess,[6] on the other hand, believed that they contained the group $\begin{array}{c}-N\\|\\-N\end{array}\rangle NH$. In 1885, Zincke and

Lawson [7] found that certain o-amidoazo-compounds of toluene and naphthalene gave on oxidation substances which contained the group N_3R, and were therefore believed to be substituted azimido-derivatives. Further investigation [8] showed that the two series of compounds are not identical, for the compound obtained by the action of nitrous acid on α-amido-β-phenyl-

amidonaphthalene $C_{10}H_6\left\langle\begin{array}{c}NH_2\ (a)\\\\NHC_6H_5\ (\beta)\end{array}\right.$ has different properties

to that obtained by the oxidation of benzeneazo-β-naphthyl-

amine, $C_{10}H_6\left\langle\begin{array}{c}NH_2\ (\beta)\\\\N=N.C_6H_5\ (a)\end{array}\right.$ The first compound like all

those obtained from nitrous acid and o-diamido-compounds, readily unites with methyl iodide, forming ammonium compounds, whilst the compounds obtained by oxidation of o-amidoazo-compounds are unacted upon by methyl iodide, and show no basic properties. It is most probable that the members of the first series have the constitution assigned to them by Kekulé and that those of the second series, which may be termed

[1] *Annalen,* **115**, 249.   [2] *Ber.* **5**, 200.   [3] *Ibid.* **9**, 219.
[4] *Ibid.* **15**, 1878.   [5] *Lehrb.* **2**, 739.   [6] *Loc. cit.*
[7] *Ber.* **18**, 3132 ; **20**, 1167.
[8] *Annalen,* **239**, 110, see also *Ber.* **20**, 2999.

pseudoazimido-compounds, correspond to the formula proposed by Griess.[1]

1117 *Azimidobenzene,* $C_6H_4\left\langle\begin{smallmatrix}N\\\\NH\end{smallmatrix}\right\rangle N$ is obtained by treating a very dilute solution of *o*-diamidobenzene sulphate with a solution of potassium nitrite,[2] or by acting on *o*-diamidobenzene hydrochloride with *p*-diazobenzenesulphonic acid.[3] It crystallizes from benzene in needles, which melt at 98·5°.

p-Bromazimidobenzene, $C_6H_4BrN_3$. To prepare this compound 20 grms. of bromonitracetanilide (1. 3. 4.) are dissolved in 150 grms. of warm 30 per cent. alcohol, 20 grms. of iron powder added, and then gradually dilute acetic acid till the reduction is complete. The hot diluted solution is treated with sodium carbonate and filtered from the precipitated iron; the filtrate on cooling deposits bromamidoacetanilide, which is at once dissolved in hydrochloric acid, and treated with nitrous acid, and the acetyl-bromazimidobenzene thus obtained hydrolyzed by boiling with concentrated hydrochloric acid. Bromazimidobenzene forms slender white needles or plates, which melt at 158—159°, are readily soluble in alcohol and acetic acid, and crystallize best from benzene. The crystals formed contain benzene of crystallization, which is only completely expelled at 100°. Its *sodium* salt, $C_6H_3Br.N_3Na$, is a white crystalline mass, and its *silver* salt, $C_6H_3BrN_3Ag$, a white amorphous precipitate. It also forms salts with acids, the *hydrochloride,* $C_6H_3Br_3N_3H,HCl$, crystallizing from hot hydrochloric acid in white needles, which are decomposed by water and alcohol. The *acetyl* compound, $C_6H_3BrN_3.C_2H_3O$, crystallizes in slender white needles melting at 117—118°.[4]

Methylbromazimidobenzene, $C_6H_3BrN_3(CH_3)$, is best obtained by distilling the compound next described under diminished pressure. It crystallizes from hot water in slender white needles which melt at 79—80°, and have an intensely bitter taste. It has slight basic properties, and gives a reddish-yellow crystalline precipitate with platinum chloride.

Dimethylbromobenzeneazammonium iodide, $C_6H_3BrH_3(CH_3)_2I$, is obtained by heating bromazimidobenzene or its acetyl compound with methyl iodide in methyl alcoholic solution. It crystallizes

[1] Zincke and Campbell, *Annalen,* **255**, 339.
[2] Ladenburg, *Ber.* **9**, 222.
[3] Griess, *ibid.* **15**, 2195.
[4] Zincke and Arzberger, *Annalen,* **249**, 360.

from alcohol in slender yellowish needles which melt at 220°
with evolution of methyl iodide. On treatment with silver
oxide in aqueous solution the corresponding hydroxide is formed.
The solution precipitates iron and copper salts, sets free ammonia
and absorbs carbon dioxide from the air, but undergoes decomposition on evaporation.

Phenylpseudoazimidobenzene, C_6H_4 ⟨N⎮N⟩ $N.C_6H_5$. In the pre-

paration of *p*-amidoazobenzene (described later on) a certain
amount of *o*-amidoazobenzene is formed, which is immediately
oxidized, forming phenylpseudoazimidobenzene. It crystallizes
in long lustrous needles, which melt at 109°, and distil without decomposition. On reduction with sodium in alcoholic
solution it yields *o*-diamidobenzene.[1]

HYDRAZINE DERIVATIVES OF BENZENE.

1118 These bodies, as well as those of the fatty series, were
discovered and carefully examined by E. Fischer.[2] The yellow
diazobenzene potassium sulphonate, $C_6H_5N{=}NSO_3K$, is formed,
as already explained, by the action of a weak alkaline solution
of potassium sulphite on diazobenzene nitrate. This is converted by reducing agents into the white phenylhydrazine
potassium sulphonate, which is also obtained by acting on diazobenzene chloride with an excess of acid potassium sulphite:

$$C_6H_5N{=}N.Cl + 3SO_3KH = C_6H_5.NH{-}NH.SO_3K +$$
$$KHSO_4 + KCl + SO_2.$$

On boiling this salt with hydrochloric acid, phenylhydrazine
hydrochloride is obtained:

$$C_6H_5.NH{-}NH.SO_3K + HCl + H_2O = C_6H_5.NH{-}NH_2.HCl$$
$$+ KHSO_4.$$

Phenylhydrazine is also obtained, together with aniline, when
an alcoholic solution of diazo-amidobenzene is treated with zinc
dust and acetic acid:

$$C_6H_5N_2.NH.C_6H_5 + 4H = C_6H_5.N_2H_3 + NH_2.C_6H_5.$$

[1] Gattermann and Wichmann, *Ber.* **21**, 1633.   [2] *Annalen,* **190**, 67.

Phenylhydrazine, $C_6H_5N_2H_3$. To prepare this compound 50 grms. of aniline are dissolved in $2\frac{1}{2}$ mol. of concentrated hydrochloric acid, and 300 grms. of water, the solution well cooled and diazotized with the theoretical quantity of sodium nitrite, and then poured into a cold nearly saturated solution of $2\frac{1}{2}$ mol. of sodium sulphite. The latter is best obtained from a commercial 40 per cent. solution of sodium hydrogen sulphite by neutralization with caustic soda. The mixed solution is warmed, treated with zinc dust and a little acetic acid till colourless, filtered hot, and at once mixed with one-third its volume of concentrated hydrochloric acid. After cooling, the magma of phenylhydrazine hydrochloride is filtered through calico, and separated from the mother-liquor as far as possible by pressing; the hydrochloride obtained is decomposed by caustic soda, the free base extracted with ether, and the ethereal solution evaporated. After drying with ignited potassium carbonate, the residue is distilled, the portion boiling from 200—240° consisting chiefly of phenylhydrazine, which may be further purified by fractionation *in vacuo*.[1]

It may also be prepared by dissolving ten grms. of aniline in 100 c.c. of concentrated hydrochloric acid, diazotizing, and pouring the solution slowly into a strongly acid solution of stannous chloride. The hydrochloride thus obtained is treated in the manner described above.

Phenylhydrazine is colourless when quite pure, but is generally obtained as a light-yellow oil, which becomes solid in the winter, forming monosymmetric tablets which melt at 23°. Its sp. gr. is $1\cdot097$ at $22\cdot7°$, and it boils at 241—242° (750 mm.) with slight decomposition, a little ammonia being evolved. Under diminished pressure it distils unchanged. It is volatile with steam but not so readily as aniline.

It is slowly attacked by reducing agents with formation of aniline and ammonia, but readily undergoes oxidation, absorbing oxygen from the air and becoming red or dark-brown; it reduces Fehling's solution in the cold with evolution of nitrogen and separation of cuprous oxide, aniline and benzene being also formed. "This property can be employed as a delicate reaction for all primary hydrazines, and also, indirectly, for the diazo-compounds; in order to detect the latter in an aqueous solution, it is treated with acid potassium sulphite in excess, heated to boiling, neutralized with caustic potash, and tested with copper solution.

[1] E. Fischer, *Org. Präparate*, p. 14.

The salts of the hydrazinesulphonic acid, which are formed in the liquid from the diazo-compounds, without exception produce the immediate precipitation of cuprous oxide; if the presence of hydroxylamine, which is readily formed by the reduction of nitrous acid, is to be feared, the alkaline solution must be boiled for some time in order to destroy it" (Fischer). On adding mercuric oxide to an ethereal solution of phenylhydrazine, nitrogen is evolved and aniline and benzene formed, as well as a considerable quantity of mercury diphenyl, $(C_6H_5)_2Hg$.[1]

While ethylhydrazine is a strong diacid base, phenylhydrazine only combines with one equivalent of an acid, and is consequently a weaker base. This is easily explained, since phenylhydrazine stands in the same relation to ethylhydrazine as aniline to ethylamine.

Phenylhydrazine is one of the most important organic reagents, and is now manufactured on the large scale, and chiefly used in the manufacture of antipyrine, a compound which will be described in a later volume. Frequent mention has been made of its property of uniting with ketones and aldehydes with elimination of water to form *hydrazones* and *osazones* (Part II. p. 14 this vol. p. 33), the more important of which compounds have already been described under the ketones and aldehydes from which they are obtained. It also acts upon organic acids, and on their chlorides, anhydrides, and ethyl salts, and also on lactones, in the same manner as ammonia, forming *hydrazides*;[2] thus phenylacetic acid, $C_6H_5.CH_2.COOH$, readily yields phenylacetophenylhydrazide, $C_6H_5.CH_2.CO.NH—NH.C_6H_5$.

The hydrazides may also be obtained by heating the amides with phenylhydrazine,[3] ammonia being evolved.

Phenylhydrazine hydrochloride, $C_6H_5.N_2H_3.HCl$, may be obtained perfectly pure by dissolving the base, freed from ammonia by distillation, in ten parts of alcohol, neutralizing with concentrated hydrochloric acid, and washing the separated crystals with alcohol and ether until they are perfectly colourless. Phenylhydrazine hydrochloride crystallizes from hot water in small, thin, lustrous plates, which sublime on cautious heating; it is almost completely precipitated from its aqueous solution by concentrated hydrochloric acid, by which means phenylhydrazine can be separated from aniline and several other amido-bases.

[1] Fischer and Ehrhard, *Annalen*, **199**, 332.
[2] *Annalen*, **190**, 125; **236**, 194; *Ber.* **19**, 1387, 1707; **20**, 401; **21**, 186 c.
[3] Just, *ibid.* **19**, 1201.

It reduces the salts of silver, mercury, gold, and platinum in the cold.

In preparing the hydrazones and osazones, it is preferable in cases where pure phenylhydrazine itself is not readily obtained, to employ the hydrochloride obtained in the manner above described. This is mixed with $1\frac{1}{2}$ times its weight of sodium acetate, dissolved in a small quantity of water and added to the ketone or aldehyde of which the hydrazone is to be prepared.

Phenylhydrazine sulphate, $(C_6H_5.N_2H_3)_2H_2SO_4$, forms small plates, readily soluble in hot water, and slightly soluble in alcohol.

Phenylhydrazine oxalate, $(C_6H_5.N_2H_3)_2H_2C_2O_4$, crystallizes from hot water in small plates, which are slightly soluble in cold water.

Phenylhydrazine picrate, $C_6H_5.N_2H_3,C_6H_2(NO_2)_3OH$, crystallizes in fine yellow needles, slightly soluble in water, readily in alcohol, and deflagrates explosively on heating.

The salts of phenylhydrazine are converted by mercuric oxide in aqueous solution into diazobenzene salts; aniline and triazobenzene being, however, simultaneously formed, as may be readily understood.

1119 Halogen-substituted anilines, on diazotizing and treatment with stannous chloride and hydrochloric acid, are converted into substituted hydrazines. The following compounds have been prepared [1]:

	Melting-point.
p-Chlorophenylhydrazine, long needles	83°
p-Bromophenylhydrazine, long needles	106°
Dibromophenylhydrazine, lustrous needles or plates . .	97°
s-Tribromophenylhydrazine	146°
Tetrabromophenylhydrazine, slender prisms	167°
p-Iodophenylhydrazine, silky needles	103°
m-Di-iodophenylhydrazine, long silky needles	112°

These all combine with ketones and aldehydes in the same manner as phenylhydrazine itself.

m-Hydrazidobenzenesulphonic acid, $SO_3H.C_6H_4.N_2H_3$, is prepared by the reduction of the diazo-derivative of *m*-amidobenzenesulphonic acid with stannous chloride and hydrochloric acid. It crystallizes from water in rhombic tablets or needles, soluble with difficulty in cold water and scarcely at all in alcohol and ether.[2]

[1] Neufeld, *Annalen*, **248**, 93.   [2] Limpricht, *Ber.* **21**, 3409.

p-Hydrazidobenzenesulphonic acid, $SO_3H.C_6H_4.N_2H_3$, is prepared from *p*-diazobenzenesulphonic acid in a similar manner to phenylhydrazine,[1] and by warming phenylhydrazine or phenylhydrazinesulphonic acid (described below) with sulphuric acid in the water-bath.[2] It crystallizes with $\frac{1}{2}H_2O$ in lustrous needles, which lose the water of crystallization at 100°. It is soluble in boiling water, but scarcely in alcohol and ether.

o-Nitrophenylhydrazine, $NO_2.C_6H_4.N_2H_3$. — Phenylhydrazine cannot be directly nitrated, as it so readily undergoes oxidation; *o*-nitrophenylhydrazine may, however, be obtained by treating *o*-nitraniline with nitrous acid and reducing the diazo-compound formed with stannous chloride and hydrochloric acid. It forms small white plates which become brown in the air.[3] A nitrophenylhydrazine has also been prepared in a somewhat complicated manner by Michael.[4]

Although phenylhydrazine is oxidized by nitric acid, the hydrazones may be readily nitrated. Thus, acetonephenylhydrazone readily yields a dinitro-compound, $(CH_3)_2.C:N.NH.C_6H_3(NO_2)_2$ which crystallizes from alcohol in yellow needles melting at 127°. Unlike the non-substituted hydrazones, these compounds are not split up by hydrochloric acid.[5]

m-Amidophenylhydrazine, $H_2N.C_6H_4.NH.NH_2$. When the diazochloride obtained from *m*-phenylene-oxamic acid (p. 271) is treated with a solution of stannous chloride in hydrochloric acid, the following reaction occurs:

$$C_6H_4 \begin{matrix} \diagup NH.C_2O_2.OH \\ \diagdown N{=}NCl \end{matrix} + 3HCl + 2SnCl_2 = C_6H_4 \begin{matrix} \diagup NH.C_2O_2\,OH \\ \diagdown NH{-}NH_2 \end{matrix}$$
$$+ 2SnCl_4.$$

The new acid separates out in white crystalline granules, which are almost insoluble in alcohol, ether, and boiling water. On boiling it with strong hydrochloric acid, amidophenylhydrazine hydrochloride is obtained:

$$C_6H_4 \begin{matrix} \diagup NH.C_2O_2.OH \\ \diagdown NH.NH_2 \end{matrix} + H_2O = C_6H_4 \begin{matrix} \diagup NH_2 \\ \diagdown NH.NH_2 \end{matrix}$$
$$+ C_2O_2(OH)_2.$$

[1] Fischer, *Annalen*, **190**, 74.   [2] Richter and Gallinek, *Ber.* **18**, 3172.
[3] Bischler, *ibid.* **22**, 240.   [4] Michael, *ibid.* **19**, 1386.
[5] Fischer and Ach, *Annalen*, **253**, 57.

On adding strong caustic potash to the solution and concentrating on the water-bath, the base separates out. At the ordinary temperature it forms a varnish-like mass, converted on warming into an oil, which is almost odourless and has a very bitter taste. It is slightly soluble in water, readily in alcohol and ether, and is very easily oxidized, reducing Fehling's solution immediately with evolution of gas.

Its salts crystallize well; the hydrochloride is precipitated by hydrochloric acid from a concentrated solution in small pointed plates. Gold chloride and platinum chloride are immediately reduced by it.[1]

1120 The hydrogen atoms in the hydrazine group may also be replaced by various radicals. Two series of compounds may be thus obtained having the general formula:

$$C_6H_5.NR.NH_2 \text{ and } C_6H_5.NH.NHR.$$

The first series are known as the a- or asymmetric, and the second as the β- or symmetric compounds.

a-Sodiumphenylhydrazine, $C_6H_5.NNa.NH_2$, is prepared by adding sodium in small quantities at a time to phenylhydrazine, aniline and ammonia being formed at the same time:

$$3C_6H_5.N_2H_3 + 2Na = 2C_6H_5.NNa.NH_2 + C_6H_5.NH_2 + NH_3.$$

The aniline and excess of phenylhydrazine are distilled off in a vacuum, when the sodiumphenylhydrazine remains as a transparent yellowish-red mass, which yields a deep yellow powder. It decomposes readily and is often spontaneously inflammable. That it has the constitution above given to it is proved by the fact that the halogen salts of alkyl-radicals convert it into asymmetric hydrazines.[2]

β-Phenylhydrazinesulphonic acid, $C_6H_5.NH.NH.SO_3H$, is only known as the potassium salt, the formation of which has been already described: it is also obtained, together with phenylhydrazine sulphate, by heating the base with potassium disulphate to 80°:

$$4C_6H_5.N_2H_3 + 2K_2S_2O_7 = 2C_6H_5.N_2H_2.SO_3K$$
$$+ (C_6H_5N_2H_3)_2SO_4H_2 + K_2SO_4.$$

It forms colourless scales, soluble with difficulty in cold water. By the action of mercuric oxide, or potassium dichromate, on its

[1] Griess, Ber. **18**, 964.   [2] Michaelis, Annalen, **252**, 266.

hot aqueous solution, it is oxidized to yellow diazobenzene potassium sulphonate, $C_6H_5.N_2.SO_3K$.

By the action of sulphur dioxide on phenylhydrazine, Fischer [1] obtained two additive compounds of the formulæ $C_6H_5.N_2H_3.SO_2$ and $2(C_6H_5.N_2H_3.)SO_2$. The first of these compounds is also obtained when sulphur dioxide is passed into a cold benzene solution of phenylhydrazine. If the solution be heated, however, this compound loses water, forming *thionylphenylhydrazone*, $C_6H_5.NH.N{=}SO$, which may also be prepared by the action of thionylchloride, $SOCl_2$, on phenylhydrazine. It forms thick sulphur-yellow prisms, which are soluble in alcohol, ether, chloroform, and carbon bisulphide, and melt at 205°. Its formation yields additional evidence of the similarity existing between sulphurous acid and the aldehydes.[2]

a-Nitrosophenylhydrazine, $C_6H_5.N(NO).NH_2$, is obtained by adding sodium nitrite to a well-cooled dilute solution of phenylhydrazine hydrochloride; the yellow flocculæ which separate are dissolved in ether and reprecipitated by light petroleum. It crystallizes in small pale yellow plates, which are very unstable. Its vapour has an extremely poisonous action, somewhat similar to that of amyl nitrite, but more intense. If even a small quantity of the vapour be inhaled, determination of the blood to the head takes place, followed by headache and nausea. On warming with dilute caustic potash it is converted into triazobenzene [3] (p. 325)

1121 *a-Methylphenylhydrazine*, $C_6H_5.N(CH_3)NH_2$, is obtained by the reduction of methylphenylnitrosamine with zinc-dust and acetic acid in aqueous solution, the temperature being kept between 10—20°. Some methylaniline is always re-formed, which is separated by the different solubility of the sulphates · in alcohol.[4] Methylphenylhydrazine is a colourless, strongly refractive oil, which boils with slight decomposition at 227° under 745 mm. pressure, and distils unchanged at 131° under 35 mm. It reduces Fehling's solution on warming, with formation of methylaniline and nitrogen.

Dimethyldiphenyltetrazone, $C_{14}H_{16}N_4$, is formed when *a*-methylphenylhydrazine is dissolved in chloroform and mercuric oxide gradually added, the mixture being kept cool:

[1] *Annalen*, **190**, 124.
[2] Michaelis, *Ber.* **22**, 2228 ; Michaelis and Ruhl, *ibid.* **23**, 474.
[3] E. Fischer, *Annalen*, **190**, 92.
[4] Fischer, *ibid.* **190**, 152 ; **236**, 198.

$$\begin{array}{l} C_6H_5N(CH_3)-NH_2 \\ + 2HgO = \\ C_6H_5N(CH_3)-NH_2 \end{array} \begin{array}{l} C_6H_5N(CH_3)N \\ || + 2Hg + 2H_2O. \\ C_6H_5N(CH_3)N \end{array}$$

The compound crystallizes in small plates, slightly soluble in cold alcohol and ether, readily in chloroform, and melts at 137° with evolution of gas. It can be boiled with water without under-going decomposition; on adding hydrochloric acid to the alcoholic solution, the tetrazone decomposes into nitrogen and methylaniline, and on adding iodine to its solution in chloroform, *dimethyldiphenyltetrazone iodide*, $C_{14}H_{16}N_4I_4$, is formed as a black crystalline precipitate, which deflagrates in the dry state spontaneously, and when suspended in carbon bisulphide and shaken with silver-dust, is reconverted into the original compound.

β-Methylphenylhydrazine or *Hydrazomethylphenyl*, $C_6H_5.NH.NH.CH_3$, is prepared from *a-β-dibenzoylphenylhydrazine* by methylation and treatment with caustic potash. It is a colourless, unstable oil, which boils at 180° and soon assumes a yellow colour in the air. It forms crystalline salts and is readily oxidized to azomethylphenyl, $C_6H_5.N:N.CH_3$ [1] (p. 345).

1122 *a-Ethylphenylhydrazine*, $C_6H_5.N(C_2H_5).NH_2$, is prepared by the reduction of phenylethylnitrosamine,[2] $C_6H_5N(NO).C_2H_5$, with zinc-dust and acetic acid, and by the action of ethyl bromide on sodiumphenylhydrazine.[3] By the action of ethyl bromide on phenylhydrazine itself, a mixture of the *a-* and *β-* compounds is obtained, together with compounds containing more ethyl groups and *diethylphenylhydrazonium bromide*,

$$H_2N - N.C_6H_5(C_2H_5)_2Br. [4]$$

*a-*Ethylphenylhydrazine forms a colourless oil, which boils at 230° (uncorr.), readily becomes brown in the air, and reduces Fehling's solution on warming. It combines with ethyl bromide to form diethylphenylhydrazonium bromide, which is readily soluble in water but is precipitated by concentrated caustic soda. It forms rhombic crystals, and is converted by moist silver oxide into the strongly alkaline hydroxide $NH_2 - N(C_6H_5)(C_2H_5)_2OH$.

On adding zinc-dust and hydrochloric acid to a warm aqueous

---

[1] Tafel, *Ber.* **18**, 1739. [2] Fischer, *ibid.* **8**, 1652.
[3] Michaelis and Phillips, *Annalen*, **252**, 270.
[4] Fischer and Ehrhard, *ibid.* **199**, 325.

solution of the bromide it is decomposed into diethylaniline, ammonia, and hydrobromic acid :[1]

$$\begin{array}{l}C_2H_5\diagdown\\C_6H_5{-}N{<}^{NH_2}_{Br}\\C_2H_5\diagup\end{array} + 2H = \begin{array}{l}C_2H_5\diagdown\\C_6H_5\diagup N\\C_2H_5\end{array} + NH_3 + HBr.$$

Diethyldiphenyltetrazone, $\begin{array}{l}C_6H_5\diagdown\\C_2H_5\diagup\end{array}N{=}N.N{=}N\begin{array}{l}\diagup C_6H_5\\\diagdown C_2H_5\end{array}$, crys-

tallizes from alcohol in monosymmetric prisms, which melt at 108° with evolution of gas.

β-Ethylphenylhydrazine or *Hydrazophenylethyl*, $C_6H_5.NH.NH. C_2H_5$. When the mixture of bases obtained as above stated by the action of ethyl bromide on phenylhydrazine is dissolved in water, decomposed with caustic soda and extracted with ether, the ethyl bases, together with the unattacked phenylhydrazine, go into solution. On adding concentrated hydrochloric acid, after the evaporation of the ether, phenylhydrazine hydrochloride separates out. The solution is then again made alkaline, the bases extracted with ether and the liquid treated with mercuric oxide; *a*-ethylphenylhydrazine is thus converted into the non-volatile tetrazone, and the β-compound into the volatile azophenylethyl, $C_6H_5N{=}NC_2H_5$, which will be subsequently described.

β-Ethylphenylhydrazine may be obtained in the pure state by the action of sodium amalgam on an alcoholic solution of azophenylethyl. It is an oily liquid, which can be distilled without decomposition. It reduces Fehling's solution in the cold, and is oxidized by the oxygen of the air. By the continued action of acetic acid and zinc-dust it is gradually split up into aniline and ethylamine.

Isopropylphenylhydrazine, $C_6H_5.N(C_3H_7)NH_2$, is obtained by the action of isopropylbromide on sodiumphenylhydrazine, and forms a colourless liquid boiling with slight decomposition at 233°. The *isobutyl, isoamyl and allyl* compounds have been prepared in a similar manner. They are all colourless liquids, and boil at 185° (172mm.), 240—245°, 260°, and 177° (109·5mm.) respectively.[2]

a-Ethylenephenylhydrazine, $\begin{array}{l}C_6H_5\diagdown\\NH_2\diagup\end{array}N.CH_2.CH_2.N\begin{array}{l}\diagup C_6H_5\\\diagdown NH_2\end{array}$, is

[1] Fischer, *Ber.* **17**, 2481.

[2] Michaelis and Phillips, *Annalen*, **252**, 278 ; *Ber.* **22**, 2233 ; see also Fischer and Knoevenagel, *Annalen*, **239**, 203.

obtained by the action of ethylene bromide on sodiumphenyl-hydrazine, and forms colourless or slightly yellow prisms or plates which melt at 90°, and are readily soluble in hot alcohol and ether, sparingly in the cold liquids. It gradually reduces Fehling's solution on boiling, and yields a *hydrochloride* $C_{14}H_{18}N_4$. 2HCl, crystallizing in silky needles which decompose on heating without melting. It unites with 2 mol. acetaldehyde, forming a *hydrazone* which melts at 82°.

Diphenylhydrazine, $(C_6H_5)_2N - NH_2$, is obtained by reducing nitrosodiphenylamine with acetic acid and zinc-dust. It is a yellow oily liquid, which on distillation partially decomposes into diphenylamine, ammonia, and resinous products. It is almost insoluble in water, and for this reason scarcely acts on Fehling's solution even on boiling. On the other hand it is readily oxidized by mercuric oxide, ferric chloride, &c., free nitrogen, diphenylamine, and a bluish violet colouring matter which contains nitrogen, being obtained, especially if the mixture be not cooled. When, however, it is shaken with a very dilute neutral and well-cooled solution of ferric chloride, the following compound is obtained as chief product, together with the above-mentioned colouring matter:

Tetraphenyltetrazone, $(C_6H_5)_2N\!=\!N.N\!=\!N(C_6H_5)_2$, separates from carbon disulphide in colourless crystals, melting at 123° without decomposition.

1123 *Dicyanophenylhydrazine*, $C_8H_8N_4$. This substance, together with the isomeric dicyan-*o*-diamidobenzene (p. 269), is obtained by passing cyanogen through an aqueous solution of phenyl-hydrazine. It is scarcely soluble in cold water, readily in dilute hydrochloric acid and alcohol, and melts above 160° with decomposition (E. Fischer). On heating with acetic anhydride, the compound $C_{10}H_8N_4$ is obtained; it is very slightly soluble in water, and crystallizes from alcohol in hard prisms; its formation is explained by the following equation:

$$\begin{array}{l}
\overset{\displaystyle C_6H_5}{\underset{\displaystyle \mathrm{CN}}{\overset{|}{\underset{|}{\overset{\textstyle N-NH_2}{\underset{\textstyle C=NH}{}}}}}}
\ \ + \ O\!\!<\!\!\begin{array}{l}CO.CH_3\\ CO.CH_3\end{array}
\ \ \overset{\displaystyle C_6H_5}{\underset{\displaystyle \mathrm{CN}}{\overset{|}{\underset{|}{\overset{\textstyle N-N}{\underset{\textstyle C=N}{}}}}}}\!\!>\!\!C.CH_3 \ + \ HO.CO.CH_3 \\
\hspace{11cm} + \ H_2O.
\end{array}$$

Its constitution is deduced from the fact that on heating with alcoholic potash it yields the acid $C_6H_5N_3(C_2H_3)C.CO_2H$, which

z 2

is slightly soluble in cold, readily in hot water and alcohol, and is converted on heating into the basic compound $C_6H_5N_3(C_2H_3)CH$ with evolution of carbon dioxide; the latter is a yellowish oily liquid, possessing a peculiar aromatic odour, and boiling at 240°.[1]

It unites with aldehydes in the same manner as phenylhydrazine, and the hydrazones obtained readily undergo condensation forming *triazole* derivatives, *i.e.* compounds containing

the ring $\begin{array}{c} N\!\!-\!\!N \\ | \qquad \diagdown \\ \quad \qquad C. \\ C\!\!-\!\!N \diagup \end{array}$[2]

β-Dicyanodiphenylhydrazine, $[C_6H_5.N(NH_2)C:NH]_2$. If cyanogen be passed through an alcoholic solution of phenylhydrazine, the reaction proceeds in a different manner to the above, 2 molecules of phenylhydrazine combining with a molecule of cyanogen in the following manner:

$$2C_6H_5.NH.NH_2 + (CN)_2 = C_6H_5.N.C(NH).C(NH)\,N.C_6H_5.$$
$$\qquad\qquad\qquad\qquad\qquad\qquad | \qquad\qquad\qquad\qquad | $$
$$\qquad\qquad\qquad\qquad\qquad\quad NH_2 \qquad\qquad\qquad NH_2$$

The compound thus obtained is sparingly soluble in alcohol, and crystallizes from it in white plates which melt at 225°. It forms a crystalline *hydrochloride* which is decomposed by water.[3]

Phenylhydrazine isonitrile, $(C_6H_5.NH.NC)_2$, is obtained by boiling an alcoholic solution of phenylhydrazine with chloroform and caustic potash. It is a white inodorous crystalline compound which melts at 180°. The molecular weight, as found by Raoult's method agrees with the above doubled moleoular weight, from which fact and from its chemical behaviour and want of smell, it follows that it cannot be a true carbamine. It yields a mono-nitro-compound, $C_{14}H_{11}N_4.NO_2$, and a monosulphonic acid, $C_{14}H_{11}N_4.SO_3H$.[4]

[1] Bladin, *Ber.* **18**, 1544.   [2] *Ber.* **22**, 796.
[3] Senf, *J. Pr. Chem.* II. **35**, 513.
[4] Ruhemann and Elliott, *Journ. Chem. Soc.* 1888, **1**, 850.

ACID DERIVATIVES OF PHENYLHYDRAZINE OR PHENYLHYDRAZIDES.

1124 The hydrogen in phenylhydrazine can be replaced by acid radicals, just as in ammonia, the amines, and the amido-bases. Phenylhydrazine also combines directly with carbon dioxide and carbon disulphide, forming compounds analogous to ammonium carbamate, &c.

β-Acetylphenylhydrazide, $C_6H_5.NH.NH.C_2H_3O$, is obtained by boiling equal molecular proportions of phenylhydrazine and acetic anhydride, or by boiling the base with glacial acetic acid.[1] It may also be obtained by warming phenylhydrazine with acetamide.[2] It crystallizes in six-sided prisms, which melt at 128·5° and are volatile with slight decomposition.

a-β-Diacetylphenylhydrazide, $C_6H_5.N(C_2H_3O)NH.C_2H_3O$. When sodiumphenylhydrazine is treated with acetyl chloride, it does not, as would be expected, yield the *a*-acetyl derivative, but forms a mixture of *β*-acetyl- and *a*-*β*-diacetylphenylhydrazine. The latter crystallizes from a mixture of alcohol and benzene in thick colourless tablets or needles, which melt at 107—108°, and are readily soluble in alcohol, chloroform, and hot water, very slightly in ether.[3]

a-Methyl-β-acetylphenylhydrazide, $C_6H_5.N(CH_3)NH.C_2H_3O$, is obtained by the action of acetic anhydride on *a*-methylphenyl-hydrazine. It forms colourless prismatic crystals which melt at 92—93.°

a-β-Dimethyl-β-acetylphenylhydrazide,

$$C_6H_5.N(CH_3).N{<}^{CH_3}_{C_2H_3O,}$$

is prepared from the foregoing compound by treatment with sodium and methyl iodide in xylene solution. It crystallizes from light petroleum in large colourless well-developed crystals which melt at 68°, and can be distilled without decomposition. It does not reduce alkaline solutions of silver and copper salts, nor can the acetyl group be removed without also decomposing the hydrazine group.[4]

---

[1] Fischer, *Annalen*, **190**, 129.
[2] Just, *Ber.* **19**, 1202.
[3] Michaelis and Schmidt, *Annalen*, **252**, 300.
[4] Fischer, *ibid.* **239**, 250.

Oxalyldiphenylhydrazide, $(C_6H_5.N_2H_2)_2C_2O_2$, is obtained by heating phenylhydrazine with ethyl oxalate; it forms a foliated crystalline mass, melting at 277—278°, and distilling at a higher temperature almost undecomposed.

Phenylhydrazine phenylcarbazide, $CO \begin{cases} NH.NH.C_6H_5 \\ O.N_2H_4.C_6H_5, \end{cases}$ may be obtained by passing carbon dioxide through a well-cooled emulsion of 1 part of phenylhydrazine with 10 parts of water. It forms a fine, soft crystalline mass, which deliquesces in the air, giving off carbon dioxide.

Diphenylcarbazide, $CO(NH.NH.C_6H_5)_2$, is formed when one molecule of urethane is heated with two molecules of phenylhydrazine. It is a crystalline substance melting at 151°, sparingly soluble in hot water, readily in hot alcohol, insoluble in ether, its solution being coloured yellow by a little copper sulphate.[1]

Phenylsemicarbazide, $C_6H_5.NH - NH.CO.NH_2$. This compound is formed on mixing solutions of potassium cyanate and phenylhydrazine hydrochloride, by warming carbamide with phenylhydrazine or phenylhydrazine hydrochloride,[2] and by heating urethane with phenylhydrazine, or carbamide with diphenylcarbazide.[3] It crystallizes from hot water in white plates which melt at 172.°

Ethylphenylsemicarbazide, $C_6H_5.NH. - NH.CO.NH(C_2H_5)$, is obtained by the direct combination of ethyl isocyanate with phenylhydrazine, and crystallizes from hot, dilute alcohol in transparent, monosymmetric tablets, melting at 151°. Its aqueous solution gives a bluish-black colouration and a precipitate of the same colour with Fehling's solution, the reaction being very delicate. This colour is rapidly converted into yellow in a closed vessel, but on shaking the liquid in the air the bluish-black colour reappears. On gently warming, complete decomposition takes place with separation of cuprous oxide. Nitrous acid converts it into *nitrosophenylethylsemicarbazide*, the fine yellow needles of which decompose on boiling with alkalis into carbon dioxide, triazobenzene and ethylamine:[4]

$$C_6H_5N \begin{cases} NO \\ NH.CO.NH(C_2H_5) \end{cases} = C_6H_5.N \begin{cases} N \\ \| \\ N \end{cases} + CO_2 + C_6H_5.NH_2.$$

[1] Skinner and Ruhemann, *Journ. Chem. Soc.* 1888, **1**, 551.

[2] Pelizzari, *Gazzetta*, **16**, 202 ; Pinner, *Ber.* **20**, 2359.

[3] Skinner and Ruhemann, *loc. cit.*   [4] Fischer, *Annalen*, **190**, 109.

Diphenylsemicarbazide, $CO\begin{cases} NH.NH.C_6H_5 \\ NH.C_6H_5 \end{cases}$, is obtained by heat-

ing phenylhydrazine with phenylisocyanate [1] or with phenylcarbamide.[2] It crystallizes from alcohol in long needles or
plates which melt at 173°.

Phenylhydrazine phenylthiocarbazide, $CS\begin{cases} NH.NH.C_6H_5 \\ S.N_2H_4.C_6H_5 \end{cases}$,

is formed by the addition of carbon disulphide to an ethereal
solution of phenylhydrazine, and crystallizes in six-sided tablets or prisms, melting at 96—97°. On dissolving it in weak
caustic potash and adding dilute sulphuric acid, *phenylthio-
carbazic acid,* $C_6H_5.N_2H_2.CS.SH$, separates out in lustrous
plates, which decompose on heating, forming carbon disulphide,
sulphuretted hydrogen, ammonia, and the following compound.

Diphenylthiocarbazide, $(C_6H_5.N_2H_2)_2CS$, crystallizes from hot
alcohol in hard, colourless, three-sided prisms melting at 150,°
which form a dark red solution in warm caustic potash. On the
addition of an acid, blue-black flakes separate out, which are
converted by precipitation with alcohol from solution in warm
chloroform into blue-black microscopic needles, which seem to
be isomeric with diphenylthiocarbazide. Its solution, which
has a slight alkaline reaction, dyes wool and silk red; the
solution of the colouring matter in chloroform is distinguished by
its fine dichroism. The colour, which is dark red when seen in
thick layers, is changed on dilution into a vivid green, which is
so intense that the formation of this body, together with the
copper test, can be employed as a delicate reaction for phenylhydrazine.

Diphenylsemithiocarbazide, $C_6H_5.N_2H_2.CS.NH(C_6H_5)$, is formed
when alcoholic solutions of phenyl mustard oil and phenylhydrazine are mixed; it crystallizes from hot alcohol in prisms, melting
at 177°. Its hot alcoholic solution at once reduces mercuric oxide.

---

[1] Kuhn, *Ber.* **17**, 2884.   [2] Skinner and Ruhemann, *loc. cit.*

COMPOUNDS OF PHENYLHYDRAZINE WITH THE CYANHYDRINS.[1]

1125 Phenylhydrazine not only forms compounds with aldehydes and ketones, but also with the cyanhydrins, or nitriles of the lactic acid group, which are obtained by the combination of the former with hydrocyanic acid.

a-Phenylhydrazidopropionitrile, $C_6H_5N_2H_2.CH(CH_3)CN$, is formed by the continued heating of ethidene cyanhydrin (lactonitrile) with phenylhydrazine to 100°. It separates from hot light petroleum in colourless crystals, which are scarcely soluble in water, but dissolve readily in alcohol, and melt at 58°. On heating with alkalis or dilute acids it is split up into hydrocyanic acid and ethidenephenylhydrazine. Cold concentrated hydrochloric acid converts it into

a-Phenylhydrazopropionamide, $C_6H_5N_2H_2.CH(CH_3)CO.NH_2$, which forms hard, readily soluble crystals, melting at 124°. On boiling with caustic soda it forms *a-phenylhydrazidopropionic acid*, $C_6H_5.N_2H_2.CH(CH_3)CO_2H$, which is slightly soluble in cold water, readily in alcohol, and crystallizes from hot dilute alcohol in lustrous white needles, melting at 187°. If the solution of the nitrile in absolute alcohol be saturated with hydrochloric acid, the *ethyl* ether, $C_6H_5.N_2H_2.(CH_3)CO_2.C_2H_5$, is obtained in crystals, readily soluble in alcohol and melting at 116°. When the acid is boiled with tin and hydrochloric acid it splits up into ammonia and *a*-anilidopropionic acid, thus indicating its constitution, which is expressed by the first of the following formulæ, while the second probably represents that of the phenylhydrazidopropionic acid obtained from pyroracemic acid:

$$C_6H_5\diagdown\atop H_2N\diagup \!\!\!\!\! >N\!-\!CH\!\!\!<\!\!{CH_3 \atop CO_2H} \qquad\qquad C_6H_5.NH\!-\!NH.CH\!\!\!<\!\!{CH_3 \atop CO_2H.}$$

a-Phenylhydrazido-isobutyronitrile, $C_6H_5.N_2H_2.C(CH_3)_2CN$, is formed when the acetone cyanhydrin obtained by the action of nascent hydrocyanic acid on acetone dissolved in ether is heated with phenylhydrazine. It is insoluble in water and crystallizes from light petroleum in needles melting at 70°. Cold concen-

---

[1] Reissert, *Ber.* **17**, 1451.

trated hydrochloric acid has no action on this compound ; when it is carefully dissolved in sulphuric acid and gently warmed, it yields the anhydride of a-phenylhydrazidobutyric acid, which forms white crystals, melting at 175°, readily soluble in acids and gradually in boiling caustic soda. On the addition of acids to the alkaline solution, the anhydride again separates out ; it has probably the following constitution:

$$C_6H_5.N—C\begin{smallmatrix}CH_3\\ \\CH_3\end{smallmatrix}$$
$$\underset{HN—CO}{|\qquad|}$$

AZO-DERIVATIVES OF BENZENE.

1126 *Azophenylethyl,* $C_6H_5.N\!=\!N.CH_3$, is prepared by treating β-methylphenylhydrazine with mercuric oxide. It is a yellow, very volatile oil, which has a peculiar smell, boils at 150° with decomposition, and is readily volatile with steam.[1]

Azophenylethyl, $C_6H_5.N_2.C_2H_5$, is formed, as already mentioned, by the oxidation of β-ethylphenylhydrazine (p. 338) and is a yellow oil which has a pungent odour, volatilizes with steam, and distils at 175—185° with slight decomposition.

Azophenylnitro-ethyl, $C_6H_5.N_2.C_2H_2(NO_2)$, is formed by the action of sodium nitro-ethane on diazobenzene nitrate[2] and crystallizes from hot alcohol in small rectangular plates melting at 136—137° with decomposition. If it be triturated with alcoholic potash, the potassium salt, $C_8H_7N_3O_2K_2+4H_2O$, is obtained; it crystallizes in small orange-coloured plates and is decomposed by hydrochloric acid with re-formation of azophenylnitro-ethyl.

Azophenylnitro-propyl, $C_6H_5.N_2.CH(NO_2)CH_2.CH_3$, has been obtained from potassium nitropropane and diazobenzene nitrate ; it crystallizes in broad, deep orange-coloured needles, readily soluble in alkalis.[3]

Azophenylnitro-isopropyl, $C_6H_5.N_2C(NO_2)(CH_3)$, is obtained in a similar manner from nitro-isopropane, and is a non-volatile, golden-yellow oil, which is insoluble in alkalis, and forms no

[1] Tafel, *Ber.* **18,** 1742.   [2] Meyer and Ambühl, *ibid.* **8,** 751, 1073.
[3] Meyer, *ibid.* **9,** 386.

metallic compounds.[1] The constitutions of azophenylnitro-ethyl and of the isopropyl-compound are expressed by the following formulæ :

$$CH_3 CH-N\!\!=\!\!N.C_6H_5 \atop | \atop NO_2$$

$$CH_3\!\!\diagdown \atop CH_3\!\!\diagup \! C-N\!\!=\!\!N.C_6H_5 \atop | \atop NO_2.$$

The fact that the latter forms no metallic compounds is readily explicable (Vol. III., Part I., p. 188), while the behaviour of the former as a dibasic acid is very singular. Several salts have been prepared, all containing two equivalents of a metal. They are probably to be looked upon as basic salts, the formula of the potassium compound being $C_8H_8N_3O_2K + KOH + 3H_2O$. It is, however, remarkable that the normal salts have not been obtained.

1127 A number of compounds have been prepared by the action of diazobenzene chloride on the sodium compounds of substances such as ethyl acetoacetate, and these were supposed to be " mixed " azo-compounds, i.e., compounds in which both fatty and aromatic radicals were combined with the azo-group. The formation of the so-called ethyl azobenzene-acetoacetate was represented by the following equation.[2]

$$CH_3.CO.CHNa \atop | \atop CO_2.C_2H_5 \quad + Cl.N\!\!=\!\!N.C_6H_5 = \quad CH_3.CO.CH.N\!\!=\!\!N.C_6H_5 \atop | \atop CO_2C_2H_5 \atop + NaCl.$$

By the action of a salt of diazobenzene on ethyl sodiomethyl-acetoacetate, the benzene-azo-group was introduced, the acetyl group being split off at the same time, and to the compound formed the constitution

$$CH_3.CH.CO_2C_2H_5 \atop | \atop N\!\!=\!\!N.C_6H_5$$

was given, the corresponding acid obtained by hydrolysis receiving the formula

$$CH_3.CH.COOH\,[3] \atop | \atop N\!\!=\!\!N.C_6H_5.$$

The latter was however found to be identical with the compound obtained by acting with phenylhydrazine on pyroracemic acid,

[1] Meyer, *Ber.* **8**, 1076.
[2] v. Richter and Münzer, *ibid.* **17**, 1928.
[3] Japp and Klingemann, *Journ. Chem. Soc.* **1**, 519.

which was supposed to be $\overset{\text{CH}_3.\text{C}.\text{CO}_2\text{H}}{\underset{\text{N}.\text{NHC}_6\text{H}_5}{\|}}$. It appears therefore
that the compounds obtained by the action of diazobenzene
chloride on the sodium compounds of certain substances
containing the CH_2 group are identical with the compounds
obtained by the action of phenylhydrazine on the corresponding ketones, containing in place of the CH_2 group the
radical CO. In order to ascertain whether the compounds are
azo-compounds or hydrazones, Japp and Klingemann treated
"benzene-azo-acetone" $CH_3.CO.CH_2.N{=}N.C_6H_5$ or $CH_3.CO.$
$CH = N - NH.C_6H_5$ with sodium ethylate and ethyl chloracetate, converted the ethereal salt into the acid, and subjected
the latter to reduction. The compound then obtained was
found to be anilidoacetic acid, and the $-CH_2.COOH$ has therefore combined with the nitrogen atom attached to the phenyl
group. Hence the second formula given for benzene-azo-
acetone must be correct, and it is really a monohydrazone of
pyruvic aldehyde. This is confirmed by the fact that on
treatment with phenylhydrazine it yields the osazone of pyruvic
aldehyde $\quad \overset{\text{CH}_3.\text{C}{=}\text{N}.\text{NH}.\text{C}_6\text{H}_5}{\underset{\text{CH}{=}\text{N}.\text{NH}.\text{C}_6\text{H}_5}{|}}$.

It is probable that many of the other "mixed" azo-compounds
prepared in a similar manner are in reality hydrazones, but the
above cases are the only ones in which any experimental proofs
of this have been brought. The compounds obtained from the
nitro-paraffins might also be hydrazones; against this view
however is the fact that dibromonitroethane $CH_3.CBr_2.NO_2$
does not combine with phenylhydrazine to form azo-phenyl-
nitroethyl, but yields a mixture of very unstable compounds.[1]

Claisen and Beyer have also shown that in certain cases
diazobenzene chloride does act on the sodium compounds with
formation of mixed azo-compounds. Thus with the sodium compound of acetoacetaldehyde, it forms *benzene-azo-acetoacetaldehyde.*
$\overset{\text{CH}_3.\text{CO}.\text{CH}.\text{CHO}}{\underset{\text{N}{=}\text{NC}_6\text{H}_5}{|}}$. This compound crystallizes in dark-red thin
prisms, which melt at 118°, and readily unite with phenyl-
hydrazine forming the *hydrazone.* $\overset{\text{CH}_3.\text{CO}.\text{CH}.\text{CH}:\text{N}-\text{NH}.\text{C}_6\text{H}_5.}{\underset{\text{N}{=}\text{N}.\text{C}_6\text{H}_5}{|}}$

[1] V. Meyer, *Ber.* **21,** 14.

This compound readily loses water forming a pyrazol derivative, which is readily understood if the constitution above given is correct, the hydrazone first passing into the tautomeric compound

$$CH_3.C(OH){:}C.CH{:}N.NHC_6H_5$$
$$\underset{N{:}NC_6H_5}{|}$$ which then loses water as follows:

$$
\begin{array}{ccc}
C_6H_5.N_2.C\text{----}CH & & C_6H_5N_2.C\text{----}CH \\
\parallel\qquad\quad\parallel & & \parallel\qquad\ \parallel \\
CH_3.\dot{C}.OH\quad N & = & CH_3\overset{|}{C}\quad N \qquad + H_2O. \\
\diagup & & \diagdown\diagup \\
\dot{N}H.C_6H_5 & & \dot{N}.C_6H_5
\end{array}
$$

If on the other hand the hydrazone had the constitution

$$CH_3.CO.C.CH{:}N\text{---}NHC_6H_5$$
$$\underset{N.NH.C_6H_5}{\|}$$ no such simple explanation of this
reaction is possible.[1]

1128 *Azobenzene,* $C_6H_5N{=}NC_6H_5$. Mitscherlich obtained this compound by distilling nitrobenzene with alcoholic potash, and named it nitrogen benzide.[2] Zinin then showed that the first product of this reaction is azoxybenzene, $C_{12}H_{10}N_2O$, which decomposes on distillation with formation of azobenzene, aniline and other products.[3] Azobenzene is also formed, together with combustible gases, aniline, &c., by heating nitrobenzene with aniline,[4] or by treating it with acetic acid and an excess of iron.[5] Alexejew has not been able to obtain it by this last method.[6] According to him and Werigo,[7] azobenzene is readily obtained by acting upon an alcoholic solution of nitrobenzene with sodium amalgam, or with zinc dust and caustic soda:[8]

$$
\begin{array}{ccc}
C_6H_5.NO_2 & & C_6H_5.N \\
& + 8H = & \parallel \qquad + 4H_2O. \\
C_6H_5.NO_2 & & C_6H_5.N
\end{array}
$$

It is also formed when nitrosobenzene is heated with aniline nitrate:[9]

$$C_6H_5.NH_2 + C_6H_5.NO = C_6H_5N{=}NC_6H_5 + H_2O.$$

[1] *Ber.* **21**, 1697.　　　　　　　[2] *Pogg. Ann.* **32**, 224.
[3] *J. Pr. Chem.* **36**, 98.　　　　　[4] Merz and Coray, *Ber.* **4**, 981.
[5] Noble, *Annalen,* **98**, 253.　　　[6] *Bull. Soc. Chim.* II. **1**, 234.
[7] *Annalen,* **135**, 176.
[8] Alexejew, *Beilst. Org. Chim.* **3**, 1129.　　[9] Baeyer, *Ber.* **7**, 1638.

It may also be obtained by the oxidation of aniline hydro-chloride with potassium permanganate (Glaser),[1] ammonia and oxalic acid being also formed if the oxidation be carried on in an alkaline solution.[2] Schmidt has obtained it by the action of bleaching powder on a solution of aniline in chloroform.[3]

In order to prepare it, caustic soda and zinc dust are added to an alcoholic solution of nitrobenzene, and, when the reaction is over, nitrogen trioxide is passed in to oxidize the hydrazoben-zene, $C_{12}H_{12}N_2$, which is always formed (Alexejew). Azoben-zene crystallizes from alcohol or light petroleum in yellowish-red plates which melt at 68° and have a faint odour of roses. It boils at 293°, and its vapour has a sp. gr. of 6·5.[4] On the spontaneous evaporation of its solution in benzene, the com-pound $C_{12}H_{10}N_2 + C_6H_6$ separates out in long, thick, yellowish-red prisms, which effloresce in the air.[5] If hydrogen chloride be passed into a solution of azobenzene in carbon disulphide, a yellow crystalline compound, $(C_{12}H_{10}N_2)_2 3HCl$, is formed; the analogous hydrobromic acid compound is a carmine-red crystal-line mass. Bromine gradually added to a solution of azobenzene in chloroform forms the additive-product $C_{12}H_{10}N_2Br_6$, which separates out in large dark-red prisms. All these compounds decompose in the air, leaving a residue of azobenzene.[6]

1129 *Substitution-products of Azobenzene.* These are either formed by direct substitution, or by the oxidation of the corre-sponding hydrazo-compounds.

			Melting-point.
p-Chlorazo-benzene,[7]	$C_{12}H_9ClN_2$	brownish-yellow needles	88—89°
m-Dichlorazo-benzene,[8]	$C_{12}H_8Cl_2N_2$	orange-red needles .	101°
p-Dichlorazo-benzene,[9]		yellow needles with a silky lustre . .	183—184°
o-Bromazo-benzene,[10]	$C_{12}H_9BrN_2$	golden plates . . .	87°
m-Bromazo-benzene,[10]		bronze tablets . . .	53—55°
p-Bromazo-benzene,[10]		orange tablets . . .	82°

[1] *Annalen*, **142**, 365.
[2] Hoogewerff and Dorp, *Ber.* **10**, 1936.
[3] *J. Pr. Chem.* II. **18**, 196.
[4] Hofmann, *Annalen*, **114**, 362.
[5] Schmidt, *Ber.* **5**, 1106.
[6] Werigo, *Annalen*, **165**, 189.
[7] Heumann and Mentha, *Ber.* **19**, 1687.
[8] Laubenheimer, *ibid.* **8**, 1625.
[9] Heumann, *ibid.* **5**, 913.
[10] Janovsky and Erb, *ibid.* **19**, 2156; **20**, 357; *Monatsh.* **8**, 51.

Melting-point.

o-Dibromazo-benzene,[1]	$C_{12}H_8Br_2N_2$	golden scales . . .	187°
m-Dibromazo-benzene,[2]		hair-like needles . .	125·5°
p-Dibromazo-benzene,[3]	$C_{12}H_8Br_2N_2$	large yellow needles.	205°
m-Di-iodazo-benzene,[2]		orange-red needles .	150°
p-Di-iodazo-benzene,[2]	$C_{12}H_8I_2N_2$	red scales	237°

p-Cyanazobenzene, $C_6H_5.N_2.C_6H_4.CN$, is obtained from *p*-amido-azobenzene by Sandmeyer's reaction. It crystallizes from benzene in short brown needles which melt at 100—101°, are insoluble in water, and yield the corresponding acid on hydrolysis.[4]

o-Nitro-azobenzene, $C_6H_5.N_2.C_6H_4NO_2$, is obtained by acting on an acetic acid solution of azobenzene with nitric acid of sp. gr. 1·52. It crystallizes from acetone in orange-yellow microscopic needles, which melt at 129·9°.[5]

p-Nitro-azobenzene, $C_{12}H_9(NO_2)N_2$, is obtained by treating azobenzene with fuming nitric acid.[6] It crystallizes from alcohol in small yellow needles, melting at 137°. It is reduced by ammonium sulphide to amido-azobenzene, but is split up by the continued action of tin and hydrochloric acid into aniline and *p*-diamidobenzene.[7]

p-Dinitro-azobenzene, $C_{12}H_8(NO_2)_2N_2$, is formed by the action of warm nitric acid, and can readily be separated from the preceding compound, since it is almost insoluble in cold acetone. It crystallizes from glacial acetic acid in fine orange-red needles, melting at 206°, and is slightly soluble in alcohol and ether. Ammonium sulphide reduces it in the cold to dinitrohydrazobenzene, on warming to diamidohydrazobenzene, while it is split up by the continued action of tin and hydrochloric acid into two molecules of *p*-diamidobenzene.

m-Dinitro-azobenzene, is a red oil, which is formed together with the para-compound, and was not further investigated by

[1] *Monatsh.* **8**, 55.   [2] Gabriel, *Ber.* **9**, 1407.
[3] Werigo, *Annalen*, **135**, 178.
[4] Mentha and Heumann, *Ber.* **19**, 3023.
[5] Janovsky and Erb, *Monatsh.* **7**, 129 ; **8**, 56.
[6] Laurent and Gerhardt, *Annalen*, **75**, 73.
[7] Janovsky and Erb, *Ber.* **18**, 1133.

Gerhardt and Laurent. It dissolves readily in alcohol, ether, and acetone, and solidifies, after standing for some weeks at 15°, to an orange-red crystalline magma. On heating with tin and hydrochloric acid it splits up into two molecules of m-diamido-benzene (Janovsky and Erb).

a-*Trinitro-azobenzene*, $C_{12}H_7(NO_2)_3N_2$, is obtained by the action of cold fuming nitric acid on p-dinitro-azobenzene. It crystallizes from acetone in chamois-coloured needles with a silky lustre, and from alcohol in long sulphur-yellow needles, fusing at 160° to a red liquid, which detonates violently on further heating.

β-*Trinitro-azobenzene* is prepared by the action of the warm acid, and forms crystals with a silky lustre, less soluble in alcohol than the a-compound, and melting at 180°.

γ-*Trinitro-azobenzene*, $C_6H_5.N_2.C_6H_2(NO_2)_3$. Fischer obtained this compound by the action of yellow mercuric oxide on a warm alcoholic solution of trinitrohydrazobenzene. It crystallizes in fine dark-red prisms, melting at 142°.[1]

Azobenzenenitrolic acid, $C_{12}H_{10}N_3O$, is formed by adding aqueous ammonium sulphide to a hot alcoholic solution of p-nitro-azobenzene until the colour becomes olive-brown and a permanent precipitate is formed. The compound separates out on cooling in yellow flakes, crystallizing from alcohol in small, brownish-red needles. It forms a deep-blue solution in cold alcoholic potash, and in warm caustic potash solution, and is reprecipitated by carbon dioxide. An alcoholic solution of potassium ferricyanide oxidizes it again to p-nitro-azobenzene.

Nitro-azobenzenenitrolic acid, $C_{12}H_9(NO_2)N_3O$, is obtained in a similar manner to the preceding compound, p-dinitro-azo-benzene being employed, and crystallizes from acetone in lustrous, amber-coloured, monosymmetric prisms, which have a blue fluorescence, and melt at 218°. It readily dissolves in aqueous alkalis and baryta water, forming splendid blue solutions.

m-Dinitro-azobenzene gives an oily nitrolic acid, which also dissolves in alkalis with a blue colour.

Janovsky and Erb consider that the following formulæ express the constitutions of these acids:

$$C_6H_5.N{=}NC_6H_4N.OH \qquad NO_2.C_6H_4.N{=}NC_6H_4N.OH$$
$$\mid \qquad\qquad\qquad\qquad\qquad \mid$$
$$C_6H_5\,N{=}NC_6H_4N.OH \qquad NO_2.C_6H_4.N{=}NC_6H_4N.OH.$$

[1] *Annalen*, **190**, 133.

Other nitro-substitution-products have been prepared, for which reference may be made to the original papers.[1]

1130 *Azobenzenesulphonic acid*, $C_{12}H_9N_2SO_3H + 3H_2O$. Griess obtained this compound by dissolving azobenzene in 5 parts of fuming sulphuric acid heated to 130°.[2] It is scarcely soluble in dilute mineral acids, slightly in alcohol and ether, tolerably in cold, and very readily in hot water, crystallizing in large, deep orange-red plates which lose their water of crystallization above 100°, and melt at 127°.[3] It forms salts which crystallize well and are only slightly soluble.

Potassium azobenzenesulphonate, $C_{12}H_9N_2SO_3K$, is formed when a solution of any potassium salt, except the sulphate, is treated with a hot solution of the acid, and separates out on cooling in large, yellowish red plates.

Azobenzenesulphonic acid is, therefore, a very strong acid, decomposing even chlorides and nitrates.

Azobenzenesulphonic chloride, $C_{12}H_9N_2SO_2Cl$, forms orange-yellow warty masses, insoluble in water, and only slowly decomposed by boiling water.[4]

Azobenzenesulphonamide, $C_{12}H_9N_2SO_2.NH_2$, is a powder, slightly soluble in boiling alcohol.

Azobenzenedisulphonic acids, $C_{12}H_8N_2(SO_3H)_2$. By dissolving azobenzene in 5 to 8 parts of fuming sulphuric acid heated to 130°, and then keeping the temperature for two hours at 150—170°, two isomeric acids are formed, and can be separated by means of their potassium salts.[5]

a-Azobenzenedisulphonic acid crystallizes with one molecule of water in deliquescent, red, concentrically grouped needles; the potassium salt, $2C_{12}H_8N_2(SO_3K)_2 + 5H_2O$, forms red prisms, slightly soluble in cold, readily in hot water.

The *chloride*, $C_{12}H_8N_2(SO_2Cl)_2$, crystallizes from ether in small brownish-red needles, melting at 220—222°; and the *amide*, $C_{12}H_8N_2(SO_2.NH_2)_2$, forms yellowish-red needles or small plates which do not melt below 300°.

If the hot concentrated solution of the potassium salt be treated with stannous chloride, *a-hydrazobenzenedisulphonic acid*, $C_{12}H_{10}N_2(SO_3H)_2$, is formed, and crystallizes in transparent, lustrous tablets containing water of crystallization.

[1] Janovsky, *Monatsh.* **7**, 124 ; Nölting, *Ber.* **20**, 2992.
[2] *Annalen*, **154**, 208.   [3] Janovsky, *Monatsh.* **2**, 221.
[4] Skandarow, *Zeitschr. Chem.* **1870**, 643.
[5] Limpricht, *Ber.* **14**, 1356.

β-Azobenzenedisulphonic acid forms a syrup; its potassium salt, $2C_{12}H_8N_2(SO_3K)_2 + 5H_2O$, is exceptionally soluble in water, and crystallizes from alcohol in deep-yellow needles.

The *chloride* crystallizes from ether in slender, red needles, melting at 123—125°, and the *amide* forms yellowish needles, melting at 258°.

No β-hydrazobenzenedisulphonic acid could be obtained by treating the potassium salt with tin and hydrochloric acid.

m-Azobenzenedisulphonic acid. Claus and Moser obtained this compound by the action of sodium amalgam on a solution of sodium *m*-nitrobenzenesulphonate;[1] it may be more readily obtained by using zinc dust and caustic potash.[2] The free acid crystallizes in pale-yellow, deliquescent, monosymmetric prisms, containing three molecules of water; it forms yellow to dark-red salts.

The potassium salt crystallizes in yellowish-red needles, which are sometimes anhydrous and sometimes contain water.

The *chloride* separates from ether in ruby-red needles, and melts at 166°, while the *amide* crystallizes from alcohol in yellow needles, melting at 295°.

On reduction the acid yields *m-hydrazobenzenedisulphonic acid*, $C_{12}H_{10}N_2(SO_3H)_2$, crystallizing in fine, colourless, monosymmetric prisms, scarcely soluble in alcohol and ether, slightly in cold, and somewhat more readily in warm water.

p-Azobenzenedisulphonic acid is formed by the oxidation of sulphanilic acid with potassium permanganate in the cold. The potassium salt, $2C_{12}H_8N_2(SO_3K)_2 + 5H_2O$, is thus obtained in red crystals, slightly soluble in water.[3] The amide crystallizes from hot water in yellow tablets, melting at 176° (Marenholtz and Gilbert).

m-Azobenzenedithiodisulphonic acid, $C_{12}H_8N_2(SO_2.SH)_2$, is obtained by the action of barium hydrosulphide on the chloride of *m*-azobenzenedisulphonic acid. It forms a pale yellow precipitate which becomes brown on heating, melts at 91—93°, and is almost insoluble in alcohol and water. Its barium salt is oxidized by potassium permanganate to *m*-azobenzenedisulphonic acid.[4]

Disazobenzene, $C_6H_5.N_2.C_6H_4.N_2.C_6H_5$, is obtained from the

---

[1] *Ber.* **11**, 762.
[2] Marenholz and Gilbert, *Annalen*, **202**, 331.
[3] Laar, *J. Pr. Chem.* II. **20** 264.
[4] Bauer, *Annalen*, **229**, 358

corresponding amidodisazobenzene (p. 360) by diazotizing and boiling with alcohol. It forms red needles resembling azobenzene, which melt at 98°, and dissolve in concentrated sulphuric acid with a yellow colour.[1]

1131 *Azoxybenzene*, $C_{12}H_{10}N_2O$, is, as already mentioned, the first product of the action of caustic potash or sodium amalgam, in presence of alcohol, on nitrobenzene :

$$C_6H_5N\!\!<\!\!\begin{matrix}O\\O\end{matrix}\!\!> \\ C_6H_5N\!\!<\!\!\begin{matrix}O\\O\end{matrix}\!\!> \quad + 6H = \quad \begin{matrix}C_6H_5.N\\ |\\ C_6H_5.N\end{matrix}\!\!>\!\!O + 3H_2O.$$

It is also obtained by oxidizing azobenzene, dissolved in acetic acid, with chromic acid,[2] as well as, together with azobenzene, by the action of potassium permanganate on aniline hydrochloride (Glaser).

Methyl alcohol is used in its preparation instead of ethyl alcohol, as the oxidation products of the latter seem to take part in the reaction and form black masses, which were found by Mitscherlich and Zinin to consist of the salts of coloured acids.

Ten parts of sodium are dissolved in 250 parts of methyl alcohol, the solution mixed with 30 parts of pure nitrobenzene, and the whole allowed to boil gently for 5 to 6 hours, in an apparatus connected with an inverted condenser. The liquid usually becomes coloured brownish-red, but, if pure nitrobenzene has been used, remains perfectly clear, the reaction proceeding according to the following equation :

$$4C_6H_5NO_2 + 3CH_3ONa = 2C_{12}H_{10}N_2O + 3HCO_2Na + 3H_2O.$$

The alcohol is then distilled off, the residue of sodium formate and azoxybenzene extracted with water, and the azoxybenzene allowed to crystallize out; the yield amounts to 90 to 92 per cent. of the theoretical.[3]

Alexejew obtained it by the action of sodium amalgam on an alcoholic solution of nitrobenzene, kept acid by acetic acid. According to Moltschanowsky, 87 per cent. of the theoretical yield is obtained.[4] It is also formed from nitrobenzene by the

---

[1] Nietzki and Diesterweg, *Ber.* **21**, 2143.   [2] Petriew, *ibid.* **6**, 557.
[3] Klinger, *ibid.* **15**, 865.   See also Moltschanowsky, *ibid.* **16**, 81.
[4] *Ibid.* **15**, 1575.

action of zinc dust and boiling aqueous solutions of salts, such as calcium, potassium, and sodium chlorides.[1]

Azoxybenzene is insoluble in water, but readily soluble in alcohol, and crystallizes in long, yellow, rhombic needles, or, on the gradual evaporation of its ethereal solution, in prisms an inch in length. It melts at 36° to a yellow, strongly refractive liquid, which solidifies on cooling to a radiating mass (Zinin). If small quantities are carefully heated, it volatilizes undecomposed; when larger quantities are distilled, a portion decomposes into azobenzene and aniline, carbonized products being formed. By the addition of three parts of iron filings, the yield of azobenzene is increased to 72·5 per cent., aniline and decomposition products being also obtained. This behaviour adapts it for the preparation of azobenzene.[2] While azoxybenzene in alkaline solution is reduced by sodium amalgam to azobenzene and hydrazobenzene, the greater portion of it is converted by stannous chloride, in an acid alcoholic solution, into aniline (Schmidt and Schultz). On heating with concentrated sulphuric acid it is converted into the isomeric hydroxyazobenzene;[3]

$$O\left\langle\begin{array}{c}NC_6H_5\\|\\NC_6H_5\end{array}\right. = \begin{array}{c}NC_6H_5\\||\\NC_6H_4.OH.\end{array}$$

Substitution-products of Azoxybenzene. The halogen compounds are obtained from the substitution-products of nitrobenzene, just as azoxybenzene is obtained from nitrobenzene. They are converted by reduction into substituted hydrazobenzenes, and these again by oxidation into substituted azobenzenes. The nitro-compounds are formed by the nitration of azoxybenzene.

			Melting-point.
m-Dichlorazoxy-benzene,[4]	$C_{12}H_8Cl_2N_2O$	flat long ochre-coloured needles	97°
p-Dichlorazoxy-benzene,[5]		light-yellow needles . .	155°
m-Dibromazoxy-benzene,[6]	$C_{12}H_8Br_2N_2O$	light yellow broad prisms	111—111·5°
p-Dibromazoxy-benzene,[7]		small yellow plates . . .	175°

[1] v. Dechener, D. R. P. 43230 ; *Ber.* **21**, 677c.

[2] Schmidt and Schultz, *Annalen*, **207**, 329.

[3] Wallach and Belli, *Ber.* **13**, 525 ; Wilsing, *Annalen*, **215**, 218.

[4] Laubenheimer, *Ber.* **8**, 1623.   [5] *Ibid.*

[6] Gabriel, *ibid.* **9**, 1405.

[7] Hofmann and Geyger, *ibid.* **5**, 919 ; Werigo, *Annalen*, **165**, 198.

			Melting-point.
m-Di-iodazoxy-benzene,[1]		flat yellow needles	—
p-Di-iodazoxy-benzene,[1]	$C_{12}H_8I_2N_2O$	light-yellow plates	199—199·5°
o-Nitro-azoxy-benzene,[2]		yellow prisms or needles	49°
p-Nitro-azoxy-benzene,[2]	$C_{12}H_9(NO_2)N_2O$	light-yellow hair-like needles	153°
γ-Nitro-azoxy-benzene,[3]		red plates	127°
Azoxybenzene-*o*-sulphonic acid,[4]		reddish-brown powder	—
Azoxybenzene-*m*-sulphonic acid,[4]	$C_{12}H_9(SO_3H)N_2O$	reddish-brown deliquescent tablets	60—70°
Azoxybenzene-*p*-sulphonic acid,[4]		citron-yellow crystals	under 100°

m-Azoxybenzenedisulphonic acid, $C_{12}H_8(SO_3H)_2N_2O$, is formed by heating *m*-nitrobenzenesulphonic acid with alcoholic potash. It is readily soluble in water and alcohol, and forms microscopic yellow needles melting at 125°. Sodium amalgam reduces it to *m*-azobenzenedisulphonic acid, and stannous chloride to the hydrazo-compound.[5]

1132 *Hydrazobenzene*, $C_{12}H_{12}N_2$. Hofmann obtained this compound in 1863 by the reduction of azobenzene with ammonium sulphide:[6]

$$\begin{array}{ccc} C_6H_5.N & & C_6H_5.NH \\ \| & + H_2S = & | \\ C_6H_5.N & & C_6H_5 NH \end{array} + S.$$

In order to prepare it, an alcoholic solution of azobenzene or azoxybenzene is boiled with zinc-dust until it has become colourless, then filtered, and precipitated with water (Alexejew). It is slightly soluble in water, readily in alcohol, and crystallizes in colourless tablets, smelling like camphor, and melting at 131°. On distillation it decomposes into aniline and azobenzene:

$$2C_{12}H_{12}N_2 = C_{12}H_{10}N_2 + 2C_6H_7N.$$

[1] Gabriel.
[2] Zinin, *Annalen*, **114**, 218.
[3] Janovsky and Erb, *Ber.* **20**, 361.
[4] Limpricht, *ibid.* **18**, 1420.
[5] Brunnermann, *Annalen*, **202**, 340.
[6] *Jahresb.* **1863**, 424.

On oxidation it is readily reconverted into azobenzene; in the damp state, therefore, it becomes coloured in the air, and its alcoholic solution readily absorbs oxygen.

Hydrochloric or sulphuric acid converts it into benzidine:

$$\begin{array}{ccc} C_6H_5.NH & & C_6H_4.NH_2 \\ | & = & | \\ C_6H_5.NH & & C_6H_4.NH_2 \end{array}$$

The product thus obtained is a mixture of two isomeric diamido-diphenylenes, and is also formed by passing sulphur dioxide through an alcoholic solution of azobenzene:

$$\begin{array}{cc} C_6H_5 N \\ || \quad + 2H_2O + SO_2 = \\ C_6H_5.N \end{array} \quad \begin{array}{c} C_6H_4.NH_2 \\ | \\ C_6H_4 NH_2 \end{array} + H_2SO_4.$$

When azobenzene is heated with a considerable excess of acid ammonium sulphite and alcohol, the ammonium salt of *amido-diphenylsulphamic acid* is obtained; this is converted by sulphuric acid into benzidine, and has, therefore, the following constitution:[1]

$$\begin{array}{c} C_6H_4.NH_2 \\ | \\ C_6H_4.NH(SO_3H). \end{array}$$

These compounds will be described under diphenyl.

Acetylhydrazobenzene, $C_6H_5.NH.N(C_2H_3O).C_6H_5$, is obtained from hydrazobenzene and acetic anhydride in the cold.[2] It crystallizes from hot alcohol in needles which melt at 159° and are insoluble in water and alkalis.

Diacetylhydrazobenzene, $C_{12}H_{10}(NC_2H_3O)_2$, is formed by boiling hydrazobenzene with acetic anhydride, and forms large, thick, yellow, rhombic crystals, slightly soluble in cold, more readily in hot water, and in alcohol (Schmidt and Schultz).

Substitution-products of hydrazobenzene are obtained by the reduction of the substituted azobenzenes or azoxybenzenes, as well as from the substituted nitrobenzenes. Monosubstitution-products have not yet been obtained.

			Melting-point.
m-Dichlorohy-drazobenzene,[3]		crystals resembling gypsum . . .	94°
p-Dichlorohy-drazobenzene,[4]	$C_{12}H_{10}Cl_2N_2$	crystals	122°

[1] Spiegel, *Ber.* **18**, 1479.   [2] Stern, *ibid.* **17**, 380.
[3] Laubenheimer, *ibid.* **8**, 1624.
[4] Hofmann, *ibid.* **5**, 918; Calm and Heumann, *ibid.* **13**, 1181.

			Melting-point.
m-Dibromohy- drazobenzene,[1] p-Dibromohy- drazobenzene,[2]	$C_{12}H_{10}Br_2N_2$	short thick prisms	107—109°
		fine needles . . .	130°
m-Di-iodohy- drazobenzene,[3] p-Di-iodohy- drazobenzene,[3]	$C_{12}H_{10}I_2N_2$	spherical aggre- gates of crystals.	89—90°
		flat needles . . .	—
Dinitrohydrazo- benzene,[4]	$C_{12}H_{10}(NO_2)_2N_2$	yellow needles . .	220°

Trinitrohydrazobenzene, $C_{12}H_9(NO_2)_3N_2$. Fischer obtained this compound by the action of picryl chloride on phenyl-hydrazine :

$$C_6H_5NH - NH_2 + C_6H_2Cl(NO_2)_3 = C_6H_5.NH - NH.C_6H_2(NO_2)_3 + HCl.$$

It crystallizes from glacial acetic acid or acetone in short, dark-red prisms, which melt at 181° with evolution of gas, and deflagrate when more strongly heated. Mercuric oxide converts it, in alcoholic solution, into γ-trinitro-azobenzene (p. 351).[5]

a-Diamidohydrazobenzene, $C_{12}H_{10}(NH_2)_2N_2$. Gerhardt and Laurent prepared this substance by boiling p-dinitro-azobenzene with alcoholic ammonium sulphide ; they gave to it the formula $C_{12}H_8(NH_2)_2N_2$, and named it *diphenine*. Julie Lermontow subsequently showed that when dinitro-azobenzene is treated with ammonium sulphide in the cold, dinitrohydrazo-benzene is formed, and that this is converted into diamidohydrazo-benzene by warming with the sulphide. It is slightly soluble in cold, more readily in hot water, and forms a yellow crystalline powder, melting at 145°. It is converted into p-diamido-benzene by heating to 100° in a sealed tube with ammonium sulphide, and into quinone by oxidation with manganese dioxide and sulphuric acid. Its *hydrochloride*, $C_{12}H_{10}N_2(NH_3Cl)_2$, crystallizes in red scales, slightly soluble in water.

β-Diamidohydrazobenzene. Haarhaus obtained this compound by the action of sodium amalgam on an alcoholic solution of m-nitraniline, and named it hydrazo-aniline.[6] It may be

[1] Gabriel, Ber. **9**, 1405.
[2] Werigo, *Annalen*, **165**, 192 : Calm and Heumann.
[3] Gabriel.   [4] Julie Lermontow, *Ber.* **5**, 231.
[5] *Annalen*, **190**, 132.   [6] *Ibid.* **135**, 164.

readily prepared by the reduction of *m*-nitraniline with zinc dust and alcoholic potash,[1] and is slightly soluble in water, readily in alcohol, and crystallizes in golden-yellow needles, melting slightly above 140°. On heating to 100° with ammonium sulphide it is converted into *p*-diamidobenzene (Lermontow). Its *hydrochloride* crystallizes in small golden-yellow plates.

HYDROXY- AND AMIDO-DERIVATIVES OF AZO-BENZENE.

THE AZO-DYES.

1133 All azo-compounds are coloured, but only those which contain an amido- or hydroxyl-group are dyes. In 1876 the only azo-dyes in use were amido-azobenzene, or aniline yellow, and triamido·azobenzene, or phenylene brown. Since that time however, a large number of azo-dyes have been discovered through the labours of Caro, Griess, Poirrier, Witt, and others, and, owing to their simple formation from the numerous amido-compounds and phenols, they have come into very general use.

The simplest compound of this class is aniline yellow or amido-azobenzene, obtained by the action of aniline on diazo-benzene chloride; diazo-amidobenzene is first formed, and is converted, in presence of aniline, into amido-azobenzene :

$$C_6H_5N{=}N.NH.C_6H_5 = C_6H_5N{=}N.C_6H_4.NH_2.$$

When potassium phenate is added to diazobenzene nitrate, hydroxyazobenzene is directly obtained :

$$C_6H_5N{=}N.NO_3 + C_6H_5.OK = C_6H_5N{=}NC_6H_4.OH + KNO_3.$$

The formation of an intermediate diazohydroxy- or diazo-amido-compound has not been observed in the preparation of this and other azo-dyes. Among the diamido-compounds only those of the meta-series form colouring matters. *m*-Diamido-benzene combines with diazobenzene chloride, diamido-azobenzene or chrysoïdine being formed :

$$C_6H_5N{=}NCl + C_6H_4(NH_2)_2 = C_6H_5N{=}NC_6H_3(NH_2)_2 + HCl.$$

---

[1] Limpricht, *Ber.* **18**, 1403.

m-Dihydroxybenzene or resorcinol yields m-dihydroxyazobenzene, $C_6H_5N{=}N.C_6H_3(OH)_2$.

According to Witt's proposal, the dyes obtained from m-diamido-compounds are called *chrysoïdines*, and those obtained from m-dihydroxy-compounds, *chrysoïns*.[1] The colour of these compounds is darker the higher the molecular weight; thus amido-azobenzene is yellow, diamido-azobenzene orange, and triamido-azobenzene brown.

Experience has shown that in the formation of azo-colours the azo-group always enters the ring in the para-position to the hydroxyl or amido-group, if this is not previously occupied. In the latter case the azo-group goes into the ortho-position. Condensations in the meta-position have not yet been observed.

If p-diazobenzenesulphonic acid be used instead of a diazo-salt, the sulphonic acids of the azo-dyes are obtained; these are called *tropaeolins*, because the shades of colour which they produce resemble those of the flowers of *Tropacolum majus*. They are distinguished in trade according to their shades as *Tropaeolin Y* (yellow), *Tropaeolin O* (orange), and the deeper redder ones as *OO,OOO*, etc.[2]

These sulphonic acids may also be obtained by the action of sulphuric acid on the azo-dyes.

When nitrogen trioxide is passed into an alcoholic solution of amido-azobenzene hydrochloride, a diazo-compound

$$C_6H_5N{=}NC_6H_4N{=}NCl,$$

is formed, and this is converted by aniline into *amidodisazobenzene*, $C_6H_5N{=}NC_6H_4N{=}NC_6H_4.NH_2$, which is a colouring matter. Similar reactions with other amido- and hydroxy-compounds, as well as with their sulphonic acids, lead to the building up of very complicated azo-dyes.[3]

Among these numerous compounds those derived from naphthalene are of special importance, but only those of the benzene group will be described at present.

At first only the *water soluble* sulphonic acids were commercially employed as dyes, since the application of the *spirit soluble* colouring matters was prevented by practical difficulties; these have now been overcome, it having been found that the azo-compounds combine with acid sodium or ammonium sulphite to

---

[1] *Ber.* **10**, 654. [2] Witt. *ibid.* **12**, 258.
[3] Caro and Schraube, *ibid.* **10**, 2230.

form soluble compounds, decomposed by steaming or by the action of alkalis, and that the dye can thus be fixed directly on the fibre.[1]

The sulphonic acids give analogous compounds, which form yellow solutions when the azo-dye has a red colour, and red solutions when they are derived from blue colouring matters. All these sulphite-compounds are characterized by a great power of crystallization, and their crystals possess in greater or less degree a dark colour and metallic lustre; as already mentioned, they readily decompose with formation of the original colouring matter, and therefore do not correspond to the *amidodiphenylsulphamic acids* (p. 357), but are addition-products of the general formula:[2]

$$X—N—H$$
$$Y—N—SO_3Na.$$

1134 *Hydroxyazobenzene*, $C_6H_5.N_2.C_6H_4.OH$. Griess obtained this substance, together with phenoldisazobenzene (p. 363), by the action of barium carbonate on diazobenzene nitrate, and named it phenoldiazobenzene.[3] He then found that the same compound is formed when azobenzenesulphonic acid is fused with caustic potash,[4] while Kekulé and Hidegh prepared it by the action of potassium phenate on diazobenzene nitrate.[5] It may also be obtained by heating the isomeric azoxybenzene with sulphuric acid (Wallach and Belli),[6] and by warming diazoamidobenzene with phenol at 100°, aniline being also formed.[7]

$$C_6H_5.N{=}N.NH.C_6H_5 + C_6H_5.OH = C_6H_5.N{=}N.C_6H_4.OH$$
$$+ C_6H_5.NH_2.$$

To prepare it, 30 grms. of potassium nitrite dissolved in 4 litres of water are added to 20 grms. of phenol and 20 grms. of aniline nitrate dissolved in 2 litres of water, the mixture allowed to stand for two hours, and the precipitate filtered off. On treatment with ammonia a resinous substance remains undissolved and is separated by filtration, the filtrate being then precipitated with hydrochloric acid.[8]

[1] Caro and Schraube, *Ber.* **17**, 452 R. ; **18**, 10 R.
[2] Spiegel, *ibid.* **18**, 1479.   [3] *Annalen*, **137**, 84.
[4] *Ibid.* **154**, 208.   [5] *Ber.* **3**, 324.
[6] *Ibid.* **13**, 525.
[7] Heumann and Oeconomides, *ibid.* **20**, 372, 904.
[8] Mazzara, *Jahresb.* **1879,** 465.

Hydroxyazobenzene is slightly soluble in water, readily in alcohol, and crystallizes in yellowish-red rhombic prisms, melting at 152—154°. It readily dissolves in alkalis and forms metallic salts. The silver compound is a yellow precipitate which detonates at 100°.

a-Hydroxyazobenzenesulphonic acid, $C_6H_4(SO_3H)N_2C_6H_4.OH$, is formed when hydroxyazobenzene,[1] or more simply azoxy-benzene,[2] is heated to 100—110° with 5 parts of fuming sulphuric acid until a small portion is found to be completely soluble in water. On dilution with water the monosulphonic acid separates out after twenty-four hours, while the di- tri- and tetra-sulphonic acids remain in solution.

It forms small, reddish plates with a metallic lustre, which are readily soluble in water and alcohol. The potassium salt $C_6H_4(SO_3K)N_2C_6H_4.OH + H_2O$. crystallizes in small, yellowish-red, lustrous plates or flat needles, tolerably soluble in cold, readily in hot water. When it is reduced with stannous chloride no aniline is formed, whence it follows that the sulpho-group and the hydroxyl are not situated in the same benzene nucleus.

β-Hydroxyazobenzenesulphonic acid, $C_6H_4(SO_3H)N_2C_6H_4.OH$, is obtained by dissolving 1 part of phenol in 10 parts of a 10 per cent. solution of caustic potash, and adding the corresponding amount of diazosulphanilic acid. The yellowish-red solution is saturated after some time with acetic acid, the potassium salt, $C_6H_4(SO_3K)N_2C_6H_4.OH$, being thus precipitated in small, an-hydrous, lustrous plates,[3] which are still less soluble than those of the isomeric compound (Wilsing). The acid, which is obtained from the salt, crystallizes in yellowish-red prisms, having a blue surface lustre, which are readily soluble in pure water, but only slightly in acidulated water. It is resolved by the action of tin and hydrochloric acid into sulphanilic acid and *p*-amidophenol.

The sodium salt, $C_6H_4(SO_3Na)N_2C_6H_4.OH$, occurs in commerce under the name of *Tropaeolin Y*, as a pale-yellow powder, which has not much value as a dye.

γ-Hydroxyazobenzenesulphonic acid, $C_6H_4(SO_3H)N_2C_6H_4.OH$. Griess obtained this substance by the action of *m*-diazoben-zenesulphonic acid on an alkaline solution of phenol. It is readily soluble in water and crystallizes in long, narrow, five-sided plates having a violet surface lustre. Tin and hydrochloric acid decompose it into *m*-amidobenzenesulphonic acid and

[1] Tschirwinsky, *Ber.* **6**, 560.   [2] Wilsing, *Annalen*, **215**, 228.

[3] Griess, *Ber.* **11** 2191.

p-amidophenol. The *potassium* salt, $C_{12}H_9N_2SO_3K$, crystallizes in long needles.

δ-*Hydroxyazobenzenesulphonic acid*, $C_6H_5N_2C_6H_3(SO_3H)OH$, is obtained by the action of diazobenzene nitrate on an alkaline solution of phenol-*o*-sulphonic acid, and is precipitated by strong hydrochloric acid from a concentrated aqueous solution, in small, lustrous yellow plates. On the gradual evaporation of its solution it is also deposited in large, cherry-red, rhombic tablets or prisms. On reduction it yields aniline and amidophenolsulphonic acid. The *potassium* salt, $C_{12}H_9N_2SO_3K$, forms small, narrow, lustrous plates or needles (Griess).

Phenoldisazobenzene, $C_{18}H_{14}N_4O$. This compound was obtained by Griess, together with hydroxyazobenzene, by the action of barium carbonate on diazobenzene nitrate. It is also formed by treating diazobenzene chloride with caustic soda, or by adding diazobenzene nitrate to an alkaline solution of hydroxyazobenzene,[1] and crystallizes from hot alcohol in brownish-red needles or small plates, which melt at 131° and have a metallic lustre. It is split up by the action of tin and hydrochloric acid into aniline and diamidophenol,[2] $C_6H_3(NH_2)_2OH$ (4 : 2 : 1). It must therefore have the following constitution:

$$HO.C_6H_3 \Big\langle {}^{N=NC_6H_5}_{N=NC_6H_5.}$$

Caro and Schraube[3] obtained an isomeric compound by the action of potassium phenate on the diazo-compound of amidoazobenzene, concerning which they only state that it is split up by nascent hydrogen into aniline, p-diamidobenzene, and p-amidophenol:

$$C_6H_5N=NC_6H_4N=NC_6H_4OH.$$

1135 *Dihydroxyazobenzenes*, $C_{12}H_8N_2(OH)_2$. According to theory twelve isomerides can exist, six containing the hydroxyls in the same benzene nucleus, and six containing one hydroxyl in each nucleus. The latter, termed azophenols, are obtained by fusing the nitrophenols with caustic potash.[4]

[1] Griess, *Ber.* **9**, 627.
[2] Percy Frankland, *Journ. Chem. Soc.* 1880, **1**, 751.
[3] *Ber.* **10**, 2230.
[4] Benedikt and Weselsky, *Annalen*, **196**, 339.

o-Azophenol, $C_6H_4(OH)N_2C_6H_4.OH$, crystallizes in small golden, lustrous plates, melting at 171°, and subliming without decomposition. It is insoluble in water, slightly soluble in alcohol and readily in ether; it forms a yellow solution in the alkalis.

p-Azophenol, is formed by adding a solution of potassium phenate to *p*-diazophenol nitrate, as well as by the action of caustic potash on quinonoxime.[1] It is slightly soluble in water, readily in alcohol, and forms light-brown asymmetric prisms, melting with decomposition at about 204°.

1136 *m-Dihydroxyazobenzene, Phenylazoresorcinol,* or *Resorcinol azobenzene,* $C_6H_5.N_2.C_6H_3(OH)_2$. By the action of diazobenzene nitrate on an alkaline solution of resorcinol a mixture of two phenylazoresorcinols is obtained,[2] which have the constitution $OH.N_2.OH = 1 : 2 : 3$, and $OH.OH.N_2 = 1 : 3 : 4$, respectively.

The first compound is only formed in small quantity. Its *ethyl* ether, $C_6H_5.N_2.C_6H_3(OH)(OC_2H_5)$, forms lustrous carmine-red plates, melting at 150°, and the *diethyl* ether, $C_6H_5.N_2.C_6H_3(OC_2H_5)_2$ crystallizes in ruby-red tablets which melt at 90°.

The second compound, generally known as *p-phenylazoresorcinol,* is obtained, in addition to the manner above described, by adding diazoamidobenzene to fused resorcinol.[3] It is best prepared by diazotizing 1 mol. aniline hydrochloride, adding 1 mol. resorcinol, pouring the mixture with constant stirring into an excess of dilute caustic soda solution, and precipitating the solution with hydrochloric acid.[4] It crystallizes from concentrated alcoholic solution on addition of water in short, slender, dark-red needles, which melt at 170°; under certain conditions, however, it separates in long reddish-yellow needles, which melt at 161°, both of which modifications may be readily converted one into the other.[5] It is insoluble in water, but dissolves readily in alcohol, ether, and benzene, and also in alkalis, forming a yellowish-red solution. On treatment with tin and hydrochloric acid it is resolved into aniline and amidoresorcinol. Its *methyl* and *ethyl* ethers have been prepared.

---

[1] Jaeger, *Ber.* **8,** 1499.

[2] Baeyer and Jäger, *ibid.* **9,** 151 ; Typke, *ibid.* **10,** 1576.

[3] Heumann and Oeconomides, *ibid.* **20,** 905 ; Fischer and Wimmer, *ibid.* **20,** 1578.

[4] Kostanecki, *ibid.* **21,** 3119.

[5] Will and Pukall, *Ber.* **20,** 1121 ; Wallach and Fischer, *ibid.* **15,** 2816.

Dihydroxyazobenzenesulphonic acid, $C_6H_4(SO_3H)N_2C_6H_3(OH)_2$, is formed by the action of fuming sulphuric acid on the preceding compound, as well as by that of p-diazobenzene-sulphonic acid on an alkaline solution of resorcinol.[1] It is slightly soluble in cold, somewhat more readily in hot water and in alcohol, and crystallizes in small, ruby-red, pointed rhombic plates with a steel-blue surface-lustre. It is split up by the action of tin and hydrochloric acid into sulphanilic acid and amidoresorcinol.

The acid potassium salt, $C_6H_4(SO_3K)N_2C_6H_3(OH)_2$, crystallizes in small, yellowish-red, rhombic plates; both it and the corresponding sodium salt occur in commerce under the names *Tropaeolin O* or *R*, *Chrysoïn*, &c.; it is chiefly used, in combination with other dyes, for olive-greens, chestnut-browns, &c.

Nitrosophenylazoresorcinol, $C_6H_5.N_2.C_6H_2.O.NOH.OH$,[$N_2$: O : NOH : OH $= 1:2:3:4$, or $1:4:3:2$], is obtained by the action of nitrous acid on p-phenylazoresorcinol.[2] It crystallizes from alcohol in lustrous brownish-red plates, which explode at **168°**. It is converted by tin and hydrochloric acid into aniline and *adj*-diamidoresorcinol.

Phenylazonitrosoresorcinol, $C_6H_5.N_2.C_6H_2.O.NOH.OH$, is prepared by adding a solution of diazobenzene chloride to a solution of nitrosoresorcinol in sodium carbonate cooled to 0°. It crystallizes from alcohol in golden-yellow plates, which melt at 225° with decomposition, and are only sparingly soluble in boiling alcohol.[3] Both these compounds dye fabrics mordanted with iron salts an intense olive-green, the second compound giving a lighter shade.

adj-Phenyldisazoresorcinol, $(C_6H_5N_2)_2C_6H_2(OH)_2$,[OH:$N_2$:OH: $N_2 = 1:2:3:4$], is obtained in small quantity in the preparation of phenylazoresorcinol. It is best prepared by pouring a solution of a mixture of 2 mol. diazobenzene chloride and 1 mol. resorcinol into a solution of sodium acetate or carbonate;[4] it is only slightly soluble in alcohol, and crystallizes in broad, red needles, which melt at 220—222°, and form a brownish-yellow solution in alkalis or sulphuric acid. On reduction with tin and hydrochloric acid it yields *adj*-diamido-resorcinol and aniline.[5]

[1] Griess, *Ber.* **11**, 2195.　　　　[2] Kostanecki, *ibid.* **21**, 3109.
[3] *Ibid.* p. 3112.　　　　　　　　　[4] *Ibid.* p. 3117.
[5] Liebermann and Kostanecki, *ibid.* **17**, 880 ; Kostanecki, *ibid.* **20**, 3136.

a- or *s-Phenyldisazoresorcinol*, $(C_6H_5.N_2)_2C_6H_2(OH)_2[OH:OH: N_2:N_2 = 1:3:4:6]$, is formed, together with the β-compound, by the action of diazobenzene chloride on an alkaline solution of phenylazoresorcinol:[1]

$$C_6H_5N{=}C_6H_3(OH)_2 + C_6H_5N{=}NCl$$
$$= \begin{matrix} C_6H_5.N{=}N \\ \\ C_6H_5.N{=}N \end{matrix}{>}C_6H_2(OH)_2 + HCl.$$

It is most readily prepared in a similar manner to the adjacent-compound, except that the mixture is poured into an excess of dilute caustic soda. It crystallizes in brownish-red needles, which melt at 213—215°, and are sparingly soluble in alcohol and ether, readily in chloroform and alkalis. On reduction with tin and hydrochloric acid it yields *s*-diamido-resorcinol.[2]

β-Phenyldisazoresorcinol is a red crystalline powder, insoluble in caustic soda; it dissolves in alcoholic potash with a brownish red, and in sulphuric acid with an indigo-blue colour.

Benzenedisazobenzeneresorcinol, or *Azo-azobenzeneresorcinol*, $C_6H_5N{=}NC_6H_4N{=}NC_6H_2(OH)_2$, is obtained in two isomeric modifications by dissolving equal molecules of amido-azobenzene and resorcinol in alcohol, adding sufficient acetic acid, and treating the well-cooled liquid with an aqueous solution of one molecule of sodium nitrite, the mixture being well agitated. The *a*-compound is a brownish-red crystalline powder, soluble in alcohol, and melting at 183—184°. It dissolves in aqueous alkalis and in concentrated sulphuric acid with a carmine-red colour.

The β-compound is also a brownish-red powder, which takes a metallic lustre when rubbed, melts at 215°, and is insoluble in alcohol and aqueous alkalis; it dissolves in alcoholic potash with a bluish violet, and in sulphuric acid with a pure deep-blue colour (Wallach and Fischer).

1137 *Amido-azobenzene*, $C_6H_5N{=}NC_6H_4.NH_2$. In the year 1863 the firm of Simpson, Maule, and Nicholson brought into the market a new dye called *aniline-yellow*, obtained by the action of nitrous acid on aniline.[3] Martius and Griess found

---

[1] Wallach and Fischer, *Ber.* **15**, 2816.

[2] Kostanecki, *ibid.* **21**, 3115.

[3] Mène, *Jahresb.* **1861**, 496.

that this was the oxalate of a base, which they named amido-diphenylimide, $C_{12}H_{11}N_3$.[1] At the same time Kekulé showed that this body is formed by a molecular change from the isomeric diazo-amidobenzene when the latter is allowed to stand in contact with an aniline salt, and must therefore be considered as amido-azobenzene;[2] this observation was confirmed by Schmidt, who obtained the same substance by the reduction of nitro-azo-benzene.[3] It is also formed by heating an aniline salt with sodium stannate.[4] To prepare it, 2 parts of aniline hydro-chloride are gradually mixed with an aqueous solution of 1 part of sodium nitrite, the temperature not being allowed to rise above 60°. After some time an excess of hydrochloric acid is added, and then water, the amido-azobenzene separating out, while the excess of aniline remains in solution and may be recovered.[5]

Amido-azobenzene is scarcely soluble in water, but readily in alcohol, and crystallizes in yellow rhombic needles or prisms, melting at 127°·4. It boils above 300° without decomposition, is oxidized by manganese dioxide and sulphuric acid to quinone, and reduced by nascent hydrogen with formation of aniline and p-diamidobenzene; its constitution is, therefore, represented by the following formula :

Amido-azobenzene is a weak monacid base, and its salts are readily decomposed by water.

Amido-azobenzene hydrochloride, $C_{12}H_{11}N_3.ClH$, crystallizes from a slightly acid, boiling, saturated solution in lustrous, bluish violet needles or scales. This salt, or the oxalate, forms the commercial aniline-yellow; it dyes wool and silk a fine yellow, but, as it is volatile in steam, is not much employed for dyeing goods; it is, however, used for the preparation of other dyes.

Amido-azobenzenesulphonic acid, $C_6H_4(SO_3H)N=NC_6H_4.NH_2$, is obtained from sulphanilic acid by converting this into p-diazobenzenesulphonic acid by means of sodium nitrite and

---

[1] *Zeitsch. Chem.* **1866**, 132. [2] *Ibid.* **1866**, 689.
[3] *Ber.* **5**, 480. [4] Schiff, *Annalen*, **127**, 346
[5] Grässler, *Dingl. Polyt. Journ.* **232**, 192; Thomas and Witt, *Journ. Chem. Soc.* 1883, **1**, 112.

hydrochloric acid, adding aniline, removing the excess of the latter by hydrochloric acid, and precipitating the sulphonic acid with common salt. It is also prepared by treating amido-azobenzene with strongly fuming sulphuric acid, and washing the product until it is free from sulphuric acid.[1] The substance thus obtained, which contains the disulphonic acid,[2] occurs commercially as a paste, under the name *Fast yellow*, and is used as a yellow dye for wool and silk; it is also employed in the preparation of Biebrich scarlet, a colouring matter which will be described under naphthalene.

Methylamido-azobenzene, $C_6H_5.N_2.C_6H_4.NH.CH_3$, is obtained by heating amido-azobenzene with methyl iodide; it crystallizes in brick-red needles melting at 180°; its *hydrochloride* forms violet needles.[3]

Dimethylamido-azobenzene, $C_6H_5.N_2.C_6H_4N(CH_3)_2$. Griess obtained this substance by the action of dimethylaniline on a diazo-salt.[4] It is also formed when equal molecules of the hydrochlorides of aniline and dimethylaniline are dissolved in water and treated with an alkaline solution of sodium nitrite :[5]

$$C_6H_5.NH_3Cl + C_6H_5.N(CH_3)_2HCl + NaNO_2 + NaHO =$$
$$C_6H_5.N_2.C_6H_4N(CH_3)_2 + 2NaCl + 3H_2O.$$

It may also be prepared by the action of methyl iodide on the preceding compound ; it crystallizes in small yellow plates melting at 115°, and its *hydrochloride* forms purple-red, hair-like needles.

The corresponding sulphonic acid, $C_6H_4(SO_3H)N_2C_6H_4N(CH_3)_2$ is obtained by the action of dimethylaniline on *p*-diazoben-zenesulphonic acid, as well as by dissolving dimethylamidoazo-benzene in fuming sulphuric acid containing 30 per cent. of sulphur trioxide, allowing the solution to stand for twenty-four hours and precipitating with water (Möhlau). Hydrochloric acid precipitates it from a boiling alkaline solution in microscopic needles, which soon change into small, strongly-lustrous, violet plates or prisms (Griess). It forms a solution in concentrated sulphuric acid with a yellowish-brown colour, which on dilution becomes a splendid red.

[1] Grässler, *Chem. Ind.* **2**, 48, 346 ; **3**, 171. [2] Nietzki, *Ber.* **13**, 800.
[3] Berju, *ibid.* **17**, 1400. [4] *Ibid.* **10**, 528.
[5] Mohlau, *ibid.* **17**, 1490.

The ammonium salt occurs in commerce under the names *Golden-orange, Helianthin, Tropaeolin D,* and *Orange III.,* and is used for dyeing wool and silk. Its aqueous solution, even when dilute, has a dark, reddish-yellow colour, which is changed by the slightest quantity of free acid into a light red; it is therefore used instead of litmus in titrations, its solution being much more stable and more sensitive towards acids.[1]

Phenylamido-azobenzene, $C_6H_5N_2C_6H_4NH(C_6H_5)$, is formed by the action of diphenylamine on diazobenzene chloride in alcoholic solution. It is readily soluble in alcohol and ether, and crystallizes in small, golden-yellow, lustrous plates or fine prisms, melting at 82°. An alcoholic solution is first coloured a splendid violet by acids, and then deposits the salts in grey crystals. This compound dissolves in concentrated sulphuric acid with a green colour, which is converted into indigo-blue, and then into reddish violet by the addition of water.

Phenylamido-azobenzenesulphonic acid, $C_6H_4(SO_3H)N_2C_6H_4NH$ (C_6H_5), is obtained by acting on *p*-diazobenzenesulphonic acid with diphenylamine, and crystallizes in grey, hair-like needles with a steel-blue lustre; it is only slightly soluble in water, forming a light-red solution, but somewhat more readily in alcohol and acetic acid. The potassium salt, $C_{18}H_{14}N_3SO_3K$. occurs in commerce as *Tropaeolin OO, Orange IV,* or *Orange N.* It is only slightly soluble in cold, but readily in hot water, and crystallizes in flat, dichroic needles, generally arranged in the shape of a fan, which appear pale- and dark-golden-yellow, and are often an inch long.[2] Small quantities of the mineral acids diminish its solubility; its slightly acid solution dyes silk and wool a fiery golden-yellow. This tropaeolin is, like helianthin, an excellent indicator for titrations, since its solution is coloured red by the slightest excess of a free acid, while carbon dioxide and sulphuretted hydrogen produce no change.[3]

1138 *as-Diamidoazobenzene,* $C_6H_5N_2C_6H_3(NH_2)_2$, was prepared almost simultaneously by Caro and Witt;[4] it is obtained by the action of *m*-diamidobenzene on diazobenzene chloride; it is slightly soluble in water, readily in alcohol, ether, &c., crystallizes in fine yellow needles, melting at 117°·5, and is

[1] Lunge, *Ber.* **11**, 1944.   [2] Witt, *ibid.* **12**, 258.

[3] v. Miller, *ibid.* **11**, 460; Danilewski, *ibid.* **14**, 115; Tropaeolin OOO or naphtholazobenzenesulphonic acid, $C_{10}H_6(OH)N_2.C_6H_4.SO_3H$, behaves in a precisely opposite manner, since it is turned red by alkalis, and is a most delicate reagent for these.   [4] *Ber.* **10**, 388.

split up by reduction with stannous chloride into aniline and
$1:2:4$ triamidobenzene. Its constitution is, therefore, expressed
by the following formula:

$$NH_2\langle\bigcirc\rangle\underset{NH_2.}{N{=}N}\langle\bigcirc\rangle$$

It is a monacid base; its hydrochloride, $C_{12}H_{12}N_4.HCl$, occurs
in commerce under the name of *Chrysoïdin*, and forms lustrous
needles, greenish-black by reflected light and deep-red by
transmitted light, which form a brownish-red solution in water.

Chrysoïdin dyes wool and silk orange-red; it absorbs the
chemically active rays of the spectrum to a very large extent,
and its solution in dilute shellac varnish is therefore used by
photographers for covering the window of the dark room; it is,
however, gradually decomposed by the action of the light.

a-Diamidodisazobenzene, $C_6H_5N_2.C_6H_2(NH_2)_2.N_2.C_6H_5$, is
formed by the action of chrysoïdin on diazobenzene nitrate,
and crystallizes from chloroform in dark-red lustrous needles
or small plates, which melt at 250°, and detonate when more
strongly heated. It is a weak monacid base, and forms salts
which are decomposed by water.[1]

β-Diamidodisazobenzene, $C_6H_5.N_2.C_6H_4.N_2.C_6H_3(NH_2)_2$. Griess
obtained this compound by the action of diazo-azobenzene
chloride on *m*-diamidobenzene. It crystallizes from chloroform
in fine, brownish-red needles, melting at 185°. It differs from
the isomeric diamidodisazobenzene in being a tolerably strong
diacid base.[2]

p-Diamido-azobenzene or *p-Azo-aniline,* $NH_2.C_6H_4.N_2.C_6H_4.NH_2$.
In order to prepare this compound, an alkaline solution of
p-nitracetanilide is treated with ammonia, zinc dust, and a
little platinum chloride; after some days, *p*-azoxyacetanilide,
$N_2(C_6H_4.NH C_2H_3O)_2O$, separates out, and on evaporation of
the mother-liquor, *p*-diamidobenzene and *p*-azo-acetanilide,
$N_2(C_6H_4.NH.C_2H_3O)_2$, are obtained; the former of these is
removed by dilute hydrochloric acid and the latter then converted into *p*-azo-aniline by heating with hydrochloric acid.[3]

This is also formed by reducing *p*-nitracetanilide with iron
and acetic acid, and then diazotizing. On adding aniline to the
neutral solution thus obtained, a bright-yellow diazo-amido-

[1] Griess, *Ber.* **16**, 2028.   [2] *Ibid.* p. 2033.
[3] Mixter, *Am. Chem. Journ.* **5**, 1 and 282.

compound separates out, and is converted by heating with
aniline and aniline hydrochloride into *acetyldiamido-azobenzene*,
$NH_2.C_6H_4.N_2.C_6H_4.N(C_2H_3O)H$; this is then decomposed by
heating with dilute sulphuric acid.[1]

p-Azo-aniline crystallizes from dilute alcohol in long, flat,
golden-yellow needles, melting at 235°. On gradually adding
hydrochloric acid to the alcoholic solution, the colour is first
changed to a dark-green, which is then converted into red; this
is caused by the formation of a normal and an acid salt, the
former of which, $C_{12}H_{12}N_4(HCl)_4$, crystallizes in almost black
needles with a green surface lustre.

p-Amidobenzeneazodimethylaniline,$C_6H_4(NH_2)N_2C_6H_4N(CH_3)_2$.
Meldola has given this name to a base which is obtained
by adding an aqueous solution of *p*-diazonitrobenzene chloride
to a well-cooled solution of dimethylaniline hydrochloride; the
p-nitrobenzene-azodimethylaniline hydrochloride is thus formed,
and crystallizes in long red needles with a splendid blue
metallic lustre. On treatment with ammonia this yields
the free base, which separates from solution in hot alcohol in
reddish-brown, microscopic needles, melting at 229—230°; and
is reduced by ammonium sulphide to the amido-compound.
This crystallizes from dilute alcohol in small brick-red needles
having a feeble metallic lustre, and melting at 182—183°; its
salts readily dissolve in water forming red solutions. Its
alcoholic solution is coloured green by acetic and oxalic acids,
and red by the addition of water. On heating the base with
aniline and aniline hydrochloride, a violet dye, which is one of
the indulines, is obtained.

Nitrous acid converts the base into a diazo-compound, the
very dilute, neutral or slightly ammoniacal solution of which
assumes a splendid blue colour in the air; this soon disappears,
brown flakes being deposited. Traces of nitrous acid can be
detected by means of this reaction more readily and with greater
certainty than by *m*-diamidobenzene (p. 270), since 1 part
of sodium nitrite in 64,000 parts of water gives a distinct
colouration.

If the base be dissolved in hydrochloric acid and treated with
zinc dust, dimethyldiamidobenzene is obtained.[2]

1139 *Azylines.* Lippmann and Fleissner[3] have given this

[1] Nietzki, *Ber.* **17**, 343.   [2] *Journ. Chem. Soc.* 1884, **1**, 106.
[3] *Monatsh.* **3**, 705; *Ber.* **13**, 2163; **16**, 1415.

name to a group of compounds obtained by the action of nitric oxide on tertiary amido-compounds:

$$2C_6H_5N(CH_3)_2 + 2NO = (CH_3)_2N.C_6H_4N{=}NC_6H_4N(CH_3)_2$$
$$+ H_2O + O.$$

No oxygen is set free but resinous oxidation products are formed.

Nölting obtained the same compound by treating a well-cooled solution of dimethyl-p-diamidobenzene in hydrochloric acid with sodium nitrite and, after some time, adding a solution of dimethylaniline in glacial acetic acid:

$$(CH_3)_2NC_6H_4N{=}NCl + C_6H_5N(CH_3)_2 =$$
$$(CH_3)_2NC_6H_4N{=}NC_6H_4N(CH_3)_2 + HCl.$$

It is also formed by the action of methyl iodide on p-amido-benzene-azodimethylaniline.[1]

Dimethylaniline-azyline, $N_2[C_6H_4N(CH_3)_2]_2$, separates from benzene in red crystals, melting at 266° to a green liquid, and forming a deep-green solution in glacial acetic acid.

Diethylaniline-azyline, $N_2[C_6H_4.N(C_2H_5)_2]_2$, is slightly soluble in cold, more readily in hot alcohol, and crystallizes in reddish-brown, monosymmetric needles which have a blue surface lustre and melt at 170°. On adding potassium nitrite to its solution in acetic acid, p-nitrosodiethylaniline is formed, and it is resolved by reducing agents into two molecules of diethyl-p-diamido-benzene.

The azylines are dyes, but they have received no application on account of their costly preparation. They are derived, like amidobenzene-azodimethylaniline, from p-azo-aniline; this may therefore be called azyline, the other compounds being most conveniently designated as follows:

Dimethyl-p-azo-aniline or dimethylazyline.
Tetramethyl-p-azo-aniline or tetramethylazyline,
Tetra-ethyl-p-azo-aniline or tetra-ethylazyline.

1140 *Triamido-azobenzene*, $NH_2.C_6H_4.N_2C_6H_3(NH_2)_2$. The hydrochloride of this base, $C_{12}H_{13}N_5(ClH)_2$, is a splendid brown dye, which was first manufactured by Roberts, Dale, and Co.,

---

[1] Nölting, *Ber.* **18**, 1143.

and brought into the market under the name of *Manchester brown*; it has also been called *Bismarck brown, Phenylene brown, Vesuvin*, &c. It is obtained as a brown crystalline mass by treating a cold dilute hydrochloric acid solution of *m*-diamido-benzene, generally called *m*-phenylenediamine, with sodium nitrite, and is much used in wool dyeing. On account of the ready formation and intense colour of this dye, *m*-diamidobenzene is, as already mentioned (p. 270), employed as a delicate reagent for nitrous acid.

The commercial product also contains other bases in small quantities. When it is decomposed with ammonia and the mixture extracted with boiling water, triamido-azobenzene goes into solution, and on cooling crystallizes in small yellowish-red plates, melting at 137°.[1]

Its constitution follows from that of **chrysoïdin**, and is expressed by the following formula:

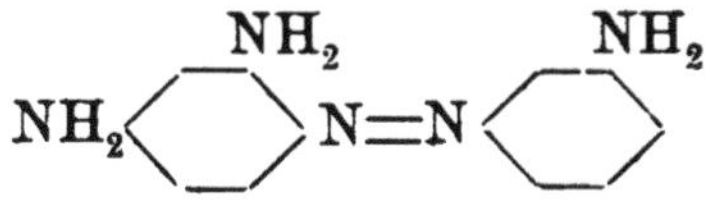

THIO-AMIDO-COMPOUNDS.

1141 *Thio-aniline* or *Diamidophenyl sulphide*, $S(C_6H_4.NH_2)_2$, is obtained by heating aniline to 150—160° with sulphur:

$$2C_6H_5.NH_2 + 2S = S{<}^{C_6H_4.NH_2}_{C_6H_4.NH_2} + H_2S.$$

The reaction proceeds more rapidly if litharge be gradually added to take up the sulphuretted hydrogen which is formed.[2]

Thio-aniline may also be prepared by dissolving phenyl sulphide in concentrated nitric acid and reducing the nitro-compound thus formed.[3] It is slightly soluble in hot water, readily in alcohol and ether, and crystallizes in long, thin, odourless needles, melting at 105°.

[1] Caro and Griess, *Zeitschr. Chem.* II. **3**, 278.
[2] Merz and Weith, *Ber.* **4**, 384.
[3] Krafft, *ibid.* **7**, 384.

Thio-aniline hydrochloride, $C_{12}H_{12}N_2S(HCl)_2 + 2H_2O$, is readily soluble in water, slightly in hydrochloric acid and alcohol, and crystallizes in lustrous prisms or needles.

Thio-aniline sulphate, $C_{12}H_{12}N_2S,SO_4H_2 + H_2O$, is slightly soluble in cold, more readily in hot water, and almost insoluble in alcohol; it forms short prisms or acute needles.

Thio-aniline oxalate, $C_{12}H_{12}N_2S,C_2H_2O_4$, crystallizes in slender needles, which are only slightly soluble in hot water.

The solutions of these salts have an acid reaction, and, even when very dilute, colour pine wood a fine orange. Ferric chloride produces a deep violet or blue colouration on warming. When thio-aniline is heated with concentrated sulphuric acid, a colourless solution is obtained, which soon becomes deep blue, and finally violet. If the blue liquid be poured into water, a splendid red solution is formed.

Thio-dimethylaniline, $S[C_6H_4.N(CH_3)_2]_2$, is prepared by the action of sulphur dichloride on dimethylaniline in light petroleum solution. It crystallizes in pale yellow stellate groups of needles, which melt at 123·5° and dissolve readily in alcohol, ether and benzene.[1] By the action of silver nitrate it loses sulphur and is converted into the corresponding *oxydimethylaniline*, $O[C_6H_4.N(CH_3)_2]_2$, which forms concentrically grouped colourless needles melting at 119°.

Thio-diethylaniline, $S[C_6H_4.N(C_2H_5)_2]_2$, is prepared in a similar manner from diethylaniline.[2] It forms yellow needles or prisms melting at 79·5—80°.

o-Amidothiophenol, $C_6H_4(NH_2)SH$, is obtained by the action of tin and hydrochloric acid on *o*-nitrobenzenesulphonic chloride.[3] It boils at 234°, and solidifies in the cold to needles melting at 26°. When cyanogen is passed through its alcoholic solution, *oxalamidothiophenol* is formed :

$$2C_6H_4{<}{\stackrel{NH_2}{SH}} + NC-CN = C_6H_4{<}{\stackrel{N}{S}}{>}C-C{<}{\stackrel{N}{S}}{>}C_6H_4.$$
$$+ 2NH_3.$$

This compound may also be prepared by adding phosphorus oxychloride to a solution of anhydrous oxalic acid in *o*-amidothiophenol, or, together with 60 per cent. of other products

---

[1] Holzmann, *Ber.* **20**, 1640.   [2] *Ibid.* **21**, 2059
[3] Hofmann, *ibid.* **13**, 19.

which have not yet been investigated, by heating acetanilide with sulphur for some time; in spite of the small yield this is the best method of preparation. Oxalamidothiophenol crystallizes in lustrous plates, which melt at about 300°, and are scarcely affected by the ordinary solvents. ˙It distils almost without decomposition; when heated to 200° with caustic potash it decomposes completely into oxalic acid and *o*-amidothiophenol, so that the latter is most readily prepared from acetanilide.[1]

m-Amidothiophenol is obtained by the reduction of *m*-nitrobenzenesulphonic chloride. It is an oily liquid, smelling of mushrooms;[2] its *hydrochloride*, $C_6H_4(SH)NH_2.ClH$, is readily soluble in water and crystallizes in lustrous, nodular masses, melting at 232°, and subliming at higher temperatures.[3]

o-Diamidophenyl disulphide, $S_2(C_6H_4.NH_2)_2$, is formed by the oxidation of *o*-amidothiophenol in the air, or more rapidly by the addition of ferric chloride to its solution in hydrochloric acid. It is insoluble in water, and crystallizes from hot alcohol in small plates melting at 93°. Its *hydrochloride* forms small plates, which are scarcely soluble in hydrochloric acid, even when very dilute. When sulphuretted hydrogen is passed through a solution of the base, amidothiophenol is regenerated (Hofmann) ˙

$$\begin{array}{l} S.C_6H_4.NH_2 \\ | \\ S.C_6H_4.NH_2 \end{array} + H_2S = 2HS.C_6H_4.NH_2 + S.$$

p-Diamidophenyl disulphide, $S_2(C_6H_4.NH_2)_2$.—When acetanilide is heated to 100° with chloride of sulphur, two compounds are formed, viz. *dithio-acetanilide*, $S_2(C_6H_4.NH.C_2H_3O)_2$, and *trithioacetanilide*, $S_3(C_6H_4.NH.C_2H_3O)_2$, both of which crystallize from glacial acetic acid in small plates. When the former is heated with dilute sulphuric acid, the sulphate of *p*-diamidophenyl disulphide, $S_2(C_6H_4.NH_2)_2SO_4H_2+2H_2O$, is obtained; it crystallizes in very fine needles, and when treated with an alkali yields the free base. This is slightly soluble in hot water, and crystallizes in long, thin, greenish needles, which have a vitreous lustre, and melt at 78—79°.

Schmidt calls this compound pseudothio-aniline, and gives the name dithio-aniline to an isomeric resinous substance, which is

[1] Hofmann, *Ber.* **13**, 1223.
[2] Glutz and Schranck, *J. Pr. Chem.* II. **2**, 223.
[3] Biedermann, *Ber.* **8**, 1674.

prepared by the action of bromide of sulphur on aniline, and possesses basic properties.[1]

Dimethylamidophenyl disulphide or *Dithiodimethylaniline,* $[S.C_6H_4.N.(CH_3)_2]_2$, is formed by acting on a benzene solution of dimethylaniline with sulphur monochloride, S_2Cl_2. It forms lustrous yellow needles or prisms, which melt at 118°, and are converted by silver nitrate into the corresponding dioxy-compound $[O.C_6H_4.N(CH_3)_2]_2$, melting at 90·5°.[2]

1142 *Thiodiphenylamine,* $S(C_6H_4)_2NH$, is obtained by heating diphenylamine with sulphur at 250—300°, and by the action of sulphur dichloride on diphenylamine in benzene solution.[3] It is also formed by heating *o*-amidothiophenol with catechol (*o*-dihydroxybenzene) and has therefore the following constitution :

Its formation is represented by the equation : [4]

$$C_6H_4{<}^{SH}_{NH_2} \; + \; {^{HO}_{HO}}{>}C_6H_4 \; = \; C_6H_4{<}^{S}_{NH}{>}C_6H_4 \; + \; 2H_2O.$$

Thiodiphenylamine crystallizes from hot alcohol in yellowish, lustrous plates, melting at 180°, distilling at a higher temperature almost without decomposition, and solidifying to a coarse radiating mass.[5] Moderately concentrated nitric acid converts it into *nitrodiphenylamine sulphoxide,* $SO(C_{12}H_7.NO_2)NH$. This substance has not yet been obtained pure, but the following compound has been obtained from it :

Amidothiodiphenylamine, $S(C_{12}H_7.NH_2)NH$, may be prepared either by reducing the nitro-compound with stannous chloride and hydrochloric acid in presence of metallic tin, or by heating amidodiphenylamine with sulphur :

$$HN{<}^{C_6H_5}_{C_6H_4-NH_2} \; + \; 2S \; = \; HN{<}^{C_6H_4}_{C_6H_3-NH_2}{>}S \; + \; H_2S.$$

[1] *Ber.* **11,** 1168. [2] Merz and Weith, *ibid.* **19,** 1571.
[3] Holzmann, *ibid.* **21,** 2064. [4] Bernthsen, *ibid.* **19,** 3255.
[5] *Ibid.* **16,** 2896.

It is slightly soluble in hot water, readily in alcohol, and crystallizes in small plates with a satin lustre.

Ferric chloride oxidizes its solution in hydrochloric acid to *imidothiodiphenylimide*, $C_{12}H_8N_2S$, a violet-red colouring matter, which will be subsequently described. Alcoholic ammonium sulphide reduces it again to amidothiodiphenylamine, which is most readily obtained pure in this manner.[1]

Fuming nitric acid converts thiodiphenylamine into two isomeric dinitrodiphenylamine sulphoxides, $SO(C_6H_3.NO_2)_2NH$.[2]

a-Dinitrodiphenylamine sulphoxide soon separates out from the solution in small, bright yellow crystals, almost insoluble in alcohol, and only slightly soluble in glacial acetic acid; it crystallizes from hot aniline in small yellowish red needles or prisms.

β-Dinitrodiphenylamine sulphoxide is precipitated by water from the nitric acid solution, and after being boiled with alcohol forms a fine, bright yellow powder.

Both these compounds are reduced by tin and hydrochloric acid to the corresponding *diamidothiodiphenylamines*, $S(C_6H_3.NH_2)_2NH$, the oxygen of the sulphoxide being eliminated; these compounds and the colouring matters which are readily obtained from them by oxidation are described later. Owing to this formation of colouring matters, even fractions of a milligram of thiodiphenylamine can be detected with ease. The substance to be tested is dissolved in a few drops of glacial acetic acid, treated with fuming nitric acid and diluted with water. The nitro-compound which is precipitated is then boiled with an acid solution of stannous chloride, the tin precipitated from the colourless solution by zinc, and an excess of ammonia added; the solution on standing in the air soon becomes coloured a deep violet. If ferric chloride be added instead of ammonia, a deep violet colouration is produced if the solution be dilute, while if it be concentrated a reddish violet precipitate is thrown down.

a-Methylthiodiphenylamine, $S(C_6H_4)_2NCH_3$, is obtained by heating thiodiphenylamine with methyl iodide and wood spirit, and crystallizes from hot alcohol in splendid long prisms, melting at 99°·3. Potassium permanganate, added to its boiling aqueous solution, oxidizes it to *a-methyldiphenylamine sulphone*, $SO_2(C_6H_4)_2NCH_3$, crystallizing in masses of small needles, and melting at about 222°. Like other sulphones this has neither

[1] Bernthsen, *Ber.* **17**, 2857.   [2] *Ibid.* p. 611.

acid nor basic properties; it forms a deep blue solution in boiling sulphuric acid, which becomes a brownish violet on dilution with water.

Dinitromethyldiphenylamine sulphoxide, $SO(C_6H_3.NO_2)_2NCH_3$, is formed by the action of fuming nitric acid on a-methylthiodiphenylamine. It is more soluble in glacial acetic acid than the dinitrodiphenylamine sulphoxides, but, unlike these, is insoluble in dilute alkalis, the hydrogen of the imido-group, which is replaceable by metals, being absent. On reduction it is converted into *diamidomethylthiodiphenylamine*, $S(C_6H_3.NH_2)_2NCH_3$, which rapidly oxidizes to a very unstable bluish green colouring matter.

β-Methylthiodiphenylamine, $S(C_6H_4)_2NCH_3$ is prepared by the action of sulphur dichloride on methyldiphenylamine.[1] It crystallizes in yellow needles melting at 78-79°. On heating with copper it yields methyldiphenylamine, whereas the a-compound yields carbazol, $\begin{matrix} C_6H_4 \\ | \\ C_6H_4 \end{matrix}\!\!\!>\!\!NH.$

Ethylthiodiphenylamine, $S(C_6H_4)_2NC_2H_5$, crystallizes from alcohol in long, thin, white prisms, melting at 102°.

Acetylthiodiphenylamine, $S(C_6H_4)_2NC_2H_3O$, is obtained in thick, colourless crystals by heating thiodiphenylamine with acetic anhydride; it is slightly soluble in alcohol and is deposited from its solution in long, thin, glittering prisms.

Dithiodiphenylamine, $S_2(C_6H_4)_2NH$, is obtained by acting on diphenylamine in light petroleum solution with sulphur monochloride, S_2Cl_2. It crystallizes from benzene in slender yellow needles, which melt at 59-60°, and readily dissolve in warm alcohol, ether, and benzene. It does not form salts.[2]

1143 *Amidodimethylanilinemercaptan*, $C_6H_3\!\!\begin{matrix} \nearrow N(CH_3)_2\ (4) \\ -NH_2\ \ \ \ (1) \\ \searrow SH.\ \ \ \ (2) \end{matrix}$

This compound was obtained by Bernthsen as the product of reduction of methylene-red, $C_8H_9N_2S_2Cl$, a colouring matter described in a later volume, which has the constitution

$$C_6H_3\!\!-\!\!N\!\!\begin{matrix} \nearrow N(CH_3)_2Cl \\ | \\ \searrow S \diagup S \end{matrix}$$. It may also be prepared by the reduction

of the corresponding thiosulphonic acid (p. 380), and of the

---

[1] *Ber.* **21**, 2065.   [2] Holzmann, *ibid.* p. 2063.

disulphide. It is best isolated in the form of its *zinc* salt, $(C_8H_{11}N_2S)_2Zn$, a white crystalline powder, which is slightly oxidized on exposure to the air. The free mercaptan has not yet been isolated owing to the ease with which it undergoes oxidation forming the disulphide. Its aqueous solution, obtained by the decomposition of the zinc salt with hydrochloric acid, is readily converted into methylene-red by the action of sulphuretted hydrogen and ferric chloride. It forms salts with the alkalis, which are immediately oxidized by the action of the air with formation of the disulphide, and also yields a hydrochloride which could not be obtained crystalline. The solutions of the latter are coloured dark-blue by the air.[1]

Ethenylamidodimethylanilinemercaptan, $C_6H_3 <^{N(CH_3)_2}_{<^N_S>C.CH_3}$.

The *hydrochloride* of this compound is obtained by treating the above zinc salt with acetyl chloride in benzene solution. It crystallizes from absolute alcohol in white or slightly green needles or prisms, and is readily soluble in water. The free ethenyl-base is a yellow oil.

Diazothiodimethylaniline, $C_6H_3 <^{N(CH_3)_2}_{<^N_S>N}$, is obtained by the action of sodium nitrite on the zinc salt of amidodimethylanilinemercaptan in dilute sulphuric acid solution. It may be obtained more readily from the thiosulphonic acid, and forms pale yellow needles, which melt at 78°. Its *hydrochloride* is readily soluble in hydrochloric acid, but is immediately dissociated by water. It yields a crystalline *platinochloride* and *aurochloride* and is reconverted into amidodimethylanilinemercaptan by zinc dust in ammoniacal solution.

The formation of these two anhydro-compounds shows that the SH group must occupy the ortho-position to the amido-group.

Amidodimethylaniline disulphide, $(C_8H_{11}N_2S)_2$, is, as already mentioned, formed by the oxidation of the mercaptan. It is also obtained by treating amidodimethylanilinemercaptanthiosulphonic acid with caustic soda and warming the solution to 40—50°. It is very difficult to obtain pure, and forms a dark-yellow oil, readily soluble in alcohol, ether and benzene, but almost insoluble in water. Its *picrate*, $(C_8H_{11}N_2S)_2.2C_6H_3N_3O_7$, is a deep yellow precipitate.

---

[1] Bernthsen, *Annalen,* **251**, 23.

Amidodimethylaniline persulphide, $C_{32}H_{40}N_8S_5$.   This compound, the constitution of which is unknown, separates out from the concentrated benzene solutions of the crude disulphide.   It crystallizes in yellow needles which melt at 97°.

$$\textit{Amidodimethylanilinethiosulphonic acid,} C_6H_3 \diagdown \begin{matrix} \diagup N(CH_3)_2 & (4) \\ NH_2 & (1) \\ \diagdown S.SO_3H & (2) \end{matrix}$$

This compound is obtained from methylene-red by treating it in very dilute solution with caustic soda or ammonia, acidifying with acetic acid, and exposing to the air for twenty-four hours in flat vessels, whereby the solution assumes a blue colouration. Ammonia is then added, any disulphide formed extracted with ether, the solution slightly acidified with acetic acid, treated with animal charcoal, and evaporated.   It is, however, most readily prepared from dimethyl-p-diamidobenzene (p-amidodimethylaniline).   This compound on oxidation yields the di-imido-derivative $C_6H_4 \diagdown \begin{matrix} \diagup N(CH_3)_2Cl \\ | \\ NH \end{matrix}$, which on treatment with a thiosulphate undergoes the following reaction : [1]

$$C_6H_4 \diagdown \begin{matrix} \diagup N(CH_3)_2Cl \\ | \\ NH \end{matrix} + HS.SO_3H = C_6H_3 \diagdown \begin{matrix} \diagup N(CH_3)_2 \\ NH_2 \\ \diagdown S.SO_3H \end{matrix} + HCl.$$

To carry out the reaction 10 grms. of amidodimethylaniline are dissolved in 100cc. of water, and a solution of 55 grms. of potassium bichromate and 4·5 grms. of acetic acid slowly added in the cold.   To the resulting crystalline magma a solution of 27 grms. of aluminium sulphate and 22 grms. of sodium thiosulphate is added with constant stirring.   The sulphonic acid quickly separates out, and is purified by solution in sodium carbonate solution, reprecipitation with acetic acid, and recrystallization from hot water.   It forms lustrous crystals, sparingly soluble in cold water and alcohol, more readily in hot water and alkalis.   It also dissolves in hydrochloric acid, forming a salt.   It melts with decomposition at 201-204° when quickly heated.

1144 *Amidodiethylanilinemercaptan,* $C_6H_3 \diagdown \begin{matrix} \diagup N(C_2H_5)_2 \\ NH_2 \\ \diagdown SH \end{matrix}$, is obtained by the reduction of the corresponding sulphonic acid

[1] Bernthsen, *Annalen,* **251,** 49.

Its zinc salt is a bluish-white pulverulent mass, and its properties closely resemble those of the dimethyl compound. The free base has not been prepared. *Diazothiodiethylaniline*,

$$C_6H_3{<}^{N(C_2H_5)_2}_{\underset{S}{-N}{>}N}$$, obtained by the action of nitrous acid,

forms slightly yellow-coloured needles, melting at 106-107°. The corresponding *disulphide*, $(C_{10}H_{15}N_2S)_2$, is a thick brownish-red oil, whose *picrate*, $(C_{10}H_{15}N_2S)_2.2C_6H_3N_3O_7$, crystallizes from hot water in green needles.

Amidodiethylanilinethiosulphonic acid, $C_6H_3{<}^{N(C_2H_5)_2}_{\underset{S.\,SO_3H}{-NH_2}}$ is obtained from *p*-amidodiethylaniline in exactly the same manner as the dimethyl-compound. It crystallizes in thin pale blue or white prisms, sparingly soluble even in hot water, more readily in acids and alkalis, and melts at 230° on quickly heating. Its dilute aqueous solution gives a beautiful purple-red colouration with ferric chloride.

Tetramethyl-p-diamidobenzenemercaptan, $C_6H_3{<}^{N(CH_3)_2}_{\underset{SH}{-N(CH_3)_2}}$,

is prepared by reducing the thiosulphonic acid next described. Its zinc salt is very similar to those previously described, is readily soluble in acids, and gives no colouration with ferric chloride.

Tetramethyl-p-diamidobenzenethiosulphonic acid, $C_6H_3{<}^{N(CH_3)_2}_{\underset{S.SO_3H}{-N(CH_3)_2}}$

is obtained just in the same manner as the thiosulphonic acids previously mentioned. It forms rhombic plates melting at about 179°, which are fairly soluble in hot water. In dilute solution it gives a beautiful reddish-violet colouration with ferric chloride.

p-Diamidobenzenemercaptan, $C_6H_3(NH_2)_2SH$. The zinc salt of this compound, obtained by the reduction of the following thio-sulphonic acid, forms white crystalline nodules. When mixed with a little hydrochloric acid and dimethylaniline hydrochloride, it gives a green colouration which becomes blue on warming. On oxidation it yields the *disulphide* as a reddish-yellow oil, whose *picrate*, $(C_6H_7N_2S)_2+2C_6H_3N_3O_7$, is a yellow crystalline compound, readily soluble in alcohol.

p-Diamidobenzenethiosulphonic acid, $C_6H_3(NH_2)_2.S.SO_3H$. This compound, obtained by the method already described, crys-

tallizes in beautiful lustrous plates, which generally assume a
pale-blue or blackish colour from surface oxidation.   It is readily
soluble in hot water, slightly in alcohol, chloroform and benzene,
and undergoes oxidation much more readily than the alkylated
compounds.   Treated with hydrochloric acid, dimethylaniline
hydrochloride and an oxidizing agent, it gives a beautiful bluish-
green colouration which changes to blue on boiling.

Many of the foregoing compounds · are closely connected
with the colouring matters known as Lauth's violet, methylene-
blue, methylene-red, etc, which will be described in a later
volume.

PHOSPHORUS DERIVATIVES OF BENZENE.

1145 *Phosphenyl chloride*, $C_6H_5PCl_2$, is formed when the
vapours of phosphorus trichloride and benzene are repeatedly
passed through a red-hot porcelain tube :[1]

$$C_6H_6 + PCl_3 = C_6H_5.PCl_2 + HCl.$$

It is also obtained by boiling 5 parts of aluminium chloride, 30
parts of phosphorus trichloride, and 50 parts of benzene for
36 hours in an apparatus connected with an inverted condenser.
According to this method, which is more especially adapted for
the preparation of small quantities, 35 grms. of the pure compound may be obtained from 500 grms. of benzene, while
according to the first method, with the necessary apparatus at
least 500 grms. can be obtained in the same time.[2]

It is also formed when mercury diphenyl, $(C_6H_5)_2Hg$, is heated
with phosphorus trichloride to 180° (Michaelis) :

$$Hg(C_6H_5)_2 + PCl_3 = Hg(C_6H_5)Cl + C_6H_5PCl_2.$$

It is a strongly refractive liquid, which fumes in the air, has a
very penetrating smell, and boils at 224°6 (Thorpe).   It is
decomposed by water with formation of phosphenylous acid.

Phosphenyl tetrachloride, $C_6H_5PCl_4$, is readily formed by the
combination of dry chlorine with the preceding compound, and

---

[1] Michaelis, *Annalen*, **181**, 280 ; where the apparatus for the preparation of
large quantities is described in detail.   [2] *Ber.* **12**, 1009.

crystallizes in white prisms, probably belonging to the mono-symmetric system. They melt at 73° and partially sublime when more strongly heated, another portion being decomposed into chlorine and phosphenyl chloride. On heating to 180° in a sealed tube, it decomposes into chlorobenzene and phosphorus trichloride. With water it first forms phosphenyl oxychloride, which is converted by the further addition of water into phosphenylic acid.

Phosphenyl oxychloride, $C_6H_5POCl_2$, is also obtained by passing air or oxygen into heated phosphenyl chloride; this must be perfectly pure, or, especially when pure oxygen is employed, very violent explosions may occur. It is also formed, together with acetyl chloride, by the action of phosphenyl tetrachloride on acetic acid; it is however best prepared by passing sulphur dioxide into the tetrachloride:

$$C_6H_5PCl_4 + SO_2 = C_6H_5POCl_2 + SOCl_2.$$

The thionyl chloride, boiling at 80°, can readily be separated from the oxychloride, which is a thick liquid having a fruit-like smell, and boiling at 258°.

It follows from the foregoing decompositions that phosphenyl tetrachloride behaves similarly to phosphorus pentachloride, and Michaelis considers that it might be advantageously employed in organic chemistry in place of the latter for the preparation of chlorides, since, on account of its high boiling point, it can be more readily separated from these than can phosphorus oxy-chloride, which boils at 110°.

Dimethylamidophosphenyl chloride, $(CH_3)_2N.C_6H_4.PCl_2$, is prepared by gradually adding 20 grms. of aluminium chloride to a mixture of 100 grms. of phosphorus trichloride and 70 grms. of dimethylaniline, boiling for eight hours, extracting three times with light petroleum, and evaporating the clear liquid which remains, on the water-bath. It forms a yellow crystalline mass consisting of thin tablets, melts at 66°, is hardly soluble in light petroleum, more readily in ether, and still more readily in benzene and phosphorus trichloride.[1]

Michaelis has also prepared several bromine derivatives of phosphenyl.

Phosphenyl bromide, $C_6H_5PBr_2$, may be obtained by heating mercury diphenyl with phosphorus tribromide, or by passing dry

[1] Schenk and Michaelis, *Ber.* **21**, 1497.

hydrogen bromide into boiling phosphenyl chloride. It is a liquid boiling at 257°.[1]

1146 *Phosphenylous acid,* $C_6H_5PO_2H_2$, is obtained by allowing phosphenyl chloride to drop gradually into water; the solution is then heated to boiling and rapidly evaporated in an atmosphere of carbon dioxide. The acid is slightly soluble in cold, very readily in hot water, and crystallizes in tablets melting at about 70°. When more strongly heated, it decomposes into triphenylphosphine, $P(C_6H_5)_3$, phosphenylic acid, $C_6H_5PO_3H_2$, and water.

It is a monobasic acid; its alkaline salts form deliquescent crystals; the *barium* salt, $(C_6H_5PO_2H)_2Ba + 4H_2O$, crystallizes in fine, oblique, rhombic prisms, and the *lead* salt, $(C_6H_5PO_2H)_2Pb$, separates out in scales having a nacreous lustre, when the solution of the sodium salt is treated with lead acetate and acetic acid. The salts of the other heavy metals are insoluble in water; the *ferric* salt, $(C_6H_5PO_2H)_3Fe$, is a characteristic, granular, white precipitate, insoluble in cold concentrated sulphuric acid but soluble in the hot acid. On cooling it separates out as a tough, white mass, which after some time becomes brittle and crystalline; when again heated, the salt fuses to an oily liquid which gradually dissolves. If the dry salt is heated to 180°, it ignites and burns with a yellowish flame.

Diethyl phosphenylite, $C_6H_5PO_2(C_2H_5)_2$, is obtained by the action of phosphenyl chloride on sodium ethylate free from alcohol; in order to diminish the violence of the reaction a quantity of anhydrous ether is added as a diluent. It is a mobile liquid, which boils at 235° and has an overpowering smell. It dissolves gradually in water with formation of *ethyl phosphenylous acid,* $C_6H_5O_2(C_2H_5)H$, which is left on the evaporation of the solution as a thick liquid possessing an aromatic odour. It is monobasic and is gradually decomposed by water into phosphenylous acid and alcohol.[2]

The action of phosphorus pentachloride on the latter acid shows that it only contains one hydroxyl group (Michaelis):

$$C_6H_5.PHO(OH) + 2PCl_5 = C_6H_5POCl_2 + POCl_3 + PCl_3 + 2HCl.$$

Phosphenyl oxychloride is thus formed, while if the acid contained two hydroxyl groups, as might be expected from its pre-

[1] Köhler and Michaelis, *Ber.* **9**, 519.
[2] *Ibid.* **10**, 816.

paration from phosphenyl chloride, the following reaction would take place :

$$C_6H_5P(OH)_2 + 2PCl_5 = C_5H_5PCl_2 + 2POCl_3 + 2HCl.$$

Michaelis has further shown by the action of phosphenyl tetrachloride on phosphorous acid itself that this only contains two hydroxyl groups :

$$PHO(OH)_2 + 3C_6H_5PCl_4 = POCl_3 + 2C_6H_5POCl_2 + C_6H_5PCl_2$$
$$+ 3HCl.$$

If it had the formula $P(OH)_3$, as was generally supposed, the following reaction would occur :

$$P(OH)_3 + 3C_6H_5PCl_4 = PCl_3 + 3C_6H_5POCl_2 + 3HCl.$$

These two acids, therefore, contain pentavalent phosphorus, and have the following constitutions :

$$O = P{\Big\langle}{\overset{\textstyle H}{\underset{\textstyle OH}{-OH}}} \qquad O = P{\Big\langle}{\overset{\textstyle H}{\underset{\textstyle OH.}{-C_6H_5}}}$$

Dimethylamidophosphenylous acid, $(CH_3)_2N.C_6H_4.P(OH)_2$, is obtained as the sodium salt by acting on the chloride (p. 383) with sodium hydroxide or carbonate, and is precipitated as the lead salt, which is washed, dried, and decomposed in presence of absolute alcohol by sulphuretted hydrogen. It crystallizes in white needles, melting at 162°, which are readily soluble in water and hot alcohol. It is stable in alkaline or alcoholic solution, but in aqueous or acid solution readily decomposes into dimethylaniline and phosphorous acid. The *sodium* salt, $(CH_3)_2N.C_6H_4.PO_2HNa + 2H_2O$, forms large well-developed crystals.[1]

1147 *Phosphenylic acid*, $C_6H_5PO(OH)_2$, is best obtained by gradually adding phosphenyl tetrachloride to water, and finally warming the solution in order to decompose all oxychloride. It crystallizes in small, oblique, rhombic plates, having a vitreous lustre, which melt at 158° and solidify to a radiating crystalline mass. On heating to 200° it is converted into diphosphenylic

---

[1] Schenk and Michaelis, *Ber.* **21**, 1498.

acid, $(C_6H_5PO)_2O(OH)_2$, which is changed at $210°$ into triphosphenylic acid, $(C_6H_5PO)_3O_2(OH)_2$. Both these compounds are tough, transparent masses, and recombine with water to form phosphenylic acid. The following formulæ explain the constitution of these substances:

Phosphenylic acid.   Diphosphenylic acid.   Triphosphenylic acid.

$$C_6H_5.PO\begin{cases}OH\\OH\end{cases} \qquad \begin{matrix}C_6H_5PO\\ \\C_6H_5PO\end{matrix}\begin{cases}OH\\O\\OH\end{cases} \qquad \begin{matrix}C_6H_5PO\\ \\C_6H_5PO\\ \\C_6H_5PO\end{matrix}\begin{cases}OH\\O\\O\\OH\end{cases}$$

When phosphenylic acid is rapidly heated to $100°$, it decomposes into benzene and metaphosphoric acid:

$$C_6H_5PO_3H_2 = C_6H_6 + PO_3H.$$

If it be fused with caustic potash, benzene is also formed, together with orthophosphoric acid.

Phosphenylic acid is a strong dibasic acid and is not precipitated by barium chloride or silver nitrate; ammonia produces a white precipitate, while ammonium molybdate does not give any precipitate.

Normal sodium phosphenylate, $C_6H_5PO(ONa)_2 + 12H_2O$, forms long pointed crystals which readily lose water when allowed to stand over sulphuric acid.

Acid sodium phosphenylate, $C_6H_5PO_3NaH$, crystallizes in prisms which contain water of crystallization, but effloresce exceedingly rapidly. The normal potassium salt only crystallizes with great difficulty and in indistinct forms, while the anhydrous acid salt forms microscopic rhombic plates.

Normal calcium phosphenylate, $C_6H_5PO_3Ca + 2H_2O$, is a precipitate consisting of small plates having a silky lustre.

Acid calcium phosphenylate, $(C_6H_5PO_3)_2CaH_2$, separates from solution in acetic acid in small lustrous plates or in lustrous, moss-like aggregates.

All the other metallic salts of phosphenylic acid are insoluble in water.

Acid ethyl phosphenylate, $C_6H_5PO_3(C_2H_5)H$, is obtained by the action of phosphenyl tetrachloride on absolute alcohol, and is a syrupy liquid which has an acid reaction and is monobasic.

Normal ethyl phosphenylate, $C_6H_5PO(OC_2H_5)_2$, is obtained by heating the silver salt with ethyl iodide. It is a thick liquid boiling at 267°, and having a peculiar smell resembling that of mustard oil. The *dimethyl* ether, which is very similar in its properties and boils at 247°, has a perfectly different smell.

Normal phenyl phosphenylate, $C_6H_5PO(OC_6H_5)_2$, is formed by the action of phosphenyl tetrachloride on phenol:

$$C_6H_5PCl_4 + 3C_6H_5OH = C_6H_5PO(OC_6H_5)_2 + C_6H_5Cl + 3HCl.$$

It is very readily soluble in alcohol and ether, and crystallizes from hot, dilute alcohol in very long thin needles, melting at 63°·5, which are not attacked by hot aqueous caustic soda, but are decomposed by the alcoholic solution. Its boiling point lies above 360°; it is also formed, together with the chloride of phosphenylic acid, $C_6H_5PO(OC_6H_5)Cl$, by the action of phosphenyl oxychloride on phenol. The acid obtained by the decomposition of the chloride with water, *i.e. acid phenyl phosphenylate*, $C_6H_5PO(OC_6H_5)OH$, is only slightly soluble in water, and crystallizes from aqueous alcohol in hair-like needles melting at 57°. It forms salts which crystallize well.

Nitrophosphenylic acid, $C_6H_4(NO_2)PO_3H_2$, is obtained by heating phosphenylic acid with seven parts of fuming nitric acid in a sealed tube for five or six hours at 100—110°. It crystallizes from ether in white, concentrically grouped needles, which deliquesce in the air and form an intense yellow solution in water. On evaporation of the solution it separates out in white, cauliflower-like masses. It melts at 132° and deflagrates explosively at 200°; on heating with soda-lime it decomposes into phosphoric acid and nitrobenzene, which is converted by the action of the alkali into aniline. It is a strong dibasic acid, the alkaline salts of which do not crystallize; the normal *barium* salt, $C_6H_4(NO_2)PO_3Ba + 2H_2O$, crystallizes from water in lustrous yellow tablets, and the more soluble acid salt, $[C_6H_4(NO_2)PO_3H]_2Ba$, forms small white plates.

Amidophosphenylic acid, $C_6H_4(NH_2)PO_3H_2$, is obtained by the action of tin and hydrochloric acid on the preceding compound. It crystallizes in fine, white lustrous needles, slightly soluble in water, more readily in hydrochloric acid, no compound being, however, formed. At 280° it decomposes without melting, and becomes coloured bluish-green. On heating with soda-lime it decomposes into phosphoric acid and aniline. Bleaching powder added to its solution in hydrochloric acid produces a dark-red colouration, which is neither destroyed by boiling nor by standing. The solutions of its salts with the metals of the alkalis and alkaline earths are coloured red on evaporation, even when exposed in vacuo over sulphuric acid. The *silver* salt, $C_6H_4(NH_2)PO_3Ag_2$, is a yellowish white precipitate.

The action of sodium amalgam on a solution of nitrophosphenylic acid does not produce azophosphenylic acid, but sodium amidophosphenylate, $C_6H_4(NH_2)PO_3Na_2 + 3H_2O$, crystallizing in white prisms.

Diazophosphenylic acid nitrate, $C_6H_4(N_2.NO_3)PO_3H_2 + 3H_2O$, is obtained by passing nitrogen trioxide into a boiling solution of amidophosphenylic acid in nitric acid. It crystallizes in colourless prisms, which are readily soluble in water and alcohol with a yellow colour, melt at 188°, and explode violently at a slightly higher temperature. It is a very stable compound, and is not decomposed by long-continued boiling with water. Its salts, which are coloured yellow to red, are explosive; those of the metals of the alkalis and alkaline earths are soluble in water and crystallize; those of most of the other metals are insoluble.

Dimethylamidophosphenylic acid, $(CH_3)_2N.C_6H_4PO(OH)_2$. This is obtained by boiling an alcoholic solution of sodium dimethyl-amidophosphenylite with mercuric chloride. After recrystallization from alcohol it melts at 133°. It is readily soluble in water and alcohol, and is decomposed on warming with the former.[2]

1148 *Phenylphosphine*, or *Phosphaniline*, $C_6H_5.PH_2$. Michaelis first prepared this compound by passing hydriodic acid into phosphenyl chloride, the hydriodide of phosphenyl iodide, $C_6H_5PI_2.HI$, being obtained as a dark mass; on treatment with absolute alcohol and subsequent distillation, this yields phenyl-phosphine, the formation of which is explained by the following equation:

$$3C_6H_5.PI_2.HI + 9C_2H_5.OH = C_6H_5.PH_2 + 2C_6H_5.PO_3H_2 + 9C_2H_5I + 3H_2O.$$

Later researches, however, have shown that the reaction is not completely represented by this equation, but that the following also takes place:

$$C_6H_5PI_2.HI + 3C_2H_5.OH = C_6H_5.PO_2H_2 + 3C_2H_5I + H_2O.$$

[1] Michaelis and Benzinger, *Annalen*, **188**, 275.
[2] Schenk and Michaelis, *Ber.* **21**, 1500.

The phosphenylous acid is then resolved, as already described, into phenylphosphine and phosphenylic acid, a decomposition which is quite analogous to that of hypophosphorus acid into phosphine and phosphoric acid. Phenylphosphine is therefore best prepared by gradually adding crude phosphenyl chloride, which has only been submitted to a few distillations, to an excess of alcohol with continual agitation, distilling off the greater portion of the alcohol in a stream of carbon dioxide, and then further heating the residue over the naked flame.

The distillation of the phenylphosphine, accompanied by violent frothing, commences at 250°; the flame should then be removed, as the distillation proceeds spontaneously, frequently with almost explosive outbursts. If it ceases, the flask must again be warmed until the formation of two non-miscible liquids is rendered evident by the turbidity in the condenser. These are water and benzene, formed by the decomposition of the phosphenylic acid. Pure phenylphosphine is readily obtained from the product by distillation in an atmosphere of carbon dioxide.[1]

It is a colourless liquid, boiling at 160—161°, and possessing a most repulsive, penetrating smell, which is so intense that the mere opening of a flask containing the compound is sufficient to contaminate the air of a tolerably large room. It rapidly absorbs oxygen with evolution of heat and formation of phosphenylous acid.

When oxygen is passed into uncooled phenylphosphine, the rise of temperature is so great that ignition takes place. It also combines with sulphur on gentle warming, to form *phenylphosphine sulphide*, $C_6H_5PH_2S$, a thick liquid having an exceedingly unpleasant smell.

Phenylphosphonium iodide, $C_6H_5PH_3I$, is formed by the combination of dry hydriodic acid with phenylphosphine. It is decomposed by heating or by the addition of water; it may, however, be sublimed in needles in a current of hydrogen iodide.

Phenylphosphonium platinochloride, $(C_6H_5.PH_2)_2H_2PtCl_6$. Phenylphosphine is only slightly soluble in concentrated hydrochloric acid; on the addition of platinum chloride, the platinochloride is obtained in yellow crystals almost insoluble in water.

1149 *Diethylphenylphosphine*, $PC_6H_5(C_2H_5)_2$, is obtained by

---

[1] Köhler and Michaelis, *Ber.* **10**, 807.

allowing zinc ethyl, diluted with benzene, to drop into a mixture of phosphenyl chloride and benzene contained in a flask which has been filled with carbon dioxide and is surrounded by a freezing mixture; a violent hissing takes place, and so much heat is evolved that the liquid soon begins to boil. When the reaction is complete, a compound of the base with zinc chloride separates out as a viscous liquid, which is then decomposed by caustic soda.

Diethylphenylphosphine is a light, strongly refractive liquid, boiling at 220°, and possessing a characteristic, disagreeable, penetrating odour, which clings to all objects and is difficult to remove.

Diethylphenylphosphine hydrochloride, $PC_6H_5(C_2H_5)_2HCl$, is formed by combination of the base with hydrochloric acid, and is a white crystalline mass which deliquesces in the air, forming a liquid smelling like the base. It readily combines with another molecule of hydrochloric acid, forming the liquid *dihydrochloride*, $PC_6H_5(C_2H_5)_2(HCl)_2$, which is decomposed on distillation into the base and free hydrochloric acid, a portion of which recombines in the condenser to form the monohydrochloride.

Diethylphenylphosphonium chloride, $P(C_6H_5)(C_2H_5)_2Cl_2$. When chlorine is passed into diethylphenylphosphine, a separation of carbon takes place accompanied by deflagration. If, however the chlorine be diluted with air and cooled by a freezing mixture, the chloride is formed as a pale yellow liquid possessing a somewhat penetrating but not unpleasant odour, It solidifies in a crystalline mass when cooled by a freezing mixture; when heated it decomposes with carbonization.

Diethylphenylphosphine oxide, $PC_6H_5(C_2H_5)_2O$, is only slowly formed when the base is kept in contact with air or oxygen, while on heating, a more complete oxidation, accompanied by deflagration, takes place. The oxide is, however, readily obtained by decomposing the chloride with water, removing the greater portion of the hydrochloric acid by evaporation, and treating the residue with silver oxide. It forms colourless, transparent needles, which are very hygroscopic, exceptionally soluble in water and have an aromatic, fruit-like odour. It melts at 55—56°, and boils above 360°.

Triethylphenylphosphonium iodide, $PC_6H_5(C_2H_5)_3I$, is formed by the combination of diethylphenylphosphine with ethyl iodide. It is soluble in water, and crystallizes from alcohol in radiating needles, melting at **115°**.

Triethylphenylphosphonium hydroxide, $PC_6H_5(C_2H_5)_3OH$, is obtained by boiling the aqueous solution of the iodide with silver oxide. It is a crystalline mass which is exceptionally soluble in water, and attracts both moisture and carbon dioxide from the air. The *platinochloride*, $[PC_6H_5(C_2H_5)_3]_2PtCl_6$, crystallizes in fine, orange-yellow tablets, slightly soluble in alcohol and readily in water.

1150 *Diphenylphosphorus chloride*, $(C_6H_5)_2PCl$, is formed when mercurydiphenyl is heated with an excess of phosphenyl chloride for one hour to 220—230° in an apparatus connected with an inverted condenser : [1]

$$C_6H_5PCl_2 + Hg(C_6H_5)_2 = (C_6H_5)_2PCl + Hg(C_6H_5)Cl.$$

It is best prepared by heating phosphenyl chloride for 4-5 days in a sealed tube at 300°,[2] and forms a colourless, oily liquid, boiling at 320° under the ordinary pressure, and at 210-215° under 57 mm. It is readily soluble in benzene, is decomposed by alcohol and water, and absorbs oxygen from the air, with formation of $C_6H_5.POCl$.

Diphenylphosphorus trichloride $(C_6H_5)_2PCl_3$, is readily formed by the combination of the preceding compound with chlorine, and is a crystalline mass.

Diphenylphosphinic acid, $(C_6H_5)_2PO(OH)$, is obtained by the oxidation of diphenylphosphorus chloride with nitric acid (Michaelis), or by the decomposition of the trichloride with water. It is, however, best prepared by warming one molecule of phosphenyl chloride to 100°, and then allowing one molecule of water to flow in gradually, heating first to 200° and then to 260°. On cooling, a very hard, bright yellow mass is obtained, from which water extracts phosphenylic acid, together with a little phosphenylous acid. The diphenylphosphinic acid is obtained from the residue by treatment with alcohol, a yield of about 30 per cent. on the phosphenyl chloride employed being obtained. A tough, yellow mass, which will be subsequently mentioned, is left behind, being insoluble in alcohol.

Diphenylphosphinic acid forms large, apparently asymmetric crystals, which melt at 190° and are insoluble in water, slightly soluble in cold, readily in hot alcohol. It dissolves in nitric

---

[1] Michaelis, *Ber.* **10**, 627 ; Michaelis and Link, *Annalen*, **207**, 208 ; Michaelis and La Coste, *Ber.* **18**, 2109.   [2] Dörken, *ibid.* **21**, 1505.

acid, and crystallizes from this solution in needles, which lose a molecule of water at 230°, and are converted into the anhydride $[C_6H_5)_2PO]_2O$.

Its soluble salts also crystallize very well, the most characteristic being the *calcium* salt, $[(C_6H_5)_2PO_2]_2Ca + 3H_2O$, which forms asymmetric crystals, and is much more soluble in cold water than in hot.

Ethyl diphenylphosphinate, $(C_6H_5)_2PO(OC_2H_5)$, crystallizes in colourless needles melting at 165°.

The yellow mass mentioned above contains the compound $C_6H_5P_4H$, which is, therefore, a phenyl derivative of solid hydrogen phosphide, and is formed in larger quantities when phosphenyl chloride is allowed to decompose gradually in damp air. It is a dark yellow, amorphous powder, having a feeble odour resembling that of phenylphosphine; it is insoluble in water, alcohol, ether and cold carbon disulphide, kindles when warmed in the air, and is oxidized by nitric acid to phosphenylic acid and phosphoric acid.

Together with this substance occurs a compound, $(C_6H_5)_2P_5O_2H$ or $C_6H_5P_4.O.PO(C_6H_5)H$, which is very soluble in carbon disulphide, crystallizes in yellow needles, and yields the same oxidation products as the insoluble compound.

The compounds described above are also formed by the action of phosphenyl chloride on phosphenylous acid:[1]

$$C_6H_5PCl_2 + C_6H_5PO_2H_2 = 2C_6H_5PO + 2HCl$$
$$5C_6H_5PO = (C_6H_5)_4P_2O_3 + P_2 + C_6H_5PO_2.$$
$$5C_6H_5PO + H_2O + 3P_2 = 2C_6H_5P_4H + 3C_6H_5PO_2.$$
$$= 2(C_6H_5)_2P_5O_2H + C_6H_5PO_2.$$

The anhydrides, $P_2(C_6H_5)_4O_3$, and, $P(C_6H_5)O_2$, then combine with water forming diphenylphosphinic and phosphenylic acids.

1151 *Diphenylphosphine*, $(C_6H_5)_2PH$, is formed together with diphenylphosphinic acid, by allowing diphenylphosphorus chloride to drop into a dilute solution of caustic soda, the operation being carried on in an atmosphere of hydrogen. Diphenylphosphinous acid is first formed in this reaction but is immediately decomposed by water:

$$2(C_6H_5)_2POH = (C_6H_5)_2PH + (C_6H_5)_2PO.OH.$$

This behaviour corresponds to the decomposition of phosphenylous acid into phenylphosphine and phosphenylic acid, which, however, only takes place on heating.

[1] Michaelis and Götter, *Ber.* **11,** 885.

Diphenylphosphine is also obtained by heating diphenylphosphorus chloride (p. 391) with zinc and decomposing the compound thus formed, which has the composition $(C_6H_5)_2.P.ZnCl$, with water.[1]

Diphenylphosphine is a liquid boiling at 280°, and possessing a very unpleasant smell, which, however, is not so penetrating as that of the primary base. It dissolves in concentrated hydrochloric and hydriodic acids, but is reprecipitated by water. It is oxidized by nitric acid or chlorine water to diphenylphosphoric acid.[2]

Methyldiphenylphosphine, $(C_6H_5)_2PCH_3$, is obtained by the action of zinc methyl on diphenylphosphorus chloride. It is a colourless, strongly refractive liquid, boiling at 284° (Michaelis and Link), and having a penetrating odour.

Methyldiphenylphosphine oxide, $(C_6H_5)_2CH_3PO$, is formed by the oxidation of the preceding compound in the air, as well as by treating an aqueous solution of triphenylmethylphosphonium iodide with silver oxide, the hydroxide, which is first formed, decomposing even in the cold, more rapidly on warming, with separation of benzene, thus showing that the methyl group is more firmly connected with the phosphorus than the phenyl:

$$P(C_6H_5)_3(CH_3)OH = P(C_6H_5)_2(CH_3)O + C_6H_6.$$

Methyldiphenylphosphine oxide crystallizes from boiling ether in prisms which melt at 110—111°, and are odourless in the pure state. In the preparation of this compound, as well as of its homologues, from the phosphonium iodides, resinous byproducts are formed, smelling like peppermint or chloral.[3]

Ethyldiphenylphosphine, $(C_6H_5)_2PC_2H_5$, is a liquid boiling at 293°; its oxide, $P(C_6H_5)_2(C_2H_5)O$, forms lustrous prisms melting at 121°.

Diethyldiphenylphosphonium iodide, $P(C_6H_5)_2(C_2H_5)_2 I$, is formed by the combination of ethyldiphenylphosphine with ethyl iodide, and forms fine colourless crystals, having a bitter taste and melting at 204°.

Derivatives of phenylphosphine and diphenylphosphine containing both methyl and ethyl, and others containing ethylene, have also been obtained.[4]

[1] Dörken, *Ber.* **21,** 1508.
[2] Michaelis and Gleichmann, *ibid.* **15,** 801
[3] Michaelis and v. Soden, *Annalen,* **229,** 315.
[4] Gleichmann, *Ber.* **15,** 198

Phenoxydiphenylphosphine, $P(C_6H_5)_2OC_6H_5$, is obtained by warming diphenylphosphorus chloride with phenol in an atmosphere of hydrogen. The mass is then heated somewhat more strongly in a current of hydrogen to remove the hydrochloric acid formed by the reaction, and the temperature is finally raised to 200° and any unaltered phenol distilled off.

Phenoxydiphenylphosphine is a thick, colourless oil, which is soluble in ether and alcohol, boils at 265—270° under a pressure of 62 mm., and becomes very viscous, without, however, solidifying when cooled in a freezing mixture. As it does not boil under the ordinary pressure without decomposition, its vapour density was determined under diminished pressure and found to be 9·97—10·07.

On boiling with water it decomposes into phenol, diphenylphosphine and diphenylphosphinic acid. This decomposition, which is analogous to that of diphenylphosphorus chloride, is more rapidly brought about by caustic soda.

Phenoxydiphenylphosphine oxide or *Phenyl diphenylphosphinate*, $(C_6H_5)_2PO(OC_6H_5)$, is obtained by the direct combination of phenoxydiphenylphosphine and oxygen. The latter also combines with bromine to form the compound $(C_6H_5)_2PBr_2(OC_6H_5)$, which has not been obtained pure, but is converted by treatment with water or boiling with caustic soda into the oxide. This crystallizes from alcohol in small needles or prisms melting at 135—136°.

Phenoxydiphenylphosphine sulphide, $(C_6H_5)_2PS(OC_6H_5)$, is readily obtained by mixing solutions of phenoxydiphenylphosphine and sulphur in carbon disulphide. It crystallizes from hot alcohol in fine needles, and from ether in small transparent prisms melting at 124°.

The analogous selenium compound has also been prepared.

Phenoxydiphenylmethylphosphonium iodide, $(C_6H_5)_2P(OC_6H_5)$ CH_3I, is obtained by the continued heating of phenoxydiphenylphosphine with methyl iodide; it forms a crystalline mass, which deliquesces in damp air, and is decomposed by boiling water with formation of hydriodic acid, phenol and diphenylmethylphosphine oxide.[1]

1152 *Triphenylphosphine*, $P(C_6H_5)_3$, is obtained by adding sodium to an ethereal solution of phosphenyl chloride and bromobenzene (Michaelis and Gleichmann). It may be obtained

---

[1] Michaelis and La Coste, *Ber.* **18**, 2109.

in a more simple manner by replacing the phosphenyl chloride by phosphorus trichloride : [1]

$$PCl_3 + 3C_6H_5Br + 6Na = P(C_6H_5)_3 + 3NaCl + 3NaBr.$$

The cheaper chlorobenzene may be employed instead of bromobenzene.[2]

Triphenylphosphine crystallizes in imperfect, short, transparent prisms or tablets, which from their optical properties must belong to the monosymmetric system; they are almost odourless and dissolve readily in alcohol and ether. It melts at 79°, but the melting-point is lowered by small traces of impurities. In an atmosphere of any indifferent gas it boils above 360° with slight decomposition, and hence the determinations of its vapour density give too high a number, viz. 9·63—10·61 instead of 9·07.

It is not attacked by dry chlorine even on heating; it readily dissolves in concentrated hydrochloric acid even at the ordinary temperature, but is reprecipitated by water. On adding platinum chloride to the hydrochloric acid solution, an amorphous yellow precipitate of $[P(C_6H_5)_3]_2H_2PtCl_6$ is obtained.

Triphenylphosphonium iodide, $P(C_6H_5)_3HI$, is obtained by dissolving the phosphine in hot hydriodic acid; on cooling it separates out in fine needles, while it crystallizes from hot glacial acetic acid in long prisms. It is resolved into its components by water, and melts at 215° with decomposition, a portion subliming unaltered.

Triphenylphosphine also combines with the elements of the chlorine group, but the compounds formed have not been obtained pure.

Triphenylphosphine oxide, $P(C_6H_5)_3O$. Oxygen is not absorbed by the phosphine even when the latter is fused. The hydroxide, $P(C_6H_5)_3(OH)_2$, may, however, be obtained by treating it with an excess of water, adding the necessary amount of bromine, warming, and then boiling with caustic soda until the oily product has become colourless, or by adding potassium chlorate to the warm solution of the phosphine in hydrochloric acid. It is slightly soluble in hot water, readily in alcohol, and crystallizes from a mixture of light petroleum and benzene in well developed prisms. It readily loses water and is converted at

---

[1] Michaelis and Reese, *Ber.* **15**, 1610.
[2] Michaelis and v. Soden, *Annalen*, **229**, 295.

100° into the oxide, melting at 153°·5 and boiling at 360° without decomposition; the vapour density was found to be 9·79, the theoretical value being 9·68.[1]

Triphenylphosphine oxide is not acted on by bromine, oxygen, sulphur, methyl iodide, &c, while the isomeric phenoxydiphenylphosphine, like triphenylphosphine, readily combines with these. This is due to the fact that the phosphorus is trivalent in the two latter, and pentavalent in their oxides:[2]

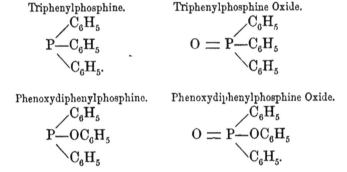

Triphenylphosphine.

Triphenylphosphine Oxide.

Phenoxydiphenylphosphine.

Phenoxydiphenylphosphine Oxide.

Triphenylphosphine nitrate, $P(C_6H_5)_3(NO_3)_2$, is obtained by dissolving the phosphine in fuming nitric acid and evaporating the solution. It forms a yellow, radiating, crystalline mass which continually gives off nitric acid, and on standing over sulphuric acid and slaked lime is changed into the basic salt $P(C_6H_5)_3NO_3.OH$, which melts at 75°, and is converted into the hydroxide on boiling with water.

Triphenylphosphine sulphide, $P(C_6H_5)_3S$, is obtained by evaporating a solution of the phosphine and sulphur in carbon disulphide; it crystallizes from hot alcohol in long needles having a silky lustre, melting at 157°·5, and boiling with slight decomposition above 360°.

Triphenylphosphine selenide, $P(C_6H_5)_3Se$, is a very similar substance, which melts at 183—184°, and is obtained by heating the phosphine with selenium.

Compounds of triphenylphosphine with the alcoholic iodides are very readily obtained; they are only slightly soluble in water, and readily in alcohol, but almost insoluble in ether, have a bitter taste and become coloured yellow in the air.

Triphenylmethylphosphonium iodide, $P(C_6H_5)_3CH_3I$. Methyl iodide combines energetically with the phosphine to form this compound, which crystallizes from water in small plates having a vitreous lustre ; the melting point of these crystals lies at first at 165—166°, but after repeated recrystallization rises gradually to 182—183°, at which point it remains constant.

On boiling its aqueous solution with silver chloride, the chloride, $P(C_6H_5)_3CH_3Cl + H_2O$, is formed and may be obtained by evaporation as a crystalline mass, losing water at 100—110°, and melting at 212—213°. The *platinochloride*, $[P(C_6H_5)_3CH_3]_2PtCl_6$, is a precipitate crystallizing from water or alcohol in yellowish red needles.

Triphenylethylphosphonium iodide,

Melting-point.

$P(C_6H_5)_3C_2H_5I$, broad, colourless tablets . . .   164—165°

Triphenylpropylphosphonium iodide,

$P(C_6H_5)_3C_3H_7I$, strongly lustrous, thick, mono-symmetrical plates . . . .   201·5°

Triphenylisopropylphosphonium iodide,

$P(C_6H_5)_3CH(CH_3)_2I + 2H_2O$, thick, opaque plates . . . .   —

Triphenylisobutylphosphonium iodide,

$P(C_6H_5)_3C_4H_9I$, small lustrous plates or needles. 176—177°

Triphenylamylphosphonium iodide,

$P(C_6H_5)_3C_5H_{11}I$, colourless prisms . . . . . .   174°

By the action of freshly precipitated silver oxide on these compounds the corresponding hydroxides are obtained ; these, as already mentioned, are rapidly resolved into benzene and diphenylphosphine oxides, containing an alcohol radical.

Methylenehexphenylphosphonium iodide, $P_2(C_6H_5)_6CH_2I_2$, is formed when the phosphine is dissolved in methylene iodide, and crystallizes in small lustrous needles.

Ethylenehexphenylphosphonium bromide, $P_2(C_6H_5)_6C_2H_4Br_2$, is a crystalline powder, slightly soluble in water.[1]

1153 *Trinitrotriphenylphosphine oxide*, $P(C_6H_4.NO_2)_3O$. One part of triphenylphosphine oxide is carefully added to a mixture of two parts of fuming nitric acid and five parts of concentrated sulphuric acid, so that the temperature does not become higher than 15—20°, and the mixture then poured into cold water. A mixture of two isomeric nitro-compounds then separates out, one of which can be easily removed by boiling alcohol. This is a bright yellow resinous body, and has not been obtained

---

[1] Michaelis and Gleichmann, *Ber.* **15**, 803 ; *Annalen*, **229**, 318.

perfectly pure. The chief product, amounting to from 85 to 90 per cent. of theoretical yield, remains behind as a yellowish white crystalline powder. It is then dissolved in boiling glacial acetic acid, and precipitated from solution in the form of long, yellowish, oblong plates by the addition of four volumes of alcohol. These can be obtained nearly colourless by re-crystallization from glacial acetic acid, with or without sub-sequent addition of alcohol. The crystals melt at 224° and deflagrate at a higher temperature.

Triamidotriphenylphosphine oxide, $P(C_6H_4.NH_2)_3O$, is obtained by reducing the nitro-compound with tin and hydrochloric acid. It crystallizes from alcohol in colourless prisms containing a molecule of alcohol, which is driven off at 100—110°. It separates from a large quantity of hot water, in anhydrous, glittering, reddish-coloured plates melting at 258°. Its salts are very soluble and remain on evaporating their solutions as gum-like or vitreous masses.

When the base is boiled with acetic anhydride a *triacetyl* compound, $P(C_6H_4.NH.C_2H_3O)_3O + H_2O$, is formed. This separates from solution in dilute acetic acid or alcohol in nodular crystals, which do not become perfectly anhydrous below 130—150°, and then fuse at 186—187° with evolution of gas.

Hexmethyltriamidotriphenylphosphine oxide, $P[C_6H_4.N(CH_3)_2]_3O$, is formed when the amido-compound is heated with methyl iodide and wood-spirit. It crystallizes from alcohol either in needles containing one molecule of alcohol, or in crystals free from alcohol, melting at 149—152°. If an excess of bromine-water be added to the hot aqueous solution of the amido-compound containing hydrochloric acid, a light reddish grey precipitate of $P(C_6H_3Br_2NH_2)_3O$ is formed, whilst chlorine water gives a brown precipitate (Michaelis and von Soden).

Diphosphobenzene hydroxide, $C_6H_5P{=}POH$, is prepared by the action of spontaneously inflammable hydrogen phosphide on phosphenyl chloride, the product being triturated and washed with alcohol. It is a yellow powder, readily soluble in carbon disulphide, which takes fire when warmed in the air, and is oxidized to phosphenylic and phosphoric acids when heated with nitric acid.[1]

Diphosphenyl, $C_6H_5P{=}PC_6H_5$, corresponds to azobenzene, as the foregoing compound does to diazobenzene hydroxide; hence it has been termed *phosphobenzene*. To prepare it, phenyl-

[1] Michaelis, *Ber.* **8**, 499.

phospine is brought into a flask through which a current of dry hydrogen is passing, and phosphenyl chloride allowed to drop in :

$$C_6H_5PCl_2 + C_6H_5PH_2 = \begin{matrix} C_6H_5P \\ \| \\ C_6H_5P \end{matrix} + 2HCl.$$

The mass is gently warmed, and the product washed first with water and then with pure ether. It is a yellowish powder readily soluble in hot benzene, melting at 149—150° and solidifying to a crystalline mass. It gradually oxidizes in the air forming *oxyphosphobenzene*, $(C_6H_5P)_2O$. On heating with concentrated hydrochloric acid it splits up into phenylphosphine and phosphenyl chloride, which latter is, however, at once converted by the water present into phosphenylous acid. This acid is also formed by oxidation with dilute nitric acid, whilst the concentrated acid converts it into phosphenylic acid (Köhler and Michaelis).

ARSENIC DERIVATIVES OF BENZENE.

1154 *Arsenphenyl chloride*, $C_6H_5AsCl_2$. This is formed together with diphenyl, $C_{12}H_{10}$, when the vapours of benzene and arsenic trichloride are passed together through a red-hot tube. The two compounds cannot be separated either by crystallization or distillation, and hence the chloride is best prepared by the action of arsenic trichloride on mercurydiphenyl.[1] It is a colourless, powerfully refracting, not very mobile liquid, boiling at 252—255° and not fuming in the air. When cold it possesses a faint unpleasant smell, but when warm its odour is penetrating; it acts on the skin as a powerful caustic. It is not attacked by water even when boiling. With chlorine it combines to form a tetrachloride, but bromine decomposes it with formation of *p*-dibromobenzene.

Arsenphenyl oxide, C_6H_5AsO, is prepared by the action of a solution of sodium carbonate on the chloride; it separates from alcoholic solution in crystalline crusts, which melt at 119—120°, and possess when cold a peculiar odour resembling that of aniseed, but when hot have a penetrating smell, the vapour attacking the mucous membrane of the nose violently. It does

[1] La Coste and Michaelis, *Annalen*, **201**, 191.

not dissolve in water but is soluble in alkalis. When strongly heated it decomposes into arsenic trioxide and triphenylarsine.

Arsenphenyl bromide, $C_6H_5AsBr_2$, is obtained by warming the oxide with an excess of concentrated hydrobromic acid. It is a liquid possessing a faint smell and boiling at 285° with decomposition. It is readily converted by bromine into bromobenzene and arsenic tribromide.

Arsenphenyl iodide, $C_6H_5AsI_2$, is a heavy, oily, red-coloured liquid. When its alcoholic solution is treated with phosphorous acid, *iodarsenobenzene*, $C_6H_5AsI.AsIC_6H_5$, is formed. This separates out in bright yellow needles, which on addition of iodine are converted into the original compound. It is a very unstable body, and deliquesces in the air.[1]

Arsenphenyl tetrachloride, $C_6H_5AsCl_4$, crystallizes in broad yellow needles melting at 45°, which fume in the air and are readily decomposed by water. It behaves towards acids in quite a different manner from the corresponding phosphorus compound; it dissolves readily in cold glacial acetic acid; on warming no acetyl chloride is formed, but chloracetic acid is obtained, whilst sulphur dioxide does not act upon it. When heated in an open vessel, or when carbon dioxide is led through the warm liquid, it is decomposed into chlorine and arsenphenyl chloride. When heated to 150° in a closed vessel, it is converted into arsenic trichloride and chlorobenzene.

Arsenphenyl oxychloride, $C_6H_5AsOCl_2$, is formed when the tetrachloride is allowed to drop into the necessary quantity of water; it may, however, be obtained more easily and in a purer condition by the combination of chlorine and arsenphenyl oxide. It is a white crystalline mass which melts at about 100° and fumes slightly in the air.

Phenylarsinic acid, $C_6H_5AsO(OH)_2$, is prepared by acting on the oxychloride or tetrachloride with water. It is slightly soluble in cold, readily in hot water, and crystallizes from alcohol in long prisms and compact masses. When heated to 140° it is converted into the anhydride, $C_6H_5AsO_2$, an amorphous powder which readily re-combines with water. It is a very stable compound which resists the action of powerful oxidizing agents, and is scarcely acted on by reducing agents; on boiling with iodine and amorphous phosphorus it yields arsenphenyl iodide. When it is heated with soda-lime, benzene is formed, which is also obtained together with phenol on fusing it with alkalis.[2]

---

[1] Michaelis and Schulte, *Ber.* **14**, 913.   [2] La Coste, *Annalen*, **208**, 9.

Phenylarsinic acid is poisonous; the symptoms begin and death follows rather more slowly than is the case with animals poisoned by arsenious acid, but hardly less rapidly than with arsenic acid. The phenylarsinates of the alkali metals are acid salts, but possess a neutral reaction, and, with the exception of the ammonium salt, they do not crystallize. The metals of the alkaline earths also readily form acid salts, which however crystallize well. The heavy metals replace two atoms of hydrogen, and their salts are either slightly soluble or insoluble in water.

Dimethylphenylarsine, $AsC_6H_5(CH_3)_2$, is formed when zinc methyl acts on the dichloride. It is a thin liquid which boils at 200° and possesses a pungent disagreeable smell.[1]

Diethylphenylarsine, $AsC_6H_5(C_2H_5)_2$, is a colourless, powerfully refracting liquid which possesses a disagreeable smell and boils at 240°. It combines with chlorine forming diethylphenylarsine chloride, $AsC_6H_5(C_2H_5)_3Cl_2$, which crystallizes well.

Triethylphenylarsonium iodide, $AsC_6H_5(C_2H_5)_3I$, is obtained by heating diethylphenylarsine with ethyl iodide to 100°. It crystallizes from aqueous solution in prisms which have a very bitter taste, melt at 112—113°, and are resolved into their constituents when heated in a current of carbon dioxide. The strongly alkaline hydroxide and the chloride obtained from the above compound are syrupy masses (La Coste and Michaelis).

1155 *Arsendiphenyl chloride,* $(C_6H_5)_2AsCl$, is prepared by boiling an excess of arsenphenyl chloride with mercurydiphenyl. It is a light yellow, oily, non-fuming liquid which when heated has a powerfully pungent smell, and attacks the skin, though less violently than arsenphenyl chloride. It boils at 280° and is converted by alcoholic potash into arsendiphenyl oxide, $[(C_6H_5)_2As]_2O$, which forms a nodular crystalline mass melting at 91—92°.

Arsendiphenyl trichloride, $(C_6H_5)_2AsCl_3$. This compound is formed by the combination of the monochloride with chlorine; it crystallizes from benzene in tablets which melt at 174°. On heating to 200° it decomposes into chlorobenzene and arsenphenyl chloride.

Arsendiphenyl oxychloride, $[(C_6H_5)_2AsCl_2]_2O$, is a white powder melting at 117°, and formed by the direct union of chlorine with diphenylarsine oxide. Water decomposes it with formation of diphenylarsinic acid.

[1] Link and Michaelis, *Annalen,* **207,** 208.

Diphenylarsinic acid, $(C_6H_5)_2AsO(OH)$, is prepared by the action of water on the trichloride. It crystallizes in long needles melting at 174°, and is a weak acid which, together with its soluble salts, acts as a strong poison; being more rapid in its action than phenylarsinic acid. Boiling concentrated nitric acid does not attack it.

Diphenylmethylarsine, $As(C_6H_5)_2CH_3$, is obtained by the action of zinc methyl on the chloride. It is an oily, powerfully refractive liquid, boiling at 206° and possessing a very penetrating smell.

Diphenylethylarsine, $As(C_6H_5)_2C_2H_5$, boils at 220°, and has a not unpleasant fruit-like smell. It combines with chlorine to form diphenylethylarsine chloride, which crystallizes from benzene in long needles melting at 137°.

1156 *Triphenylarsine*, $As(C_6H_5)_3$, is formed when arsenphenyl oxide is heated to 180—200° :

$$3C_6H_5AsO = (C_6H_5)_3As + As_2O_3.$$

It can however be prepared more readily by acting on a mixture of arsenic trichloride, bromobenzene and ether with sodium.[1] It crystallizes from arsendiphenyl chloride in thin, brittle plates having a vitreous lustre, and melting at 58—59°. In a current of carbon dioxide it boils above 360° and does not combine with ethyl iodide even on heating.

Triphenylarsine chloride, $(C_6H_5)_3AsCl_2$, crystallizes from hot benzene in plates which fume slightly in the air and melt at 171°. When heated in a sealed tube it decomposes into arsendiphenyl chloride and chlorobenzene.

Triphenylarsine hydroxide, $(C_6H_5)_3As(OH)_2$, is formed by boiling the chloride with water, or better with dilute ammonia. It crystallizes on evaporation in plates or white needles which melt at 108°. Over sulphuric acid it effloresces at 105—110°, forming *triphenylarsine oxide*, $(C_6H_5)_3AsO$, which melts at 189°.

Triphenylarsine sulphide, $(C_6H_5)_3AsS$, is obtained by fusing triphenylarsine with sulphur, or more easily by boiling triphenylarsine chloride with yellow ammonium sulphide. It is insoluble in water and the alkali-sulphides, but crystallizes from hot alcohol in silky needles melting at 162° (La Coste and Michaelis).

p-Trianisylarsine, $(CH_3O.C_6H_4)_3As$. This compound is pre-

---

[1] Michaelis and Reese, *Ber.* **15**, 2876.

pared by adding 20 grms. of sodium to a mixture of 50 grms. of p-bromanisoïl, 30 grms. of arsenic trichloride and a little ethyl acetate, the whole being diluted with four times its volume of ether. p-Trianisylarsine separates from a mixture of benzene and alcohol in cubical transparent crystals which melt at 156°, and are soluble with difficulty in alcohol and ether, readily in benzene. It may be readily converted into derivatives containing one or two p-anisyl groups.[1]

p-Triphenctylarsine, $(C_2H_5O.C_6H_4)_3As$, is prepared in a similar manner from p-bromophenetoïl. It melts at 88-89° and is readily soluble in ether.[2]

Arscnobenzene, $C_6H_5As{=}AsC_6H_5$. This body is formed by the action of most reducing agents on an alcoholic solution of arsenphenyl oxide, C_6H_5AsO; the best method is to boil it with crystallized phosphorous acid :

$$2C_6H_5AsO + 2PO_3H_3 = \begin{matrix} C_6H_5As \\ || \\ C_6H_5As \end{matrix} + 2PO_4H_3.$$

It crystallizes in yellowish needles, melting at 196°, slightly soluble in alcohol, but readily in benzene; the solution soon becomes resinous. On heating strongly it decomposes into triphenylarsine and metallic arsenic; chlorine converts it into arsenphenyl chloride.[3]

ANTIMONY DERIVATIVES OF BENZENE.

1157 *Diphenylstibinc chloridc*, $(C_6H_5)_2SbCl_3$. This compound is formed together with triphenylstibine and its dichloride by adding 17 grms. of sodium to a solution of 40 grms. of antimony trichloride and 40 grams. of chlorobenzene in benzene, and allowing the mixture to stand. After evaporating off the benzene the residue is warmed with alcoholic hydrochloric acid, filtered, the alcohol evaporated off, and the dark oily compound remaining behind extracted with hot dilute hydrochloric acid. On cooling long lustrous needles separate from the filtered extract, which have the composition $(C_6H_5)_2SbCl_3 + H_2O$. Diphenylstibine

---

[1] Michaelis and Weitz, *ibid.* **20**, 49.
[2] *Ibid.* p. 52.
[3] Michaelis and Schulte, *Ber.* **14**, 912.

chloride loses its water of crystallization at 110°, and then melts at 180°. It is insoluble in water, but is slowly decomposed by it, and dissolves in dilute hydrochloric acid, more readily in alcohol. It cannot be obtained by the methods employed in the case of the corresponding phosphorus and arsenic compounds.[1]

Diphenylstibic acid, $(C_6H_5)_2SbO.OH$ is obtained by treating an alcoholic solution of the chloride with ammonia. It is a light white powder, insoluble in water, alcohol, and sodium carbonate solution, but soluble in acetic acid and caustic soda. It decomposes on heating above 250° without melting, and is converted by hydrochloric acid into the above chloride.[2]

Triphenylstibine, $(C_6H_5)_3Sb$. This compound is best prepared by dissolving 40 grms. of antimony trichloride and 40 grms. of chlorobenzene in four times their volume of benzene, and adding 50 grms. of sodium. After allowing to stand for twenty-four hours, the mixture is warmed and filtered, the insoluble matter being well washed with benzene, and the combined filtrates evaporated. The residue is warmed with alcoholic hydrochloric acid, when all diphenylstibine chloride passes into solution, and triphenylstibine oxide is converted into the chloride. The insoluble portion is washed with alcohol, dried, mixed with light petroleum, and chlorine passed on to the surface of the solution as long as any chloride is precipitated. This is then filtered off, washed with light petroleum, re-crystallized from alcohol, dissolved in alcoholic ammonia, and decomposed by sulphuretted hydrogen. The separated triphenylstibine is filtered off, washed with water, and recrystallized from alcohol, ether, or light petroleum. It forms transparent asymmetric tablets, which melt at 48°, and boil with slight decomposition at 360°, are insoluble in water and dilute acids, but dissolve readily in alcohol, ether, benzene, and light petroleum. It combines directly with halogens, but not with sulphur.[3]

Triphenylstibine hydroxide, $(C_6H_5)_3Sb(OH)_2$, is obtained by the action of alcoholic potash on the chloride or bromide. It is a light white powder which melts at 212°, and is insoluble in ether and light petroleum, but readily soluble in alcohol. It is converted by haloid acids into the corresponding salts.

Triphenylstibine dichloride, $(C_6H_5)_3SbCl_2$, is obtained in the preparation of triphenylstibine, and forms long, thin needles,

[1] Michaelis and Reese, *Annalen,* **233,** 57.
[2] *Ibid.* p. 59.   [3] *Ibid.* p. 42.

melting at 143°. It is readily soluble in carbon bisulphide and benzene, and is quickly converted by alcoholic potash into the hydroxide, though only slowly attacked by aqueous alkalis.

Triphenylstibine dibromide, $(C_6H_5)_3SbBr_2$, and *triphenylstibine di-iodide*, $(C_6H_5)_3SbI_2$, are obtained by the action of bromine and iodine on triphenylstibine. The first forms glassy crystals melting at 216°, and the second lustrous white tablets melting at 153°.

Triphenylstibine nitrate, $(C_6H_5)_3Sb(NO_3)_2$, is prepared by dissolving triphenylstibine hydroxide in fuming nitric acid. It melts at 156° with evolution of red fumes.

BISMUTH DERIVATIVES OF BENZENE.

1158 *Diphenylbismuth bromide*, $(C_6H_5)_2BiBr$, is prepared by mixing ethereal solutions of bismuthtriphenyl and bismuth bromide. It crystallizes from chloroform in yellow nodules, which melt at 157—158°, dissolve readily in chloroform, but not in ether. It is decomposed by alcohol with formation of an oxybromide, and is converted by ammonia into bismuth hydrate and bismuthtriphenyl.[1]

Bismuthtriphenyl, $(C_6H_5)_3Bi$, is prepared by heating a mixture of 500 grms. of bromobenzene, and 500 grms. of a bismuth sodium alloy (1 part of Na and 10 parts of Bi) at 160° for 50 hours. Water is then added, and after distilling off the benzene with steam, the solution is extracted with chloroform, the extract evaporated, and alcohol added. After recrystallization from alcohol and chloroform with addition of animal charcoal, bismuthtriphenyl is obtained in long thin monosymmetric prisms melting at 78°, and is sometimes obtained in monosymmetric plates melting at 75°. It is sparingly soluble in alcohol, readily in ether and light petroleum, and even more readily in chloro-[2]

Bismuthtriphenyl dichloride, $(C_6H_5)_3BiCl_2$, is formed hy the action of chlorine on bismuthtriphenyl dissolved in ligbt petroleum. It forms long lustrous prisms wbich melt at 141·5° are sparingly soluble in ether and cold alcohol, readily in benzene.

---

[1] Michaelis and Marquardt, *Annalen*, **251**, 327.
[2] Michaelis and Polis, *Ber.* **20**, 55; Michaelis and Marquardt, *Annalen*, **251** 324.

Bismuthtriphenyl dibromide, $(C_6H_5)_3BiBr_2$, is obtained from bismuthtriphenyl and bromine and forms long needles which melt when quickly heated at 122°.

Bismuthtriphenyl nitrate, $(C_6H_5)_3Bi(NO_3)_2$, is prepared by the addition of silver nitrate to an alcoholic solution of the bromide. It forms colourless needles, which decompose with explosive violence on heating.

BORON DERIVATIVES OF BENZENE.

1159 *Phenylboron chloride*, $C_6H_5BCl_2$. When boron trichloride is heated to 180—200° with mercurydiphenyl, this body is formed. It is a colourless liquid which fumes in the air, and is violently decomposed by water. It boils at 175°, solidifying at the ordinary temperature to a crystalline mass which melts about 0°. At the ordinary temperature it does not absorb chlorine, but when placed in a freezing mixture it takes up about two atoms, becoming liquid. If the freezing mixture be removed, a portion of the compound produced is resolved into chlorine and phenylboron chloride, another part yielding chlorobenzene and boron trichloride:

$$C_6H_5BCl_4 = C_6H_5Cl + BCl_3.$$

This decomposition corresponds exactly to that of arsenmethyl tetrachloride into methyl chloride and arsenic trichloride.

Phenylboric acid, $C_6H_5B(OH)_2$, is obtained by slowly dropping the chloride into water. It crystallizes from warm water in aggregates of needles melting at 204° and volatilizing slightly in a currrent of steam. The acid reddens litmus feebly and yields crystalline salts.

Sodium phenylborate, $C_6H_5B(ONa)_2$, is soluble in water and crystallizes in large quadratic tablets.

Acid calcium phenylborate, $(C_6H_5BO_2)CaH_2$, forms nodular druses.

Acid silver phenylborate, $C_6H_5.BOAg(OH)$, is a yellow precipitate which rapidly undergoes change on exposure to light. It is formed by adding silver nitrate and some ammonia to a solution of the acid. Heated with water it decomposes into benzene, boric acid, and silver oxide:

$$2C_6H_5BO_2AgH + 3H_2O = 2C_6H_6 + 2B(OH)_3 + Ag_2O.$$

Ethylphenyl borate, $C_6H_5B(OC_2H_5)_2$, is formed by acting on the chloride with absolute alcohol. It is a colourless, pleasantly smelling liquid, boiling at 176° and readily passing into the acid on standing in moist air.

A characteristic reaction for phenylboric acid is that, even in very dilute solutions, it gives a precipitate of phenylmercuric chloride with corrosive sublimate solution :

$$C_6H_5B(OH)_2 + HgCl_2 + H_2O = C_6H_5HgCl + B(OH)_3 + HCl$$

By this test one part of the acid in 25,000 parts of water can be detected.

Phenylboron oxide, C_6H_5BO. This anhydride of phenylboric acid is obtained when the latter compound is heated above its melting point. It forms a colourless crystalline mass which melts at 190° and boils without decomposition above 360°. It is soluble in alcohol and ether, but not in water, although it combines with water on long continued boiling, yielding phenylboric acid.[1]

Boric acid possesses well-known antiseptic properties, and it was natural to expect that its phenyl compound would exhibit similar properties in even a more marked degree. The experiments of Filehne and Rothaas seem to promise a future for phenylboric acid in therapeutic practice, as it possesses a mild aromatic taste, and even when taken in large doses does not produce any corrosive action, only giving rise in the human subject to singing in the ears, giddiness, and headache, which soon disappear, though it acts as a powerful poison on the lower animals. Thus, 2 mgm. acts fatally on a frog, whilst 0·5 grm. can be given to a rabbit without producing any serious effects. Its antiseptic action is from five to ten times as powerful as that of its sodium salt ; a solution of the acid of the strength of 1 to 100,000 diminishes the rate of putrefactive decomposition, and meat can be preserved fresh in a solution of 1 to 5,000 provided the quantity of preservative solution employed is large enough to prevent the acid from being neutralized by the alkaline salts contained in the flesh. The formation of bacteria is prevented by a solution containing 1 to 1,000, and greatly retarded by one of from 1 to 10,000.[2]

---

[1] Michaelis and Becker, *Ber.* **13**, 58 ; **15**, 180.   [2] *Ibid.* **15**, 182.

SILICON DERIVATIVES OF BENZENE.

1160 *Siliconphenyl chloride*, $C_6H_5SiCl_3$, is obtained by heating silicon tetrachloride with mercurydiphenyl to 300°. It is a powerfully refractive liquid which fumes in the air and possesses a smell similar to that of silicon chloride. It boils at 197°, and on ignition burns. with a strong, smoky, green-mantled flame, leaving a residue of silica.

Phenylsilicic acid, or *Silico-benzoic acid*, $C_6H_5.SiO.OH$, is formed when the chloride is added drop by drop to an excess of dilute ammonia. The acid which separates out is dissolved in ether, and on evaporation remains as a transparent mass resembling solidified beads of molten glass. From alcoholic solution it separates out as a syrup which dries to a solid mass and is then insoluble in alcohol though dissolving in ether. It melts at 92° and dissolves in caustic potash, from which it is not reprecipitated by hydrochloric acid, but if this solution be allowed to stand in the air, or if ammonia be added to the acid solution, the acid separates out. On evaporating the alkaline solution and heating the residue, benzene distils over :

$$C_6H_5SiO_2K + KOH = C_6H_6 + SiO_3K_2.$$

This decomposition corresponds to that of benzoic acid into benzene and carbon dioxide.

The salts of this acid have not been prepared. If the acid be dried at 100° it is converted into silicobenzoic anhydride, $(C_6H_5SiO)_2O$, an amorphous powder which, like the acid, burns with a smoky flame, silica coloured black by carbon remaining behind.

Phenylsilicon ether, or *Ethyl orthosilicobenzoate*, $C_6H_5Si(OC_2H_5)_3$, is formed by the action of the chloride on absolute alcohol, and is a pungent smelling ethereal liquid boiling at 235°. When heated with concentrated hydriodic acid, silicobenzoic acid is formed :

$$C_6H_5Si(OC_2H_5)_3 + 3HI = C_6H_5.SiO_2H + 3C_2H_5I + H_2O.$$

Silicontriethylphenyl, $C_6H_5Si(C_2H_5)_3$ is obtained when the chloride is heated with zinc ethyl at 150—165°. It is a colourless liquid boiling at 230°, its vapour possessing a faint smell of oil of cloves. The chloride, $SiC_{12}H_{19}Cl$, is formed by passing chlorine into the above compound at a low temperature; it is a thick liquid boiling at 260—265°, and possessing a faint aromatic odour. It is not converted into an acetate when treated with an alcoholic solution of potassium acetate even at 250°, and it is therefore probably silicon chlorophenyltriethyl, $C_6H_4ClSi(C_2H_5)_3$.[1]

Silicondiphenyl chloride, $(C_6H_5)_2SiCl_2$, is obtained by acting on 1 molecule of silicontetraphenyl with 2 molecules of phosphorus pentachloride. It is a colourless liquid, which boils at 230—237° under 90 mm. pressure.[2]

Silicontriphenyl chloride, $(C_6H_5)_3SiCl$, is obtained by heating silicontetraphenyl and phosphorus pentachloride in equimolecular proportions to 180° for $1\frac{1}{2}$ hours, and after allowing the hydrochloric acid formed to pass off, again heating at 240°, and then fractionating under 90 mm. pressure. It separates from light petroleum in colourless crystals, which fume slightly in the air, melt at 88—89°, and are readily soluble in light petroleum, absolute ether, and chloroform.

Silicontriphenyl carbinol, $(C_6H_5)_3SiOH$, is best prepared by boiling the chloride with water containing ammonia. It forms colourless transparent crystals, which melt at 139—141°, distil without decomposition, and dissolve readily in alcohol, ether, chloroform and benzene. Its *ethyl* ether is obtained by acting on the chloride with alcohol.

Silicontetraphenyl, $Si(C_6H_5)_4$. To prepare this substance 23 grms. of sodium are added to an ethereal solution of 20 grms. of silicon tetrachloride, 50 grms. of chlorobenzene and a little ethyl acetate. After the reaction is over, the ether is poured off, the unaltered sodium separated, and the residue, after washing with water, recrystallized from benzene, and finally from ether or chloroform. It forms prismatic or pyramidal quadratic crystals, which melt at 233°, and boil above 530° without decomposition. It is insoluble in water, sparingly soluble in alcohol and ether, more readily in chloroform, and still more easily in benzene, and also dissolves unaltered in hot sulphuric acid. It burns in the air with separation of light flakes of silica which are carried away by the other products of combustion, leaving no residue. It is acted on by bromine forming the compound $C_{18}H_{16}Si_4O_7$ (?), an infusible and non-volatile powder.[3]

[1] Ladenburg, *Annalen*, **173**, 151.   [2] Polis, *Ber.* **19**, 1019.
[3] Polis, *Ber.* **18**, 1541 ; **19**, 1012.

Tetranitrosilicontctraphenyl, $Si(C_6H_4.NO_2)_4$, is obtained by adding silicontetraphenyl to a mixture of sulphuric and fuming nitric acids. It is a powder, which melts at 93—105°.

1161 *Dichlorosilicondiphenyldiamide,* $SiCl_2(NHC_6H_5)_2$, is obtained by the addition of silicon tetrachloride to a benzene solution of aniline. The separated aniline hydrochloride is filtered off in a current of dry air, and the filtrate evaporated over sulphuric acid in a vacuum. A thick viscous liquid remains which solidifies to a white amorphous mass. This could not be obtained crystalline, and is at once decomposed by moisture.[1]

Silicotetraphenylamide, $Si(NH.C_6H_5)_4$. This substance is prepared by gradually adding a solution of 100 grms. of silicon tetrabromide in 2 vols. benzene to a solution of 438 grms. of aniline in a similar quantity of benzene. After filtering from aniline hydrochloride the filtrate is distilled as far as possible on the water bath, and then in a current of hydrogen at 100—105°. The warm syrup is mixed with carbon bisulphide, filtered, and evaporated at as low a temperature as possible to the crystallizing point. After recrystallization from carbon bisulphide it forms colourless transparent monosymmetric crystals, which melt at 137—138°. It is very soluble in benzene, less in carbon bisulphide, and is precipitated by light petroleum from both solutions. It is slowly decomposed by water, forming silicic acid and aniline; rapid decomposition takes place if acids or alkalis are present especially in alcoholic solution.[2]

TIN DERIVATIVES OF BENZENE.

1162 *Tintriethylphenyl,* $C_6H_5Sn(C_2H_5)_3$, is formed when a mixture of tintriethyl iodide and bromobenzene diluted with ether is heated with sodium:

$$C_6H_5Br + (C_2H_5)_3SnI + 2Na = (C_2H_5)_3SnC_6H_5 + NaBr + NaI$$

It is a colourless, powerfully refractive liquid possessing rather a pleasant smell, and boiling at 254°. Its vapour undergoes partial oxidation on exposure to air, and on ignition it burns with a luminous smoky flame leaving a residue of tin oxide. If

---

[1] Harden, *Journ. Chem. Soc.* **1887**, I. 40.
[2] J. Emerson Reynolds, *Journ. Chem. Soc.* **1889**, 1. 475.

its alcoholic solution be warmed with silver nitrate, a fine silver mirror is deposited and diphenyl, $C_{12}H_{10}$, and tintriethyl nitrate are formed. When acted on by iodine it decomposes into iodobenzene and tintriethyl iodide, while fuming hydrochloric acid converts it into benzene and tintriethyl chloride.

Tinethylphenyl chloride, $C_6H_5(C_2H_5)SnCl_2$, is formed by the action of tin tetrachloride on the last named compound :

$$C_6H_5(C_2H_5)_3Sn + SnCl_4 = C_6H_5(C_2H_5)SnCl_2 + (C_2H_5)_2SnCl_2.$$

It crystallizes from ether in scales which melt at 45°, and are only slightly soluble in water and hydrochloric acid, but dissolve readily in absolute alcohol.[1]

Tindiphenyl chloride, $(C_6H_5)_2SnCl_2$. In order to prepare this compound, a mixture of equal parts of tin tetrachloride and mercurydiphenyl with light petroleum is boiled in a flask connected with a reversed condenser for twelve hours, and the liquid then distilled at a temperature below 160°. The product obtained from 300 grms. of mercurydiphenyl is next poured into 500 to 750 cc. of cold water and the mixture well shaken. The aqueous solution contains hydrochloric acid, stannic chloride, and tindiphenyl chloride; it is warmed on the water-bath from 85—90°, when, after standing for a short time, a part of the tindiphenyl chloride separates out as a heavy oil. The aqueous solution is then poured off and again warmed for two hours; a gummy powder consisting of a mixture of the chloride and hydroxychloride is deposited. After its removal and further heating for two hours almost pure hydroxychloride separates out, and then a mixture of this compound with tindiphenyl oxide and stannic oxide. The pure chloride is obtained from this last-named mixture and from that of tindiphenyl chloride and hydroxychloride by passing hydrogen chloride over the substance and finally warming to 45°.[2] The product is then extracted with light petroleum.

Tindiphenyl chloride is readily soluble in alcohol and light petroleum, crystallizing from the latter solvent in compact, transparent prisms often an inch in length, which have a diamond lustre and belong to the asymmetric system. It possesses a penetrating sweetish taste, melts at 42° and boils with partial decomposition at 333—337°. When heated with concentrated

[1] Ladenburg, *Annalen*, **159**, 251.
[2] Aronheim, *Annalen*, **194**, 145.

hydrochloric acid to 100° it decomposes into benzene and stannic chloride.

Tindiphenyl hydroxychloride, $(C_6H_5)_2Sn(OH)Cl$, is formed when the chloride is gently warmed with water, or when it is exposed to moist air. As stated above, it is also obtained in the preparation of the chloride, and it is prepared in the pure state from the product, which contains some chloride, by washing it with alcohol and then allowing the residue to stand for some time in contact with water.

It is an amorphous powder, melting at 187° and insoluble in the ordinary solvents. Concentrated hydrochloric acid converts it into the chloride.

Tindiphenyl oxide, $(C_6H_5)_2SnO$, is obtained by the decomposition of the chloride or hydroxychloride by alkalis. It is a white powder which after drying does not fuse, and possesses analogous properties to the hydroxychloride.

Tindiphenyl chlorobromide, $(C_6H_5)_2SnClBr$. When hydrogen bromide acts upon the chloride or oxychloride, this compound is formed. It closely resembles the chloride but possesses a more powerful odour, and separates from its solutions as an oily liquid which is converted into a mass of crystals on bringing into it a fragment of the crystallized compound. These crystals melt at 37°.

Tindiphenyl chloriodide, $(C_6H_5)_2SnClI$, is formed in a similar manner to the above-mentioned compound. It crystallizes in transparent, glistening monosymmetric prisms melting at 69°.

A fact worthy of note is that by the action of hydrobromic or hydriodic acids on the chloride, the weaker halogen displaces the stronger, just as phosphenyl chloride is converted by hydrobromic acid into phosphenyl bromide. This shows that tindiphenyl chloride does not act like the chloride of an organo-metallic radical but like that of a non-metal.

Tindiphenyl bromide, $(C_6H_5)_2SnBr_2$, is formed by the action of hydrobromic acid on the oxide, and of bromine on tin tetraphenyl. It is a thick oily liquid, which does not solidify, even on standing for some days, but does so at once when a crystal of the chloro-bromide is placed in the liquid, showing that this latter compound is isomorphous with the bromide, which it otherwise closely resembles. It melts at 38°, and boils at 230° under 42 mm. pressure. Hydriodic acid acts in a similar way on the oxide, but the iodide has not yet been isolated as it at once undergoes the further change into benzene and stannic iodide :

$$(C_6H_5)_2SnI_2 + 2HI = 2C_6H_6 + SnI_4.$$

Tindiphenyl di-ethyl ether, $(C_6H_5)_2Sn(OC_2H_5)_2$. This compound is prepared by adding sodium to a solution of the chloride in absolute alcohol. It separates from this solution in very bright cube-shaped crystals or in long prisms which melt with decomposition at 124° and are readily decomposed by water into the oxide and alcohol.

Tintriphenyl chloride, $(C_6H_5)_3SnCl$, is prepared by the action of sodium amalgam on an ethereal solution of tindiphenyl chloride, or by heating the latter in a current of ammonia at 100—200°. Alcohol extracts from the product tintriphenyl chloride and stannic chloride, whilst tindiphenyl hydroxychloride remains behind. This latter compound is probably derived from $(C_6H_5)_3SnCl(NH_2)$, which is first formed. The reaction, doubtless takes place in two separate stages :

(1) $3(C_6H_5)_2SnCl_2 = 2(C_6H_5)_3SnCl + SnCl_4.$

(2) $(C_6H_5)_2SnCl_2 + 2NH_3 = (C_6H_5)_2SnCl(NH_2) + NH_4Cl.$

The best mode of preparing tintriphenyl chloride is to dissolve one part of tindiphenyl chloride in from four to five parts of glacial acetic acid, and to add one molecule of sodium nitrite for every molecule of the former compound. Tintriphenyl chloride is here formed in the same way as above, but in this case it is the chief product. At the same time nitrosobenzene[1] is produced :

$$(C_6H_5)_2SnCl_2 + N_2O_3 = 2C_6H_5NO + SnOCl_2.$$

Tintriphenyl chloride separates from solution in large crystals melting at 106°. Its alcoholic solution yields a gelatinous precipitate with ammonia, and this on drying forms a white, strongly electric powder which melts at 117—118°, and has the formula $2(C_6H_5)_3SnOH + 3H_2O$, or $[(C_6H_5)_3Sn]_2O + 4H_2O$. It is soluble in hot water and yields stable, crystalline salts.

Tintetraphenyl, $Sn(C_6H_5)_4$. For the preparation of this compound 500 grms. of an alloy of 25 per cent. of sodium and 75 per cent. of tin are boiled with 600 grms. of bromobenzene and 25

---

[1] Aronheim, *Ber.* **12**, 509.

cc. of ethyl acetate for 30 hours in an oil bath. The product is extracted with benzene, and the brown crystals which separate from the cooled extract recrystallized from benzene and chloroform, with addition of animal charcoal. It forms thin colourless quadratic prisms, which melt at 225—226°, and boil without decomposition above 420°. It is readily soluble in hot benzene, chloroform, and acetic acid, sparingly in the cold liquids and in alcohol and ether, insoluble in light petroleum. It burns in the air with formation of tin dioxide, and is converted by bromide into tindiphenyl bromide.

LEAD DERIVATIVES OF PHENYL.

1163 *Leaddiphenyl oxide*, $(C_6H_5)_2PbO$. To prepare this substance the nitrate is gradually added to a boiling caustic soda solution. It is a white powder which does not melt, and is not volatile without decomposition. It is insoluble in neutral solvents, but dissolves in acids forming the corresponding salts. It does not form a hydroxide.

Leaddiphenyl chloride, $(C_6H_5)_2PbCl_2$, is obtained by passing a current of dry chlorine on to the surface of a carbon bisulphide solution of leadtetraphenyl, or by treating the nitrate with potassium chloride. It is a white powder, insoluble or sparingly soluble in ordinary solvents, which decomposes on heating without melting.

Leaddiphenyl bromide, $(C_6H_5)_2PbBr_2$, obtained by precipitating the nitrate with potassium bromide, is a powder which also decomposes on heating without melting.

Leaddiphenyl iodide, $(C_6H_5)_2PbI_2$, is prepared in a similar manner to the bromide, or by acting on leadtetraphenyl with iodine in chloroform solution. It forms golden yellow plates resembling lead iodide, and melting at 101—103°.

Leaddiphenyl nitrate, $(C_6H_5)_2Pb(NO_3)_2$, is prepared by adding leadtetraphenyl to boiling nitric acid of sp. gr. 1·4. It crystallizes from water in small colourless plates, which explode on heating. It is partially converted by boiling water into the basic salt $(C_6H_5)_2Pb(OH)NO_3$.

Besides these compounds a number of other salts have been prepared. The *chromate*, $(C_6H_5)_2PbCrO_4$, forms a yellow precipitate, the *phosphate*, $[(C_6H_5)_2Pb]_3(PO_4)_2$, a flocculent precipitate, and the *carbonate*, $(C_6H_5)_2PbCO_3$, a white powder. The *sulphide*, $(C_6H_5)_2PbS$, is obtained by the action of sulphuretted hydrogen ou an aqueous solution of leaddiphenyl acetate containing acetic acid. It separates from a mixture of alcohol and benzene in small, yellowish prismatic crystals, which melt at 80—90°, and decompose on further heating into diphenyl and lead sulphide. The *formate*, $(C_6H_5)_2Pb(CHO_2)_2$, obtained by boiling leadtetraphenyl with concentrated formic acid, crystallizes in small lustrous needles, melting above 200° with decomposition. The *acetate*, $(C_6H_5)_2Pb(C_2H_3O_2)_2 + 2H_2O$, forms long prisms which lose their water of crystallization at 110° and then melt at 195°, and is converted by hydrochloric acid into leaddiphenyl chloride and acetic acid, wherein it differs from mercuryphenyl acetate (p. 418). The *basic cyanide*, $(C_6H_5)_2Pb(OH)CN$, and the *thiocyanate*, $(C_6H_5)_2Pb(CNS)_2$, are both insoluble precipitates.

Leadtetraphenyl, $(C_6H_5)_4Pb$, is obtained by boiling 500 grms. of bromobenzene and 500 grms. of a lead sodium alloy containing 8 per cent. of Na for 60 hours, 20 cc. of ethyl acetate being also added. The bromobenzene is distilled off from the filtered liquid under diminished pressure, and the residue recrystallized from benzene. Leadtetraphenyl crystallizes in quadratic prisms, which melt at 224—225°, and commence to decompose at 270°. It is also sometimes obtained in small needles, closely resembling those of mercurydiphenyl. It is sparingly soluble in alcohol, ether, acetic acid, more readily in chloroform and benzene.[1]

MERCURY DERIVATIVES OF BENZENE.

1164 *Mercurydiphenyl*, $Hg(C_6H_5)_2$, was prepared by Dreher and Otto by acting on bromobenzene with sodium amalgam.[2]

The best yield is obtained by boiling a mixture of bromobenzene with its own volume of anhydrous coal-tar naphtha boiling between 120° and 140°, one-tenth of its weight of cthyl acetate and an excess of 2·7 per cent. sodium amalgam,[3] for some hours in connection with a reversed condenser. The product is filtered whilst hot, and the crystals which separate out purified by recrystallization from benzene and absolute alcohol.

---

[1] Polis, *Ber.* **20**, 716, 3331.　　[2] *Annalen*, **154**, 93.　　[3] *Ibid.* **194**, 148.

This compound crystallizes from its saturated benzene solution in small, brilliant needles, which have a strong lustre and resemble asbestos. It separates from dilute solutions in long white prisms, which melt at 120°, and with care can be sublimed. It is somewhat volatile in a current of steam, but boils far above 300°, being then partially converted into mercury, benzene, diphenyl, and charcoal, and completely converted into these products at a red-heat.

It is odourless, perfectly insoluble in water, slightly soluble in cold, more readily in hot alcohol and ether, but very soluble in benzene, carbon disulphide, and chloroform.

Ladenburg, when working with this compound, suffered severely from the irritant power of its vapour, especially on the eyes, and became so sensitive to its action that he found himself unable to remain in the room where it was being prepared.

When hydrogen chloride is passed over it, this compound splits up into benzene and mercuric chloride ; the concentrated aqueous acid acts in a similar way, as does strong hydriodic or moderately concentrated sulphuric acid, &c. The halogens decompose it, according as an excess is employed or not, either into monochloro-substitution-products of benzene and salts of mercury, or into those of mercurydiphenyl. On heating it with sodium, benzene and sodium amalgam are obtained. It combines with sulphur trioxide forming mercuric benzenesulphonate $(C_6H_5SO_3)_2Hg$.

p-Mercurydimethylaniline, $Hg[C_6H_4.N(CH_3)_2]_2$, is prepared by boiling 100 grms. of dimethylaniline, 70 grms. of xylene, a little ethyl acetate, and an excess of sodium amalgam for 24 hours, extracting the product with benzene, and evaporating the extract at 100° in vacuo. The yellowish needles thus obtained are recrystallized from benzene, from which it is deposited in lustrous transparent crystals containing 1 mol. benzene, which they lose in the air, falling to a white powder. It crystallizes from chloroform in colourless crystals, the surface of which becomes green in the air. It is very slightly soluble in alcohol and ether, readily in benzene and chloroform. It behaves as a base, and dissolves in dilute hydrochloric acid, but the solution is decomposed on warming. Phosphorus trichloride converts it into dimethylamidophosphenyl chloride [1] (p. 383).

Mercuryphenyl hydroxide, C_6H_5HgOH, is readily formed by the decomposition of the haloid salts by alkalis, but is best

[1] Michaelis and Schenk, *Ber.* **21,** 1501.

prepared by acting on the alcoholic solution of the chloride with moist oxide of silver.[1] It forms small rhombic crystals slightly soluble in cold but readily in hot water, benzene, and alcohol. It acts as a powerful base; its aqueous solution has a strong alkaline reaction, precipitating alumina from its salts, decomposing ammonium salts and absorbing carbon dioxide from the air.

Mercuryphenyl chloride, C_6H_5HgCl, is best obtained by heating equal molecules of mercurydiphenyl and mercuric chloride together with alcohol in a sealed tube at 100—110°. It crystallizes in rhombic scales melting at 250° and subliming when gently heated; these are insoluble in water and slightly soluble in cold alcohol.

Mercuryphenyl bromide, C_6H_5HgBr, is formed, together with bromobenzene, when equal molecules of mercurydiphenyl and bromine dissolved in carbon disulphide are brought together. It crystallizes from a hot mixture of alcohol and benzene in glistening rhombic scales melting at 275—276°.[2]

Mercuryphenyl iodide, C_6H_5HgI, is prepared in a similar manner to the bromide, and also forms rhombic scales melting at 265—266°. Sodium amalgam decomposes it into mercury, mercurydiphenyl, and sodium iodide.

Mercuryphenyl nitrate, $C_6H_5HgNO_3$. This body is prepared by boiling the chloride with alcohol and silver nitrate. It is insoluble in water, but dissolves slightly in cold and readily in boiling alcohol and benzene, crystallizing in fine rhombic tablets possessing a silky or pearly lustre.

Mercuryphenyl carbonate, $(C_6H_5Hg)_2CO_3$, is obtained by the action of silver carbonate on the chloride. It crystallizes in small white needles slightly soluble in boiling water, readily in alcohol.

Mercuryphenyl cyanide, C_6H_5HgCN. This is formed when mercurydiphenyl is heated with mercuric cyanide and alcohol to 120°. It crystallizes in long, glistening, rhombic prisms, melting at 203—204°. When heated to 120° with strong hydrochloric acid, it decomposes into benzene, formic acid, ammonium chloride, and mercuric chloride, whilst when heated with alcoholic potash it yields mercury, benzene, and potassium cyanate:

$$C_6H_5HgCN + KOH = Hg + C_6H_6 + CNOK.$$

[1] Otto, *J. Pr. Chem.* II. **1**, 179.
[2] *Ibid.* p. 186,

Mercuryphenyl formate, $C_6H_5HgCHO_2$. When mercurydiphenyl is heated with concentrated formic acid this compound is produced together with benzene. It crystallizes on cooling in small plates having a vitreous lustre, and melting at 171°.

Mercuryphenyl acetate, $C_6H_5HgC_2H_3O_2$, is obtained by a similar method, and crystallizes from boiling water in small, oblique, rhombic prisms having a vitreous lustre, which are usually grouped in stellate masses and melt at 148—149°.[1] It is also easily prepared by boiling mercuryphenyl iodide with silver acetate. or by heating mercuric acetate with mercury-diphenyl and alcohol to 120° (Otto).

Mercuryphenyl myristate, $C_6H_5HgC_{14}H_{27}O_2$, is formed when equal molecules of mercurydiphenyl and myristic acid are heated with alcohol to 120°. It is deposited in rhombic scales possessing an unctuous touch.

[1] Otto, *J. Pr. Chem.* II. **1**, 186.

INDEX

INDEX

A.

THE END.

Lightning Source UK Ltd.
Milton Keynes UK
UKHW010003090219
336872UK00005B/254/P